Catalysis for Low Temperature Fuel Cells

Special Issue Editors

Vincenzo Baglio

David Sebastián

MDPI • Basel • Beijing • Wuhan • Barcelona • Belgrade

MDPI

Special Issue Editors
Vincenzo Baglio
CNR-ITAE Institute
Italy

David Sebastián
CNR-ITAE Institute
Italy

Editorial Office
MDPI AG
St. Alban-Anlage 66
Basel, Switzerland

This edition is a reprint of the Special Issue published online in the open access journal *Catalysts* (ISSN 2073-4344) from 2016–2017 (available at: http://www.mdpi.com/journal/catalysts/special_issues/catalysis_for_low_temperature_fuel_cells).

For citation purposes, cite each article independently as indicated on the article page online and as indicated below:

Author 1; Author 2. Article title. *Journal Name*. **Year**. Article number/page range.

First Edition 2018

ISBN 978-3-03842-658-5 (Pbk)
ISBN 978-3-03842-659-2 (PDF)

Table of Contents

About the Special Issue Editors

Vincenzo Baglio is a researcher at CNR – Institute for Advanced Energy Technologies "Nicola Giordano" of Messina, Italy. He obtained a B.Sc. degree in chemistry (1998) from University of Messina (Italy) and a PhD in "Materials for Environment and Energy" (2005) from University of Rome "Tor Vergata" (Italy). His current research is focused on energy conversion and storage, in particular in the field of fuel cells, batteries and electrolysers. He published more than 150 articles (H-index: 40 in SCOPUS) in international journals, 8 book chapters, 1 book, 1 international patent and he had more than 200 contributions in national and international conferences.

David Sebastián received his PhD in Chemical Engineering in 2011 from the University of Zaragoza, Spain, and the Spanish National Research Council (CSIC). Then he worked as a postdoctoral researcher at the Italian National Research Council (CNR). Currently, he is working at the Institute of Carbochemistry, CSIC, Spain. His main research interests include design of novel carbon materials for efficient and cost-effective energy conversion and storage devices like fuel cells and water electrolyzers. He has co-authored about 60 peer-reviewed papers.

![catalysts logo] **catalysts**

MDPI

Editorial

Catalysis for Low-Temperature Fuel Cells

David Sebastián [1,2] and Vincenzo Baglio [1,*]

[1] Istituto di Tecnologie Avanzate per l'Energia "Nicola Giordano", CNR, Via Salita S. Lucia Sopra Contesse 5, 98126 Messina, Italy; sebastian@itae.cnr.it
[2] Instituto de Carboquímica, CSIC, Miguel Luesma Castán 4, 50018 Zaragoza, Spain; dsebastian@icb.csic.es
* Correspondence: baglio@itae.cnr.it; Tel.: +39-090-624237

Received: 23 November 2017; Accepted: 28 November 2017; Published: 1 December 2017

1. Introduction

Today, the development of active and stable catalysts still represents a challenge to be overcome in the research field of low-temperature fuel cells. Operation at low temperatures offers great advantages, such as quick start-up and longer stability, but demands the utilization of highly active catalysts to reduce the activation energy of the electrochemical reactions involved in the electrodes, and thus obtain practical performances and high efficiencies. At present, the best-performing catalysts in low-temperature fuel cells are based on highly-dispersed Pt nanoparticles [1]. However, these present several drawbacks, such as high cost, limited earth resources, sensitivity to contaminants, low tolerance to the presence of alcohols, and instability due to carbon support electrochemical oxidation and Pt dissolution [2–4].

In the search for alternative catalysts, researchers have pursued several strategies: increase of the utilization of Pt catalysts by means of novel structures (metal/support) [5,6]; alloying with earth-abundant and cost-effective transition metals [7–9]; new carbon and non-carbon supports [10–13]; cheaper platinum-group-metals like Pd [14–16]; non-platinum-group metals catalysts (Fe-N-C, Co-N-C, etc.) [17–19]; among others.

This Special Issue (SI) is intended to cover the most recent progress in advanced electro-catalysts, from the synthesis and characterization to the evaluation of activity and degradation mechanisms, in order to gain insights towards the development of high-performing fuel cells.

2. This Special Issue

This SI includes 10 high-quality contributions related to recent works in catalysis for low-temperature fuel cells. It comprises four review papers covering aspects such as palladium-based catalysts, non-precious catalysts, the catalyst/support interaction, and the effects related to cell reversal. Six full research articles are also included, dealing with several strategies aimed to maximize catalytic activity through innovative and advanced synthesis techniques.

Calderón-Gómez et al. reviewed the progress for the last ten years concerning palladium-based catalysts as electrodes for direct methanol fuel cells (DMFCs) [20]. Palladium, although belonging to the precious platinum group metals, is cheaper and more abundant than platinum. In the review, Pd catalysts are categorized in two typologies: carbon-supported and non-carbon-supported palladium alloys. Recent studies of these catalysts for the methanol oxidation reaction (MOR) and for the oxygen reduction reaction (ORR) point to a great potential of palladium-based catalysts for replacing platinum as active phase in the electrodes of DMFCs. The alloying of Pd with appropriate metals appears to be crucial for increasing the catalytic activity. It is also concluded that carbon supports aid in dispersing noble metal nanoparticles while favoring the electronic transference, and non-carbon supports (perovskites, titanium oxide, tungsten carbide, etc.) provide modifications in the electronic structure of Pd active sites.

In the review article of Mora-Hernández et al., some of the best-performing catalyst formulations are covered, involving precious and non-precious catalysts [21]. It is divided in three main sections concerning the design of catalytic materials (precious metals, chalcogenides, and non-precious catalysts), the fuel cell reactions of interest (hydrogen oxidation and oxygen reduction), and the important role of catalyst supporting materials (low and high graphitic carbon as well as oxide and composites), and finally a review of micro fuel cells. Novel synthetic routes including photo-deposition, the use of ligands or chalcogenides appear as promising strategies to improve the activity and stability of catalysts for ORR through strong metal-support interactions, nanoalloys with some transition metals, and enhanced tolerance to the presence of small organic molecules like methanol.

Non-precious catalysts are becoming increasingly attractive from sustainable and economical points of view. Brouzgou et al. review the most recent achievements in novel carbon materials for the oxygen reduction in alkaline environment [22]. In particular, the attention is focused on three-dimensional hybrid catalysts, of which nitrogen-doped reduced graphene oxides present the most promising electrochemical activity but short stability over time. Dual-doped ORR catalysts based on mesoporous carbon structures with co-doping metals embedded throughout the carbon matrix are also reviewed, with great prospects in terms of activity and still room for improving stability. More recent three-dimensional interconnected structures in which two metals interact with a graphene-based support are also reviewed, concluding that the main challenge remains to be the enhancement of stability.

The fourth review paper by Qin et al. regards the effects of fuel cell reversal (i.e., anode voltage being superior to cathode and demanding energy rather than supplying it) on irreversible damage: causes, consequences and mitigation strategies [23]. Carbon support corrosion, catalyst and sinter agglomeration, and membrane degradation can permanently harm the fuel cell operation. Two main strategies can be adopted: material modification to resist cell reversal conditions and proper system management. The proper choice of catalyst features for high activity appears as a crucial parameter in addition to operating conditions and carbon supports to mitigate the cell damaging.

This SI also includes six full research papers. Four of them deal with Pt-based catalysts, highlighting the importance that this noble metal still represents in this field. Three of these papers concern the MOR. At the anode of DMFCs, the latter is still a sluggish reaction conditioning the performance of the global cell together with the ORR. Caballero-Manrique et al. investigated the effect of Cu electrodeposition time of PtCu and PtRuCu catalysts [24]. They observed an optimum charge associated with Cu electrodeposition, together with Ru-decoration, to maximize carbon monoxide and methanol oxidation electro-catalytic activity. The enhancement was attributed to the formation of a Pt shell completely covering the Cu core, leading to maximum electrochemical surface areas. The works of Xu et al. and Wang et al. regard the utilization of graphene-supported Pt-based catalysts for the electro-oxidation of methanol [25,26]. Xu et al. propose a one-pot synthetic method to produce composites based on Pt and graphene mediated by a template (polyamidoamine or chitosan) [25]. The use of templates leads to a stronger interaction between Pt and the support resulting in improved MOR activity, whereas the addition of a second metal in Pt-Ag catalysts supported on graphene also resulted in improved performance for the MOR [26], particularly in the presence of UV irradiation. Qayyum et al. explored the synthesis of Pt nanoparticles by pulsed laser deposition [27]. This preparation technique permits the synthesis of high-performing electrodes characterized by low loading of platinum for the ORR, being demonstrated at the cathode of a single cell.

Among Pt-free related papers, the work of Feng et al. describes the fabrication of PdAg-nanochain-based catalysts for the ethanol oxidation reaction (EOR) [28]. The use of a simple one-step hydrothermal method with dimethyl formamide resulted in a particular nanochain structure with enhanced EOR activity when compared to Pd/C catalysts in alkaline medium. They associated this to the improved electrochemical surface area of PdAg nanoparticles. Regarding the efforts for completely removing platinum-group metals in the catalysts, new catalyst structures are being pursued with great results. In this SI, the development of active and stable non-precious metal catalyst for the ORR was reported by Park et al. [29]. They reported a novel Fe_xN_y/NC nanocomposite obtained by pyrolysis of carbon black and an iron precursor in ammonia atmosphere. They report the essential role of Fe_3N species as the active site for the ORR, obtaining excellent activity and stability.

Acknowledgments: We would like to acknowledge Keith Hohn, Editor-in-Chief, Shelly Liu, Managing Editor, and all the staff of the Catalysts Editorial Office (Assistant Editors Candice Zhuo, Jamie Li, Adela Liao, Zu Qiu, Jenny Li, Philip Guo and Camile Wang) for their great support during the preparation of this Special Issue. We would also like to thank all the authors for their great contributions and the reviewers for their time dedicated at reviewing the manuscripts.

Conflicts of Interest: The authors declare no conflict of interest.

References

1. Gasteiger, H.A.; Kocha, S.S.; Sompalli, B.; Wagner, F.T. Activity benchmarks and requirements for Pt, Pt-alloy, and non-Pt oxygen reduction catalysts for PEMFCs. *Appl. Catal. B Environ.* **2005**, *56*, 9–35. [CrossRef]

2. Ferreira, P.J.; la O', G.J.; Shao-Horn, Y.; Morgan, D.; Makharia, R.; Kocha, S.; Gasteiger, H.A. Instability of Pt/C Electrocatalysts in Proton Exchange Membrane Fuel Cells: A Mechanistic Investigation. *J. Electrochem. Soc.* **2005**, *152*, A2256. [CrossRef]

3. Sebastián, D.; Serov, A.; Matanovic, I.; Artyushkova, K.; Atanassov, P.; Aricò, A.S.S.; Baglio, V. Insights on the extraordinary tolerance to alcohols of Fe-N-C cathode catalysts in highly performing direct alcohol fuel cells. *Nano Energy* **2017**, *34*, 195–204. [CrossRef]

4. Borup, R.; Meyers, J.; Pivovar, B.; Kim, Y.S.; Mukundan, R.; Garland, N.; Myers, D.; Wilson, M.; Garzon, F.; Wood, D.; et al. Scientific aspects of polymer electrolyte fuel cell durability and degradation. *Chem. Rev.* **2007**, *107*, 3904–3951. [CrossRef] [PubMed]

5. Du, S.; Lu, Y.; Malladi, S.K.; Xu, Q.; Steinberger-Wilckens, R. A simple approach for PtNi-MWCNT hybrid nanostructures as high performance electrocatalysts for the oxygen reduction reaction. *J. Mater. Chem. A* **2014**, *2*, 692. [CrossRef]

6. Sebastián, D.; Alegre, C.; Gálvez, M.E.; Moliner, R.; Lázaro, M.J.; Aricò, A.S.; Baglio, V. Towards new generation fuel cell electrocatalysts based on xerogel-nanofiber carbon composites. *J. Mater. Chem. A* **2014**, *2*, 13713. [CrossRef]

7. Mani, P.; Srivastava, R.; Strasser, P. Dealloyed binary PtM_3 (M = Cu, Co, Ni) and ternary $PtNi_3M$ (M = Cu, Co, Fe, Cr) electrocatalysts for the oxygen reduction reaction: Performance in polymer electrolyte membrane fuel cells. *J. Power Sources* **2011**, *196*, 666–673. [CrossRef]

8. Baglio, V.; Stassi, A.; Di Blasi, A.; D'Urso, C.; Antonucci, V.; Aricò, A.S. Investigation of bimetallic Pt-M/C as DMFC cathode catalysts. *Electrochim. Acta* **2007**, *53*, 1360–1364. [CrossRef]

9. Antolini, E. Alloy vs. intermetallic compounds: Effect of the ordering on the electrocatalytic activity for oxygen reduction and the stability of low temperature fuel cell catalysts. *Appl. Catal. B Environ.* **2017**, *217*, 201–213. [CrossRef]

10. You, P.Y.; Kamarudin, S.K. Recent progress of carbonaceous materials in fuel cell applications: An overview. *Chem. Eng. J.* **2017**, *309*, 489–502. [CrossRef]

11. Antolini, E. Carbon supports for low-temperature fuel cell catalysts. *Appl. Catal. B Environ.* **2009**, *88*, 1–24. [CrossRef]

12. Sebastián, D.; Suelves, I.; Moliner, R.; Lázaro, M.J.; Stassi, A.; Baglio, V.; Aricò, A.S. Optimizing the synthesis of carbon nanofiber based electrocatalysts for fuel cells. *Appl. Catal. B Environ.* **2013**, *132–133*, 22–27. [CrossRef]

13. Cavaliere, S.; Subianto, S.; Savych, I.; Jones, D.J.; Rozière, J. Electrospinning: Designed architectures for energy conversion and storage devices. *Energy Environ. Sci.* **2011**, *4*, 4761. [CrossRef]

14. Shao, M. Palladium-based electrocatalysts for hydrogen oxidation and oxygen reduction reactions. *J. Power Sources* **2011**, *196*, 2433–2444. [CrossRef]

15. Lo Vecchio, C.; Alegre, C.; Sebastián, D.; Stassi, A.; Aricò, A.S.; Baglio, V. Investigation of supported Pd-based electrocatalysts for the oxygen reduction reaction: Performance. *Materials* **2015**, *8*, 7997–8008. [CrossRef] [PubMed]

16. Rivera Gavidia, L.; Sebastián, D.; Pastor, E.; Aricò, A.; Baglio, V. Carbon-Supported Pd and PdFe Alloy Catalysts for Direct Methanol Fuel Cell Cathodes. *Materials* **2017**, *10*, 580. [CrossRef] [PubMed]

17. Lefèvre, M.; Proietti, E.; Jaouen, F.; Dodelet, J.-P. Iron-based catalysts with improved oxygen reduction activity in polymer electrolyte fuel cells. *Science* **2009**, *324*, 71–74. [CrossRef] [PubMed]
18. Sebastián, D.; Serov, A.; Artyushkova, K.; Atanassov, P.; Aricò, A.S.; Baglio, V. Performance, methanol tolerance and stability of Fe-aminobenzimidazole derived catalyst for direct methanol fuel cells. *J. Power Sources* **2016**, *319*, 235–246. [CrossRef]
19. Sebastián, D.; Serov, A.; Artyushkova, K.; Gordon, J.; Atanassov, P.; Aricò, A.S.; Baglio, V. High Performance and Cost-Effective Direct Methanol Fuel Cells: Fe-N-C Methanol-Tolerant Oxygen Reduction Reaction Catalysts. *ChemSusChem* **2016**, *9*, 1986–1995. [CrossRef] [PubMed]
20. Calderón Gómez, J.; Moliner, R.; Lázaro, M. Palladium-Based Catalysts as Electrodes for Direct Methanol Fuel Cells: A Last Ten Years Review. *Catalysts* **2016**, *6*, 130. [CrossRef]
21. Mora-Hernández, J.; Luo, Y.; Alonso-Vante, N. What Can We Learn in Electrocatalysis, from Nanoparticulated Precious and/or Non-Precious Catalytic Centers Interacting with Their Support? *Catalysts* **2016**, *6*, 145. [CrossRef]
22. Brouzgou, A.; Song, S.; Liang, Z.-X.; Tsiakaras, P. Non-Precious Electrocatalysts for Oxygen Reduction Reaction in Alkaline Media: Latest Achievements on Novel Carbon Materials. *Catalysts* **2016**, *6*, 159. [CrossRef]
23. Qin, C.; Wang, J.; Yang, D.; Li, B.; Zhang, C. Proton Exchange Membrane Fuel Cell Reversal: A Review. *Catalysts* **2016**, *6*, 197. [CrossRef]
24. Caballero-Manrique, G.; Nadeem, I.; Brillas, E.; Centellas, F.; Garrido, J.; Rodríguez, R.; Cabot, P.-L. Effects of the Electrodeposition Time in the Synthesis of Carbon-Supported Pt(Cu) and Pt-Ru(Cu) Core-Shell Electrocatalysts for Polymer Electrolye Fuel Cells. *Catalysts* **2016**, *6*, 125. [CrossRef]
25. Wang, Y.; Li, Z.; Xu, S.; Lei, F.; Lin, S. One Pot Synthesis of Pt/Graphene Composite Using Polyamidoamine/Chitosan as a Template and Its Electrocatalysis for Methanol Oxidation. *Catalysts* **2016**, *6*, 165. [CrossRef]
26. Xu, S.; Ye, L.; Li, Z.; Wang, Y.; Lei, F.; Lin, S. Facile Synthesis of Bimetallic Pt-Ag/Graphene Composite and Its Electro-Photo-Synergistic Catalytic Properties for Methanol Oxidation. *Catalysts* **2016**, *6*, 144. [CrossRef]
27. Qayyum, H.; Tseng, C.-J.; Huang, T.-W.; Chen, S. Pulsed Laser Deposition of Platinum Nanoparticles as a Catalyst for High-Performance PEM Fuel Cells. *Catalysts* **2016**, *6*, 180. [CrossRef]
28. Feng, Y.; Zhang, K.; Yan, B.; Li, S.; Du, Y. Hydrothermal Method Using DMF as a Reducing Agent for the Fabrication of PdAg Nanochain Catalysts towards Ethanol Electrooxidation. *Catalysts* **2016**, *6*, 103. [CrossRef]
29. Park, M.; Lee, J.; Hembram, K.; Lee, K.-R.; Han, S.; Yoon, C.; Nam, S.-W.; Kim, J. Oxygen Reduction Electrocatalysts Based on Coupled Iron Nitride Nanoparticles with Nitrogen-Doped Carbon. *Catalysts* **2016**, *6*, 86. [CrossRef]

catalysts

MDPI

Review

What Can We Learn in Electrocatalysis, from Nanoparticulated Precious and/or Non-Precious Catalytic Centers Interacting with Their Support?

Juan Manuel Mora-Hernández, Yun Luo and Nicolas Alonso-Vante *

IC2MP, UMR-CNRS 7285, University of Poitiers, 4 rue Michel Brunet, 86022 Poitiers, France;
jmmora@live.com.mx (J.M.M.-H.); yun.luo@univ-poitiers.fr (Y.L.)
* Correspondence: nicolas.alonso.vante@univ-poitiers.fr; Tel.: +33-054-945-3625

Academic Editors: Vicenzo Baglio and David Sebastián
Received: 12 July 2016; Accepted: 9 September 2016; Published: 21 September 2016

Abstract: This review is devoted to discussing the state of the art in the relevant aspects of the synthesis of novel precious and non-precious electrocatalysts. It covers the production of Pt- and Pd-based electrocatalysts synthesized by the carbonyl chemical route, the synthesis description for the preparation of the most catalytically active transition metal chalcogenides, then the employment of free-surfactants synthesis routes to produce non-precious electrocatalysts. A compilation of the best precious electrocatalysts to perform the hydrogen oxidation reaction (HOR) is described; a section is devoted to the synthesis and electrocatalytic evaluation of non-precious materials which can be used to perform the HOR in alkaline medium. Apropos the oxygen reduction reaction (ORR), the synthesis and modification of the supports is also discussed as well, aiming at describing the state of the art to improve kinetics of low temperature fuel cell reactions via the hybridization process of the catalytic center with a variety of carbon-based, and ceramic-carbon supports. Last, but not least, the review covers the experimental half-cells results in a micro-fuel cell platform obtained in our laboratory, and by other workers, analyzing the history of the first micro-fuel cell systems and their tailoring throughout the time bestowing to the design and operating conditions.

Keywords: low temperature fuel cells; electrocatalysts; catalytic centers; micro laminar flow fuel cells

---

1. Introduction

Fuel cell technology is a promising strategy to provide power for areas where there is no access to the public or where the cost of wiring and electricity transfer network is important. These devices are usually employed as an energy source for purposes that require non-intermittent power stations such as power generation and energy distribution systems. The use of hydrogen as a fuel makes them clean energy systems with less noise, and pollution generation [1]. Currently, fuel cells are used in both small and large systems such as combined-heat-power (CHP) systems, mobile power, portable computers and communication equipment [2]. Despite all these advantages, there are some limitations for the use of fuel cells. For example, the short lifetime caused by impurities in the gas stream inlets, degradation and poisoning of the catalysts used as electrodes, slow kinetics to perform multi-electron charge transfer reactions, low density output power per unit volume and low accessibility are, to name a few, challenges to overcome. However, in the last years, a great progress in the development of this technology using novel catalysts and fuels aiming at optimizing the operation of fuel cells [2] has been made. Among the main fuel cells operation issues, we can find the high cost and availability of electrocatalysts, the degradation of the proton exchange polymeric membrane, and the oxygen reduction reaction (ORR) kinetics. This reaction takes place on the most active element, Pt, at an overpotential of around 0.3 to 0.4 V [3]. For this reason, fuel cells operate, in practice, at a much lower potential relative to the theoretically possible and with a very low efficiency [4]. The ORR is, in fact, one of the most important reactions in electrochemistry because of its central action in metal-air batteries, electro-synthesis of hydrogen peroxide and the manufacture of cathodes for fuel cells [4]. The molecular oxygen reduction is ideally performed by a transfer mechanism of four electrons. A process involving a single step in which molecular oxygen is reduced to water [5]. Any new ORR electrocatalyst should be investigated to determine the ORR kinetics, since the H_2O_2 produced often causes the degradation of the polymer electrolyte membrane in the heart of proton exchange cells [5]. Nowadays, an important variety of precious metals are used as ORR electrocatalysts.

Platinum supported on carbon is usually used as an anode to conduct the oxidation reaction of hydrogen (HOR) as well as at the ORR at the cathode. It is well stablished that Pt is costly and very sensitive to the presence of contaminants [6]. Numerous strategies have been explored to reduce the mass loading and improve its activity. One of these strategies is the formation of Pt-alloys with 3d transition metals. These kind of alloys can also inhibit the formation of adsorbed species such as hydroxyl, Pt-OH [7]. The transition metal alloys of Pt with, e.g., Ir, Ru, Rh, Re, Os, increase substantially the electrochemical activity, reduce the mass loading, induce lower overpotentials, and promote the reaction of four-e$^-$ [8]. A study carried out by Min et al. [9] considers that the electrochemical property of Pt and its alloys is a result of electronic factors. They suggested that the ORR kinetics increases with the decreasing vacancies in the 5d band of Pt, since such hybridization is less favorable as there is an increase in particle size, implying a decrease of the adsorption strength of the oxygen species. Mukerjee et al. [7] concluded that, on PtCo/C, PtCr/C, PtMn/C, PtFe/C, and PtNi/C alloys, the catalytic activity is increased by an electronic combination which produces changes in the vacancies of the d band, and alloy network contractions. Thus, better ORR properties were obtained on PtCr/C. Markovic et al. [10] reported the synthesis of two Pt-based materials; PtFe and PtCo nanowires. The PtFe electrocatalysts presented higher activity than PtCo toward the ORR due to a greater Co leaching from the alloy. The mass activity and specific surface of the PtFe nanowires were higher than the commercial platinum electrocatalyst. Besides the effect obtained by alloying with early or late transition metals, the electronic properties of Pt NPs can also be modulated by interaction with the support. Indeed, in recent years, exhaustive research activities have been conducted for searching active Pt structures to reach the Department of Energy (DOE) targets [11], which led to the development of many different types of alloys, often identified as skin-layer-, core-shell-, and thin film-electrocatalysts. Greeley et al. [12] has identified, by a density functional theory (DFT) computational screening, a skin-type Pt_xY catalyst as a promising cathode material for the ORR. In this connection, Pt_3Y alloys were the most stable Pt_3M alloys able to bind oxygen species slightly less than pure Pt, therefore showing a higher ORR activity [13]. The high stability of Pt_3Y alloy was attributed to the almost half-filling of the metal-metal d-band of the two alloyed metals, which corresponds to the bonding states filling, whereas the antibonding states are empty.

Another kind of electrocatalysts are the transition metal chalcogenides. Indeed, the substitution in a cluster compound of Ru by Mo atoms in Mo_6Se_8 led to a ternary chalcogenide, $Ru_2Mo_4Se_8$, with interesting properties toward the ORR in acid media [14]. These type of materials are metal clusters with octahedral centers where the delocalization of electrons provides a high electronic conductivity acting as reservoirs of electrical charges, while it remains kinetically stable [15]. Transition metal chalcogenides have also the ability to provide binding sites for the reactants neighbors and intermediaries, they possess also the property to change the volume and distance during the electronic transfer. As a result of these studies, the Ru has been positioned as one of the main electrocatalytic alternatives for the ORR: it is less noble than Pt and is oxidized at a potential 0.2 V more negative than Pt, and the oxides and also hydrides formed on its surface can act as a chemical oxidant for the CO adsorbed at a potential between 0.3 and 0.4 V more negative than pure Pt, this prevents its poisoning as occurs in the case of Pt [16]. Furthermore, it has been shown that Ru possesses good electrocatalytic activity and that combining it with other elements they show greater stability [17–20]. In the Ru_xSe_y material, the Se modifies the electrical properties of the active sites by synergistic effects facilitating the electron transfer [21–24]. Other studies demonstrated that the presence of Mo and Ru atoms decreases the percentage of peroxide produced [25], the Sn reduces the ORR overpotential [26] and finally, the W facilitates the oxygen adsorption [27].

Another strategy to tailor the electrochemical ORR activity and reduce the amount of Pt is by replacing it with Pd, resulting in cheaper electrocatalysts with high ORR activity [28]. Pd has a valence electron configuration and a lattice parameter similar to Pt [29]. Furthermore, it is suggested that the oxygen reduction activity on Pt can be enhanced if Pd is used in combination with other elements which can improve the rate of the electro-reduction of oxygen [30] and the desorption kinetics of the ORR products. Pd is an element more abundant and also cheaper than Platinum; in addition, its stability in acid medium can be comparable with Pt. However, its performance to carry out the ORR is several times smaller in comparison to Pt and its alloys [31]. In order to enhance the catalytic activity of Pd catalysts for PEMFC (proton exchange membrane fuel cell), the particle size control [32], the structure [33], alloying metals [34] and supports [31] are crucial parameters to be considered. In general, the reactivity and selectivity of Pd catalyst for a certain catalytic reaction can be tailored by controlling their shape and thus the type of facets. Pd NPs perform the ORR in acid media and proceed preferentially via four-e$^-$ [35], Pd-Ag, Pd-Cu alloys have shown catalytic activity for the ORR in acid medium with 2- and 4-electrons [36]. Recent reports have shown that Pd alloys such as $Pd_3Fe(111)$ have excellent electrochemical properties [37,38]. Similar studies prove that PdCo/C alloys display comparable activity to Pt/C towards the ORR in acid medium [39]. It has been suggested that the increased oxygen reduction activity of PdCo alloys might be attributed to a Pd electronic stabilization due to the addition of this last element [40]. The electrochemical activity of PdCo/C containing 10%–30% of Co presented a half-wave potential value of 0.66 V/RHE (reversible hydrogen electrode) while Pt/C showed an $E_{1/2}$ = 0.68 V/RHE. Both electrodes were evaluated in a 0.1 M $HClO_4$ solution [41]. Several studies have shown that the catalysts formed by the Pt-Pd alloy present a comparable activity comparable with pure Pt [42] due to the facility to modify the length of the lattice parameter depending on the Pt amount in the alloy in order to favor the highest oxygen adsorption over the catalyst surface [43]. Among the electrocatalysts with different Pt-Pd ratios, the $Pt_{30}Pd_{70}$ presented an onset potential 17 mV more positive than Pt indicating a faster kinetics. The $Pt_{30}Pd_{70}$ electrocatalyst is thermodynamically more active for the ORR than pure Pt as revealed recently by the density functional theory (DFT) studies [44]. This scenario, aside from ORR, another interesting reaction exists where Pd-based catalysts play an important role in the oxidation of organic molecules (methanol (CH_3OH) [45], and formic acid (HCOOH)) [46]. Pd-based electrocatalysts have been employed to oxidize methanol; nevertheless, besides their good electro-activity, those catalysts can overcome the effect of carbon monoxide-poisoning [47] with a high performance as anode catalysts in alkaline DMFC's. Pd-based catalysts show a higher tolerance than Pt to CO species however, the disadvantage of Pd-based catalysts is their lack of durability [48].

Under this scenario, many efforts have been made tailoring and improving the electrochemical activity and durability of Precious and Non-precious based catalysts to perform two of the most important electrochemical reactions in fuel cells; the hydrogen oxidation and the oxygen reduction reactions. In this work we emphasize the catalytic response of precious catalytic centers like Pt-, Pd-based electrocatalysts, and chalcogenides, synthesized via the carbonyl chemical route. We also present the catalytic performance of non-precious electrocatalysts produced via a free-surfactant chemical heating route. All these materials were tested towards the ORR and HOR. The selection of such materials was based on the experience that our research group has acquired in the synthesis methods and electrochemical evaluations. In addition, the electrocatalysts modification by the hybridizing process of some catalytic centers with some carbon and ceramic supports is discussed. Taking into account the most active materials synthesized in our laboratory, they were evaluated in a micro-fuel platform and the results were compared with the state of the art of the micro fuel cells domain analyzing the design, operating arrangements and conditions.

2. Materials Design

2.1. Precious Catalysts

Electrocatalytic materials in the nanodivided form can be obtained by the pyrolysis, in an organic solvent, of transition metal complexes $[M_y(CO)_x]$ that contain a metallic cluster in their base structure, or using metallic salts and employing the carbonyl chemical route through the reactivity of carbon monoxide to achieve the chemical coordination of CO with a metal center in organic or aqueous solvents [49]. The physicochemical and catalytic properties of electrocatalysts depend directly on the chemical nature of the surface but also on their size and shape [50]. Due to its high feasibility and great potential to tailor electrocatalysts, the carbonyl chemical route has been extensively employed to produce active, and chemically stable electrocatalysts [50]. The carbonyl chemical route [51] involves the reactivity of metal carbonyls in solvents to obtain metallic clusters. It has been employed to synthesize Ru [49,52], Pt [49,53–55] and recently Pd-based [56,57] catalysts for electrocatalytic purposes.

2.1.1. Pt- and Pd- Based Electrocatalysts

The synthesis of carbonyl complexes was initiated by Longoni et al. [58]. These authors developed a process to produce the dianions $[Pt_3(CO)_6]_n{}^{2-}$ (n = 10, 5, 4, 3) from the reduction of several platinum precursors ($Na_2PtCl_6 \cdot 6H_2O$, $H_2PtCl_6 \cdot xH_2O$, Na_2PtCl_4, $Pt(CO)_2Cl_2$) using various reducing agents (alkali hydroxides under carbon monoxide, $Fe(CO)_5$, cobaltocene, alkali metals). The n value of the final product depends greatly on the nature of the reagents and the experimental conditions, the reductive carbonylation proceeds as follows [58]:

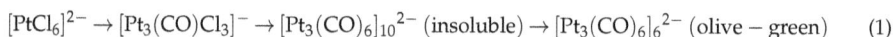

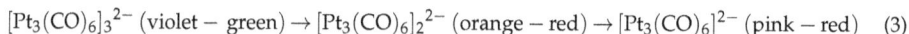

$$[PtCl_6]^{2-} \rightarrow [Pt_3(CO)Cl_3]^- \rightarrow [Pt_3(CO)_6]_{10}{}^{2-} \text{ (insoluble)} \rightarrow [Pt_3(CO)_6]_6{}^{2-} \text{ (olive − green)} \quad (1)$$

$$[Pt_3(CO)_6]_5{}^{2-} \text{ (yellow − green)} \rightarrow [Pt_3(CO)_6]_4{}^{2-} \text{ (blue − green)} \quad (2)$$

$$[Pt_3(CO)_6]_3{}^{2-} \text{ (violet − green)} \rightarrow [Pt_3(CO)_6]_2{}^{2-} \text{ (orange − red)} \rightarrow [Pt_3(CO)_6]^{2-} \text{ (pink − red)} \quad (3)$$

The reactivity of these complexes depends on the value of n and generally, the trend was that increasing n results in an increase in the reactivity towards nucleophilic and reducing agents, while decreasing n results in an increase in reactivity towards electrophiles and oxidizing agents.

For Pt-based electrocatalysts, this method was modified in our group and can be shortly described as follows: the platinum carbonyl $[Pt_3(CO)_6]_5{}^{2-}$ was prepared by mixing specific quantities of $Na_2PtCl_6 \cdot 6H_2O$ with sodium acetate (mol ratio of NaAc/Pt = 6) in methanol solution under CO atmosphere, for 24 h. Activated carbon Vulcan XC-72 was aggregated to the previous solution and stirred for another 12 h under nitrogen atmosphere. The product, nanostructured Pt/C, was washed, filtered and dried [50]. The above process is shown in the reaction (4) [51].

$$[Pt_3(CO)_6]_{10}{}^{2-} + M_yX_n \rightarrow [Pt_xM_y(CO)_z] \rightarrow Pt_xM_y \quad (4)$$

where M = Sn, Ni, Cr, Co, and X = Cl^-.

Based on these previous studies, CO tolerant carbon supported anode bimetallic electrocatalyst Platinum-Tin (Pt_3Sn) was developed [59]. Figure 1a,b schematize the process established by Longoni and Chini [58], and by Alonso-Vante's group [59], respectively. It is important to mention that the process followed by Longoni and Chini stops at step 4. The advantage of the reactivity of the cluster complex is that careful handling the platinum complex, can be taken to deposit the Pt nanoparticles over several supports (step 5). In this case, by depositing carbon black (Vulcan XC-72) Pt/C, can be obtained.

As a result of this synthesis modification, it was found that commercial Etek catalysts presented a broader particle size than the homemade Pt-Sn produced also by the carbonyl chemical route, Figure 2.

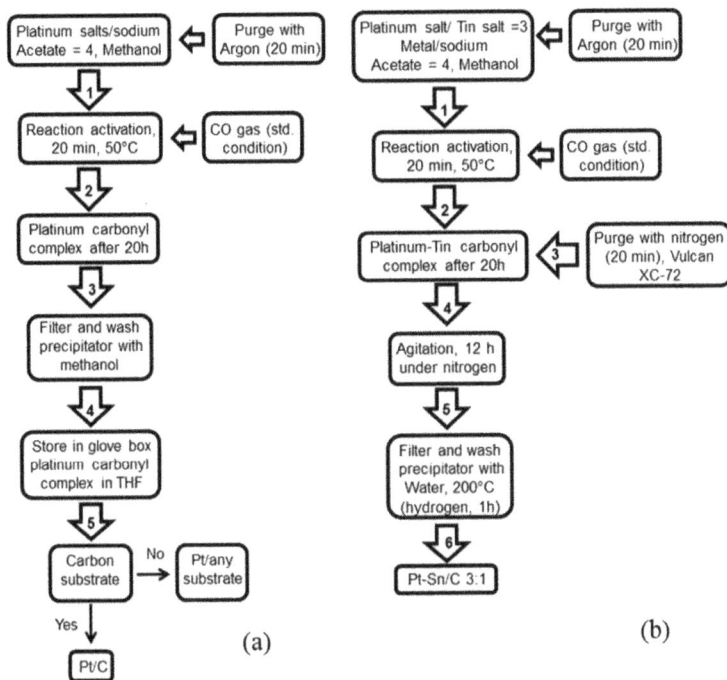

Figure 1. Pathways to obtain Pt (**a**); and Pt-based electrocatalysts (**b**) by carbonyl chemical route.

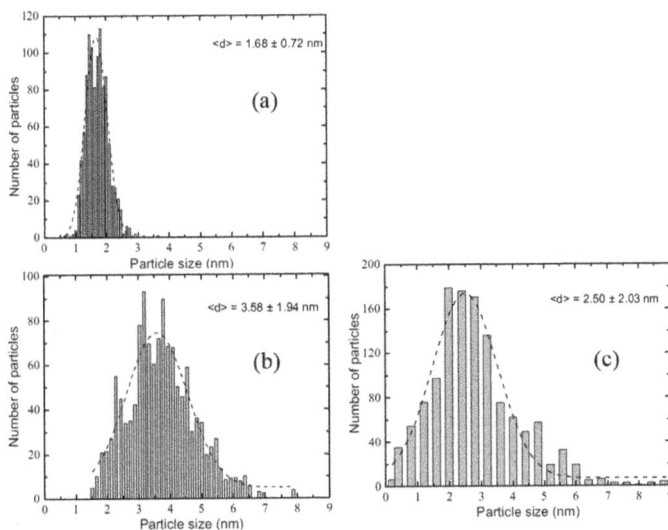

Figure 2. Particle size of the electrocatalysts (**a**) Pt-Sn/C (carbonyl route); (**b**) Pt-Sn/C Etek; and (**c**) Pt/C Etek [59] obtained by TEM (transmission electron microscopy). The *<d>* parameter was determined by a Gaussian distribution. Reprinted with permission from [59]. Copyright John Wiley and Sons, 2002.

In the same line, further bimetallic nanoalloys, e.g., carbon-supported well dispersed Pt-Ni; and Pt-Cr alloys were obtained [54,55]. In this latter, $CrCl_3$ was used with a final heat treatment from 200 °C to 500 °C. The lattice parameter presents an inversely proportional relation with the content of

Cr (Pt/C Etek 0.3923 nm, Pt:Cr (3:1) 0.3895, Pt:Cr (2:1) 0.3863 and Pt:Cr (1:1) 0.3817) [54]. Pt-Cr showed an enhancement factor of 1.5–3 in the mass activity (MA), and in the specific activity (SA), Figure 3; because of the changes in this lattice parameter and the Pt-Cr composition.

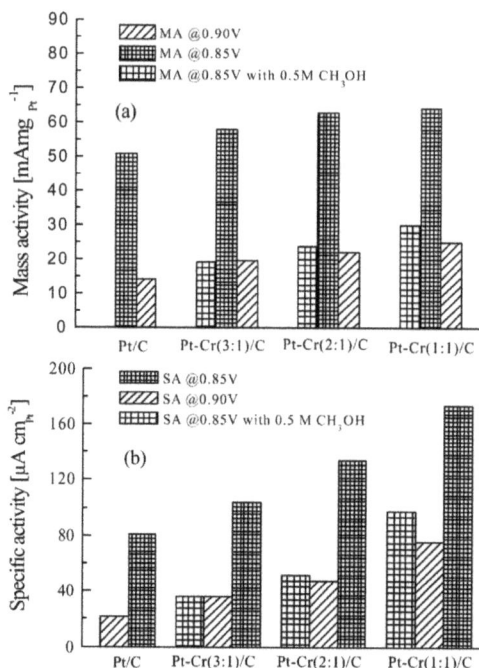

Figure 3. (a) Mass activity (MA) and (b) specific activity (SA) histograms for the Pt-based materials towards the oxygen reduction reaction (ORR). 0.5 M $HClO_4$ was used as electrolyte [54]. Reprinted with permission from [54]. Copyright American Chemical Society, 2004.

For Pt-Ni, the wide angle X-ray spectroscopy patterns, and the Debye function analysis showed that the as-prepared Pt-Ni electrocatalysts presented face-centered cubic structure and that the lattice parameter is reduced with the increase of Ni [55]. More important, both Pt-Cr and Pt-Ni alloys showed high tolerance to methanol than the Platinum electrocatalyst, Figure 4.

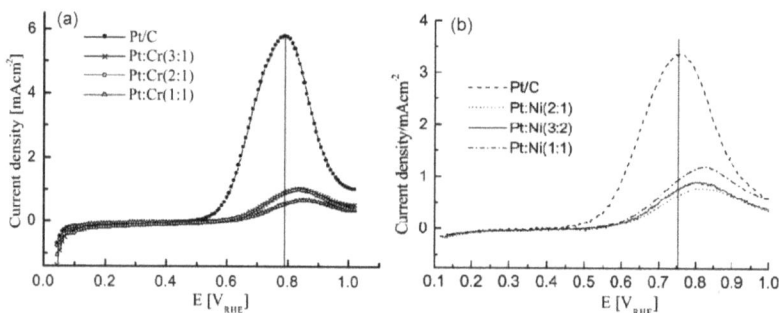

Figure 4. Methanol oxidation comparison for (a) Pt-Cr (reprinted with permission from [54]. Copyright American Chemical Society, 2004 and (b) Pt-Ni (reprinted with permission from [54]. Copyright Elsevier, 2005 in a 0.5 M $HClO_4$ + 0.5 M CH_3OH outgassed solution. The test conditions were: Scan rate $\upsilon = 5$ mV/s, and a rotating disk rate of 2000 rpm [54,60].

To explore the effect of the support material onto Pt NPs the synthesis of platinum supported on Vulcan-carbon (Pt/C) and multi-walled carbon nanotubes (Pt/MWCNTs), via the carbonyl chemical route, was done [61]. After 20 V cycles (CVs) from 0.05 to 1.2 V/RHE, an improved stability was obtained with Pt/MWCNT in comparison to Pt/C. This kinetics and stability improvement were attributed to a higher graphitization degree of MWCNT support, leading to an increased strength of the π-sites with the MWCNTs acting as anchoring centers for Platinum nanoparticles [62]. Furthermore, this interaction also limited the agglomeration of Pt nanoparticles.

Through a modification of the classic carbonyl chemical route, the synthesis of carbon supported Pt-Ti nanoparticles was performed [63]. For the chemical composition, a post heat treatment process on the distribution of the particle size, and also on the electronic properties of Ti and Pt atoms on the performance on the oxygen reduction, and the methanol tolerance were done. The catalytic activity and selectivity towards methanol showed a remarkable enhancement for the nanoalloyed structure. The improvement was caused by the annealing at 875 °C as a result of the alteration of the density of electrons present in the platinum *d*-orbitals modifying the strength of the oxygen adsorption. In consequence, the arrangement of a surface alloy with a high order degree is the responsible to mediate the methanol poisoning of platinum sites [63].

The carbonyl chemical route was further explored to synthesize, for the first time ever, carbon supported Pd NPs supported for the ORR [57]. Herein, K_2PdCl_6 and sodium acetate were used as precursors reacting with CO. All of them were dissolved in CH_3OH, a step which was the key in the modified process because this reaction has to be performed for 2 h in an ice-bath, in this way the reaction rate was decreased avoiding the agglomeration of particles during the complex formation. Thereafter, the process continued following the classical route. The synthesis process comparison for platinum- and palladium-based nanomaterials via carbonyl chemical route is contrasted in Figure 5.

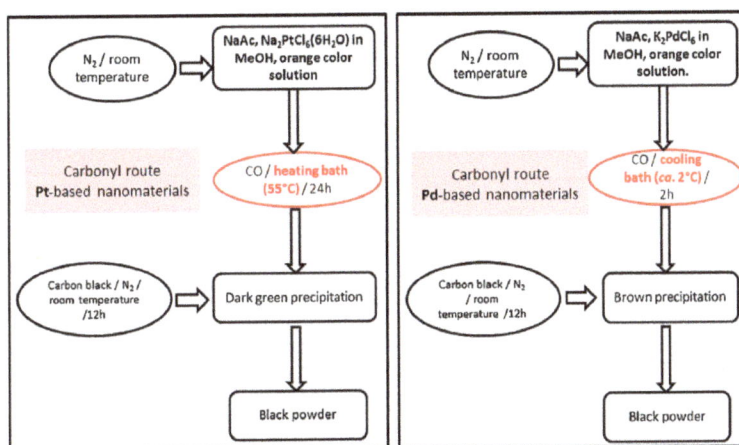

Figure 5. Process diagrams for the synthesis via the carbonyl route of platinum- and palladium-based electrocatalysts [57]. Reprinted with permission from [57]. Copyright Elsevier, 2015.

As a result of this process, Pd samples showed metal mass-loading dependent morphologies. In 10 wt % Pd/C, Pd NPs were highly agglomerated on the carbon support. In Figure 6a, we can find many particles with an average diameter higher than 10 nm. For 20 wt % Pd/C, large amounts of Pd nanowires could be observed, see Figure 6b. In Figure 6c, it is possible to observe Pd nanorods which present an average diameter of 3 nm, these nanorods were well dispersed on the carbon support presenting a total composition of 30 wt % Pd/C. In Pd/C-4, the nanoparticles were almost agglomerated, Figure 6d. In comparison with platinum-based catalysts obtained by the carbonyl chemical route, the morphology of the palladium electrocatalysts changed depending on the metal mass loading [57].

The importance of this study lies in the description to control the morphology of palladium nanoparticles depending of the metal loading following the carbonyl route.

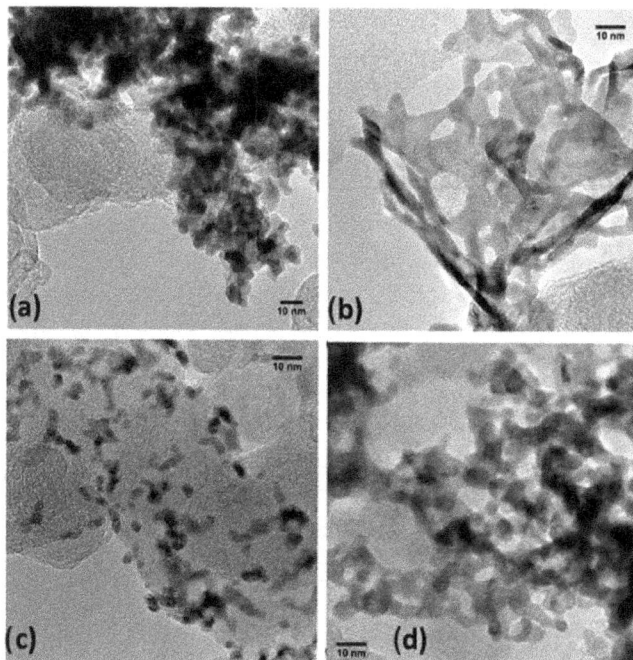

Figure 6. TEM images for (**a**) Pd/C 10%; (**b**) Pd/C 20%; (**c**) Pd/C 30% and (**d**) Pd/C 40% electrocatalysts.

2.1.2. Transition Metal Chalcogenides

Transition metal chalcogenides are promising materials in the field of catalysis and photo-electrocatalysis, since they combine the properties of electrically conducting materials with the catalytic advantage of molecular metal-clusters [64]. In the 70s, various research groups developed the so-called Chevrel phases such as M_6X_8 (M = Mo; Re, X = S, Te and Se), having exceptional superconducting properties [65]. Perrin et al. [66] reported the synthesis and properties of $Mo_xRe_ySe_z$ and $Mo_xRe_yTe_z$ but also $Mo_xRu_ySe_z$, $Mo_xRu_yTe_z$ and $Mo_xRh_yTe_z$ [67]. The authors found that molybdenum sulfide phases and those with mono or divalent elements, a rhombohedric structure with a triclinic deformation predominated. They assumed that Mo and Re d-electron bands were responsible for the electrical superconductivity in these materials. All these samples were prepared by a direct synthesis from the elements in evacuated quartz ampoules, heated up to 1200 °C. The generated samples were ground and annealed at the same temperature for 24 h. To obtain the sulfide compounds, it was necessary to let the reaction take place near 1300 °C, and several subsequent heat treatments were necessary in order to get a pure sample. For selenide and the telluride clusters the synthesis was easier. In order to evaluate and manipulate the samples, the powders were converted into pellets and finally sintered [66,67].

The use of such transition metal chalcogenides ($Mo_{4.2}Ru_{1.8}Se_8$) as electrocatalysts for the oxygen reduction in acid medium was reported in 1986 [68]. The authors reported the first test of such materials to perform the ORR with an appreciable cost advantage with respect to platinum [68]. It was further assessed that the $Mo_{4.2}Ru_{1.8}Se_8$ cluster, in powder form fixed on supports with the help of of Nafion® ionomer thin film, onto glassy carbon, Figure 7a, and on p-type gallium phosphide (p-GaP) semiconductor, Figure 7b, favored the ORR, and hydrogen evolution reaction (HER) under Visible illumination [69]. $Mo_{4.2}Ru_{1.8}Se_8$ clearly enhanced both processes. On p-GaP surface the photo-induced hydrogen evolution was maintained. The limiting currents were reduced by only 3% at -0.56 V (vs. SHE). This was quite remarkable and suggested (now a phenomenon well established) that hydrogen, besides protons, can penetrate the Nafion layer easily, curve 3, in Figure 7b. The flat-band potential of the unmodified electrode was determined to be 0.94 V vs. SHE and was not shifted as a consequence of attachment of the Nafion film and the cluster particles.

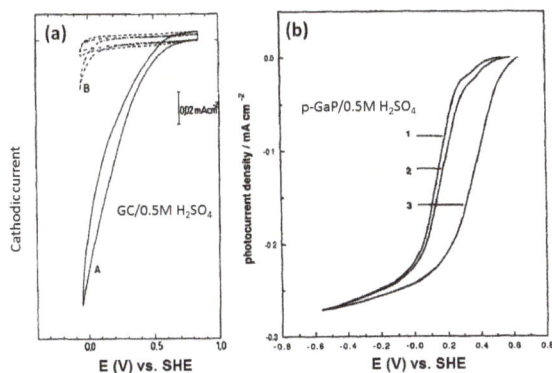

Figure 7. Linear scan voltammograms for the ORR (oxygen reduction reaction) and photo-evolution of hydrogen of the Nafion-attached cluster chalcogenides. (**a**) Glassy Carbon at 50 mV/s; and (**b**-curve (3)) p-GaP. Curve (B), in (**a**), is the unmodified GC surface. In (**b**), curves (1, 2) represent Nafion unmodified, and modified p-GaP surfaces. Scan rate = 5 mV·s^{-1}. Adapted from [69].

Summing up, the $Mo_{4.2}Ru_{1.8}S_8$ cluster employed as a model catalyst for multi-electron reduction processes for the ORR, was not suitable for multi-electron oxidation processes—the oxygen evolution reaction (OER), due to insufficient kinetic stabilization [69]. Based on previous works [70,71], a certain degree of kinetic stabilization can be accomplished with chalcogenides, which can permit a photoreaction of holes via transition-metal d-states which are not significantly mixed with chalcogen states. Regarding the electrochemical activity towards the ORR, $Mo_{4.2}Ru_{1.8}S_8$, Figure 8, a pseudo ternary cluster, outperformed the ternary clusters, e.g., $Ni_{0.85}Mo_6Te_8$, where Ni atoms occupies the free sites in the channels between the Mo-clusters. Two main positive effects on the pseudo ternary compound were claimed, namely, the increase of metal-metal distance in the cluster, thus possibly facilitating the breaking of the –oxygen–oxygen– bond liberating OH^- ions. The expected shift of the electronic levels upwards was caused by the increase of the charge in the cluster.

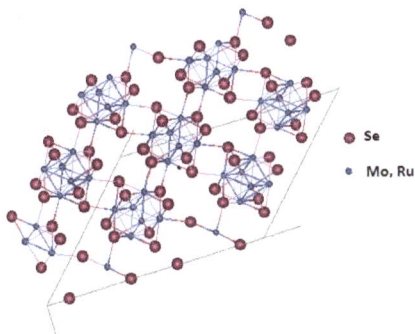

Figure 8. Structure of a pseudo ternary cluster $(Mo_4Ru_2)Se_8$.

Because the catalytic properties and the high density of d states at the Fermi level can be related, this relation on $Mo_4Ru_2Se_8$, $Mo_2Re_4Se_8$, Mo_6Se_8, and the Mo metal was undertaken [72]. Indeed, Mo_6Se_8 showed metallic properties, whereas $Mo_4Ru_2Se_8$ and $Mo_2Re_4Se_8$ showed semiconducting properties. The explanation lies on the fact that the substitution of Mo by Re and/or Ru atoms, that contain a high valence electron number, causes an increment of the electrons from 20 to 24 in the cluster M_6X_8, resulting in a filling of the e_g band. This effect shifts upwards the Fermi level, as schematized in Figure 9.

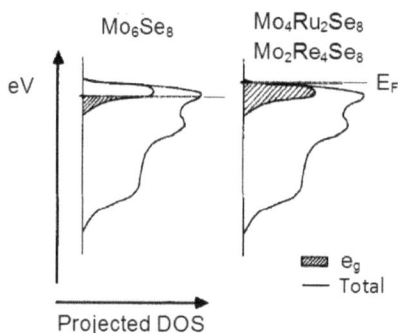

Figure 9. Band structure scheme of chalcogenide clusters. Adapted from [72,73]. Reprinted with permission from [72]. Copyright Elsevier, 1988.

These compounds were evaluated towards HER and ORR. As a result of the narrow energy band caused by the d-state density, some improvement was observed on the HER, with a significant influence on ORR. This study revealed that hydrogen species solely interact with individual metal atoms, the molecular oxygen interacts with the atoms M, and M' in the $(MM')_6X_8$ cluster [74], as shown in Figure 10, thus forming a surface transition metal complex. Based on the statistics of the calculated possible combinations and taking again into account the density of d states, the complex IV must play a major role in the ORR electrocatalytic materials activity.

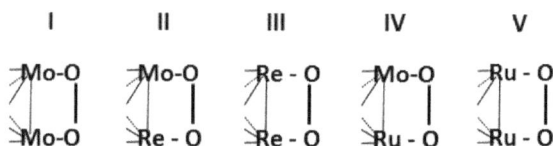

Figure 10. Surface transition metal complexes and their bridge-type, interaction with the chemisorbed oxygen, modified from [72]. Reprinted with permission from [72]. Copyright Elsevier, 1998.

Based on these Chevrel phase model compounds, prepared in quartz ampoule at high temperature [67], Mo-Ru-Se, in powder and/or in thin layer form, was synthesized via a soft chemical method; the carbonyl chemical route by reacting molybdenum hexacarbonyl, tris-ruthenium dodecacarbonyl and selenium powder in xylene [75]. The XPS (X-ray photoelectron spectroscopy) characterization revealed the presence of SeO_2 and MoO_3, Figure 11. After cathodic electrochemical treatment, in presence or absence of oxygen at 0.35 V during 5 min in 0.5 M H_2SO_4, the emission line for SeO_2, disappears and the MoO_3, emission line is strongly reduced, see curves (2) in Figure 11, indicating that the catalysts still contain a small amount of Mo. Exposed to air for more than 100 days, curves (3), the MoO_3 species is further reduced to a minimum in comparison to the shortly treated electrode, curves (2). The absence of SeO_2, species demonstrates a higher stability than Chevrel phase compounds against oxidation under oxygen atmosphere [13].

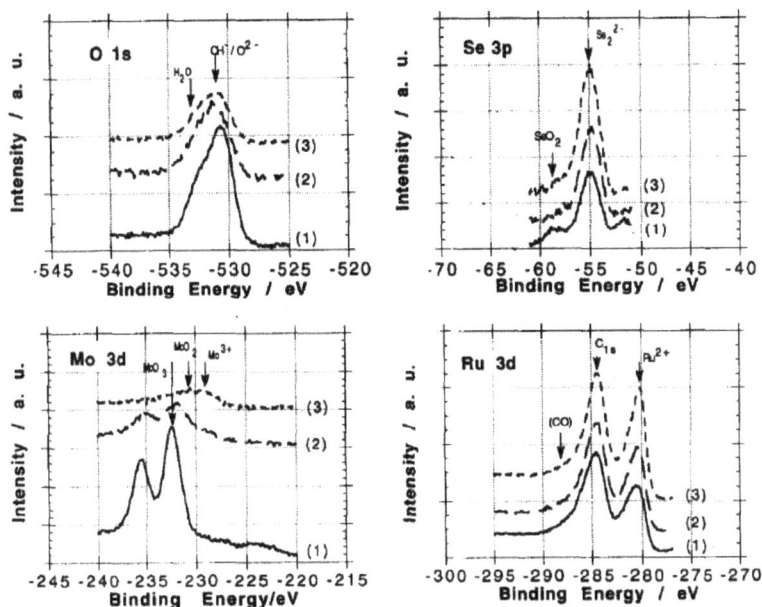

Figure 11. XPS (X-ray photoelectron spectroscopy) spectra of $(Ru_{1-x}Mo_x)_ySeO_z$ at different conditions; (1) as prepared sample in xylene solution; (2) polarized sample at +0.35 V/NHE (5 min) in 0.5 M H_2SO_4; and (3) after ORR and kept in air approximately 100 days. Reprinted with permission from [75]. Copyright Elsevier, 1994.

The Extended X-ray absorption fine structure spectroscopy (EXAFS) disclosed that chemically generated Mo-Ru-Se electrocatalysts were, structurally speaking, different to that of the Chevrel phase compound [64]. It was concluded that the Mo and Ru atoms, in Mo-Ru-Se, were located as independent phases: $MoSe_x$, MoO_x and $RuSe_x$ [76]. They, however, showed high selectivity to perform the ORR, in a solution containing methanol, with a four-electron reaction pathway, with an activity nevertheless, lower than that of the standard Pt catalysts in methanol-free electrolyte [77].

A further characterization by X-ray diffraction at wide angles (WAXS) of Ru_x and Ru_xSe_y was carried out [78]. Indeed, chalcogenide particles produced from the reaction between $Ru_3(CO)_{12}$ and Se were more resistant to oxidation than monometallic Ru_x nanoparticles with similar size. This phenomenon was associated to the coordination of Selenium atoms over the ruthenium atoms surface. As revealed by EXAFS analysis [79,80] the closest Ru-Ru averaged distance was 2.64 Å (corresponding to the *hcp* phase of metallic ruthenium) and that of the metal chalcogenide Ru-Se was 2.43 Å. It is worth noting that the calculated Debye-Waller parameters were a factor of 2 higher for the Ru-Se coordination distance than for Ru-Ru. Considering a particle diameter of 17 Å a pure ruthenium cluster-like model particle was developed containing 153 atoms (Ru_{153}). This is the third species in a series of *hcp* closed shell clusters with "magic" numbers 13, 57, 153, 323, etc. Therefore, within the *hcp* cluster model (17 Å particle size, 153 atoms), the number of ruthenium atoms available for electrocatalytic purposes at the surface of such cluster can be $153 - 57 = 96$. Because $x = 2$ and $y = 1$ in the chalcogenide material, Ru_xSe_y, the cluster model contains 99 Ru atoms and 54 Se atoms, with an increased disorder of selenium atoms, deduced from EXAFS calculations [79], almost all selenium atoms are coordinated to ruthenium atoms on the cluster in a statistical way, as shown in Figure 12. In this way, the Ru-clusters surface coverage by Se atoms induces the necessary free active sites (Ruthenium sites) to perform an efficient coordination with oxygen/water [78].

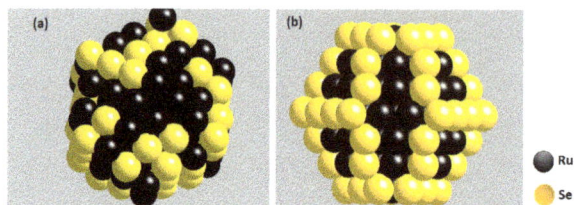

Figure 12. *hcp*-Cluster Model structures for $Ru_{99}Se_{54}$; (**a**) Se statistically bonded onto the Ru clusters; (**b**) Se replacing Ru surface atoms in an ordered positioning with 5-, 6-, and 7- fold coordinated.

The use of selenium atoms to coordinate catalytic center surface atoms, as well as impinging on it electronic modification was demonstrated for ruthenium in Ru_xSe_y [81]. This concept was further applied onto platinum, generated via the carbonyl chemical route [82]. The surface Pt NPs in Pt/C was modified with Se (selenization process). Platinum supported on carbon and SeO_2 were mixed in isopropanol solution with specific quantities and stirred for 12 h at room temperature. This mixture (20 wt % Pt_xSe_y/C) was heat-treated (200 °C) for 1 h under a nitrogen atmosphere. In order to account the optimal amount of Se atoms, to allow the adsorption and subsequent reduction of O_2 molecules, Se was concomitantly stripped from the Pt surface using Chronoamperometry at different stripping times, at an established anodic potential. Employing a stripping time of 20 min at 1.1 V/RHE, the ORR half-wave potential ($E_{1/2}$) shifted to positive values, whereas the OCP (open circuit potential) remained constant. The optimal Se coverage (see arrow in Figure 13a) kept to a maximum the ORR in the presence of methanol [83]. The partial leaching of Se atoms from the Pt surface improved the activity of Pt_xSe_y/C towards the oxygen reduction, with a high tolerance as compared to Pt/C, Figure 13.

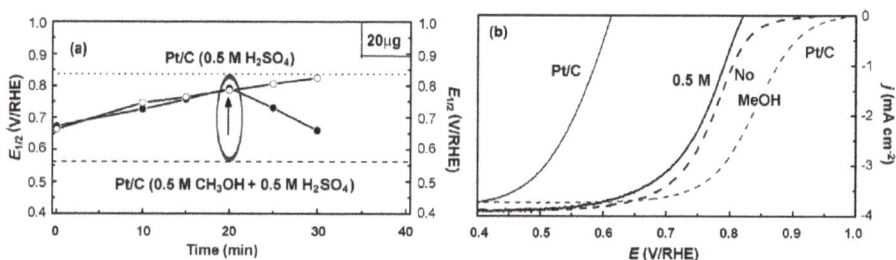

Figure 13. (**a**) Pt_xSe_y/C (20 mg of 20 wt %) ORR half-wave potentials compared to Pt/C (dotted lines). Both electrocatalysts were evaluated at 900 rpm in oxygen-saturated 0.5 M H_2SO_4 + 0.5 M CH_3OH (filled circles, left axes) and 0.5 M H_2SO_4 (empty symbols, right axes). (**b**) Linear scan voltammetries for the same test [83]. Reprinted with permission from [83]. Copyright John Wiley and Sons, 2013.

2.2. Non-Precious Catalysts

Despite the efforts carried out in the production of novel and efficient electrocatalyst, platinum-based materials are still one of the most efficient oxygen reduction catalysts employed in PEMFCs [84]; however, their high cost, low abundance and poor durability are a main obstacle in the development of PEMFCs [85]. To overcome these issues, the search of non-precious metal ORR catalysts has been extensively conducted in recent years [86]. Pt-free catalysts such as transition-metal chalcogenides have shown promising activities towards fuel cell reactions [51]. Co and Fe are also promising cathodic catalyst centers [87]. In recent studies, researchers have focused their efforts on two kinds of non-precious metals as catalytic centers: Co and Fe, and transition-metal chalcogenides with application in the ORR [88].

The development of new non-precious electrocatalyst dates back to the seventies with Co_3S_4 and $Me^{(a)}Me_2^{(b)}X_4$ ($Me^{(a)}$ = Mn, Fe, Co, Ni, Cu or Zn, $Me^{(b)}$ = Ti, V, Cr, Fe, Co or Ni, X = O, S, Se or Te). This material showed the highest catalytic activity for the oxygen reduction in a 1 M H_2SO_4 solution, and presented an open circuit potential of ca. 0.8 V vs. HRE [87]. However, Co_3S_4 was not close to the catalytic activity of Pt. Based on this work other Co-based materials [89–93] came into light. Feng et al. [94] synthesized three Co_3S_4 (-01, -02 and -03) compounds following different methodologies. Co_3S_4-01 was produced from a $Co_2(CO)_8$ decomposition in a mixture solution which contained tri-octylphosphane oxide and oleic acid as surfactants, just after sulfur and cobalt particles were reacted and deposited over the carbon substrate [91]; Co_3S_4/C-02 was prepared following the same synthesis procedure for Co_3S_4-01 but without the use of surfactants. Co_3S_4/C-03 was synthesized by means of a microwave-assisted process [94]. Co_3S_4/C-02 and Co_3S_4/C-03 synthesized without surfactant showed the highest activity toward the ORR, Figure 14. This work suggested that the surfactants, used in the synthesis of Co_3S_4/C-01, block the electrocatalyst active sites. These results demonstrated that routes without the surfactants were effective methods to produce non-precious catalysts supported onto carbon [94].

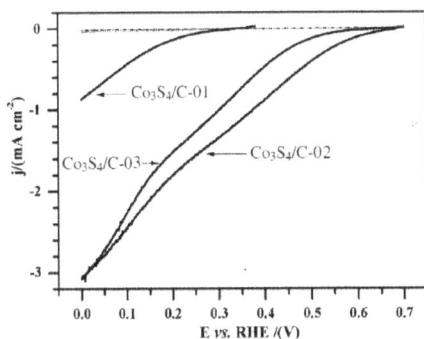

Figure 14. Linear scan voltammograms of 20 wt % Co_3S_4/C-01, -02, and -03. Rotation rate = 1600 rpm in an oxygen saturated 0.5 M H_2SO_4 solution at 25 °C. υ = 5 mV s^{-1} [94]. Reprinted with permission from [94]. Copyright American Chemical Society, 2008.

In modifying and upgrading the Co-Se electrocatalysts synthesis, $CoSe_2$ nanoparticles with different nominal loading from 20 wt % to 50 wt % were produced [95] via conventional heating process [94]. In short, for the synthesis of 20 wt % $CoSe_2$/C, $Co_2(CO)_8$ and Vulcan carbon were well dispersed in *p*-xylene under a nitrogen atmosphere at 25 °C for a half-hour. This suspension was heated and refluxed, and then cooled down. Thereafter, a specific amount of selenium was dispersed in *p*-xylene and stirred half-hour. Both suspensions were mixed, heated, and refluxed for 10 min. The final product was filtered, washed, and dried under vacuum at 25 °C. All $CoSe_2$/C electrocatalysts with different nominal ratio were heat-treated at 300 °C under high purity N_2 for 3 h.

The oxygen reduction cathodic current obtained for 20 wt % $CoSe_2$/C was in agreement to the results reported in previous works for non-precious metallic electrocatalysts (ca. 2.5 mA·cm^{-2} at 0.4 V at 1600 rpm) [96–98]. Some data from the literature were 2.0 mA·cm^{-2} for $Co_{1-x}Se$/C (50 wt % Cobalt on Vulcan carbon) [96], and 2.5 mA·cm^{-2} for (Co, Ni)S_2 thin films [98] at 2000 rpm. The OCP value of 0.81 V in oxygen-saturated 0.5 M H_2SO_4 surpassed the $Co_{1-x}Se$/C [96], and FeS_2 (0.78 V) [97]. The OCP was quite similar to that of NiS_2 (0.80 V), and CoS_2 (0.82 V) [98], but not better than that of (Co, Ni)S_2 (0.89 V) [98].

The electrode loading effect of $CoSe_2$ (from 11 µg·cm^{-2} to 44 µg·cm^{-2}) was investigated via the rotating ring disk electrode (RRDE) technique, Figure 15a. In this experiment, the disk current increased at an electrode potential of 0.5 V from 2.5 to 3.8 mA·cm^{-2} by the increase of the catalyst loading, with a concomitant decrease of H_2O_2% generation (at ca. 0.32 V). The variation of the mass loading as a function of the peroxide production can be interpreted in terms of an increase of the active sites in the electrocatalyst. In other words, the H_2O_2 has additional time to react to water in a thicker layer. As a result of this effect, an increase in the total number of electrons, and a decrease in the hydrogen peroxide was detected at the ring electrode. Similar results on galena and pyrite materials were observed by Ahlberg et al. [99]. The authors found 25%–30% H_2O_2 production at the disk electrode. On the other hand, in Figure 15b we can observe the influence of 20 wt %, 35 wt %, and 50 wt % $CoSe_2$/C systems on the oxygen reduction activity and the hydrogen peroxide generation. Since the hydrogen peroxide generated on the surface of the ring electrode decreases from 30% to 15% when the loading rate is changed from 20 wt % to 50 wt %, there must exist a limiting catalytic site density. A similar trend was observed in other systems, e.g., 20 wt % Ru_xSe_y/C, which overpasses the current density of 70 wt % Ru_xSe_y/C [100], including Pt/C [101].

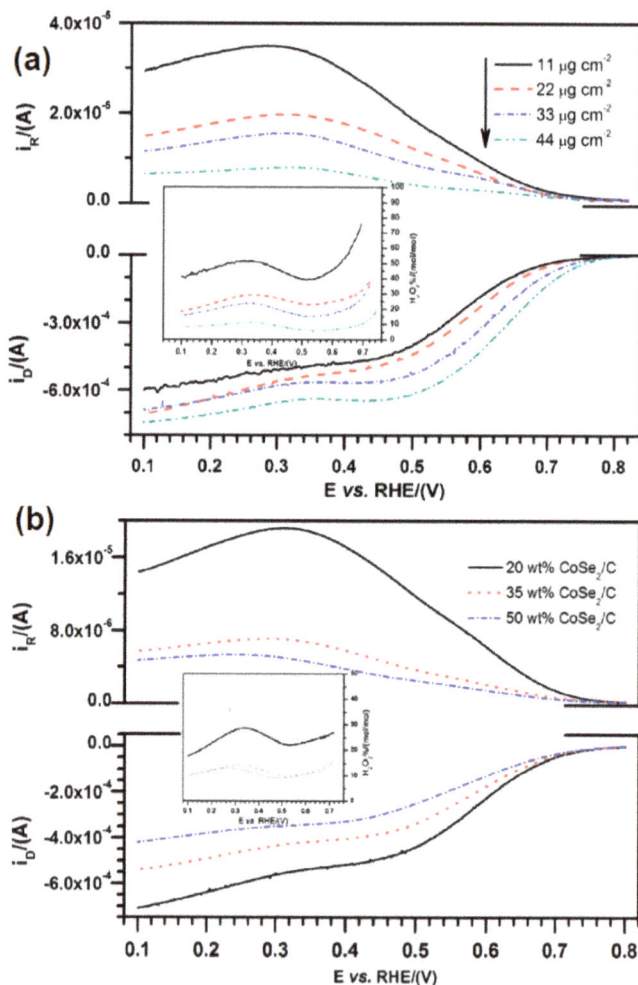

Figure 15. Linear scan voltammograms for the ORR and the formation of peroxide hydrogen (insets); (**a**) Electrode loading effect of $CoSe_2$ (from 11 $\mu g \cdot cm^{-2}$ to 44 $\mu g \cdot cm^{-2}$); (**b**) Influence of 20 wt %, 35 wt %, and 50 wt % $CoSe_2$/C systems on the ORR activity and the H_2O_2 production [95]. Reprinted with permission from [95]. Copyright Elsevier, 2009.

Besides the ORR activity on the Co-Se system, the methanol tolerance was further investigated in alkaline medium [102]. Without surprise, the $CoSe_2$ nanocatalyst performance in alkaline medium is better than that in acid medium revealing further that the carbon-supported cubic phase of $CoSe_2$ is a promising non-precious electrocatalyst for alkaline fuel cells [103]. Furthermore, the oxygen reduction activity of 20 wt % $CoSe_2$/C in 0.1 M KOH was slightly higher than that of Co_3O_4/rmGO (reduced mildly oxidized graphene oxide) and comparable to that of Co_3O_4/NrmGO (nitrogen-doped mildly oxidized graphene oxide) as reported by Liang et al. [104]. Apropos methanol tolerance, Figure 16 contrasts the ORR curves of 20 wt % $CoSe_2$/C with that of 20 wt % platinum supported on carbon in oxygen-saturated 0.1 M KOH containing methanol. For Pt/C, the onset potential shifts negatively from 1.0 to 0.6 V, with an oxidation peak observed at ca. 0.65 V when the methanol concentration increased from 0 to 5 M. For 20 wt % $CoSe_2$/C, the onset potential remains unchanged and no oxidation peak is observed testifying immunity to methanol. At $j = -3$ mA·cm^{-2}, however, the potential shifts from 0.68 V to 0.62 V for 20 wt % $CoSe_2$/C catalyst, suggesting that methanol species in alkaline medium perturb the ORR. Nevertheless, the results herein show that 20 wt % $CoSe_2$/C presented an improved selectivity to perform the oxygen reduction reaction in presence of methanol than Pt/C electrocatalyst in alkaline medium. However, $CoSe_2$ is not hundred percent tolerant to methanol.

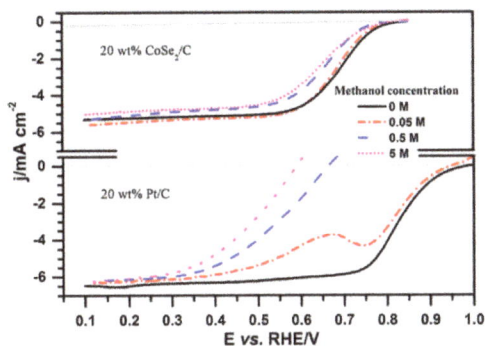

Figure 16. Methanol effect comparison $CoSe_2$/C and commercial Pt/C from E-TEK towards the oxygen reduction at concentrations of 0, 0.05, 0.5 and 5 M CH_3OH and a rotating speed conditions of 2500 rpm in an oxygen saturated 0.1 M KOH solution at 25 °C [102]. Reprinted with permission from [102]. Copyright Elsevier, 2012.

3. Fuel Cell Reactions

3.1. Hydrogen Oxidation Reaction—HOR

The electrocatalysis of hydrogen is a central issue in low temperature membrane fuel cells (PEMFC). H_2 is oxidized at the anode electrode:

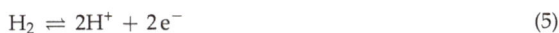

$$H_2 \rightleftharpoons 2H^+ + 2e^- \tag{5}$$

The HOR largely depends on the strength of the intermediate interaction and bonding of metal and adsorbed hydrogen atoms [105]. As a consequence, the binding energies of the adsorbed hydrogen to the surface should always be in a range of an adsorption energy. A stronger bond would hamper the catalysis process by reducing surface dangling states and active sites, while a weaker bond could possibly indicate an inert electrode surface character with low number of dangling states and unsuitability of electrode for the catalysis [106]. Equation (5) can be represented by the following microscopic steps:

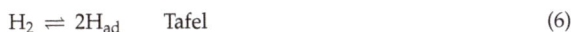

$$H_2 \rightleftharpoons 2H_{ad} \qquad \text{Tafel} \tag{6}$$

$$H_2 \rightleftharpoons H_{ad} + e^- + H^+ \qquad \text{Heyrovsky} \qquad (7)$$

$$H_{ads} \rightleftharpoons H^+ + e^- \qquad \text{Volmer} \qquad (8)$$

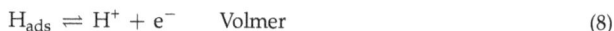

So, that the overall reaction can be revealed knowing some parameters, such as the exchange current density, activation energy, charge transfer coefficients, and Tafel slope. In PEMFCs the mass loading reduction of the expensive Pt is required to catalyze the HOR. Significant efforts have been made on the potential use of partial or complete replacement of Pt electrocatalyst at the cathode [107–109].

For the replacement of the most common based on platinum group metals (PGMs = Pt, Pd, Ru, Ir, Rh and Os [110]) catalysts employed so far for the HOR in acidic media [43,111–117], anionic exchange membrane fuel cells (AEMFCs) are also very interesting because their applicability of non-Pt metal catalysts in alkaline medium, where most of the transition non-precious metals presents high stability [118,119]. While research on oxygen ORR catalysts in alkaline medium has been ongoing for many years [120], studies on hydrogen HOR catalysts for AEMFCs constitute practically a new field of investigation. In contrast to the fast kinetics of the HOR on PGMs catalysts in acidic media, it was found the HOR activity of several PGMs in alkaline electrolyte is around two orders of magnitude slower than the activity of these materials in acid electrolyte, all this as a result of an intrinsic HOR kinetic barrier [121,122].

Contrary, to the simple Volmer reaction performed in acidic media, Equation (8), in alkaline medium, Equation (10) is required to complete the HOR that comprise two different sub-processes of hydrogen adsorption on the metallic surface and water formation in alkaline medium [123].

$$H_2 + OH^- \rightleftharpoons H_{ad} + H_2O + 1e^- \qquad \text{Heyrovsky} \qquad (9)$$

$$OH^- + H_{ad} \rightarrow H_2O + 1e^- \qquad \text{Volmer} \qquad (10)$$

Recent attempts to reduce the amount of Pt in the design and production of electrocatalysts for the HOR were carried out by Gu et al. [124]. Using a hydrothermal method, the synthesis of Pt@Ru in the face-centered cubic (*fcc*) phase with a metastable phase for Ru under ambient conditions was performed. The *fcc* Pt@Ru nano crystals were produced by a one-step hydrothermal procedure, in which K_2PtCl_4 and $RuCl_3$ were reduced by formaldehyde (HCHO) at 160 °C in the presence of sodium oxalate ($Na_2C_2O_4$), HCl, and poly(vinyl-pyrrolidone) (PVP). The samples were referred to as $Pt_x@Ru_{100-x}$, in which x was the molar percentage of Pt in the feedstock.

As revealed by density functional theory (DFT) calculations, a preferentially epitaxial growth of Ru atom layers on the non-closest-packed facets of hetero *fcc* metal seeds led to the formation of *fcc* Ru shells to form Pt@Ru tetrahedrons/C, Figure 17; these morphological arrangements showed enhanced electrocatalytic activity with an order of magnitude more efficient toward the hydrogen oxidation reaction (HOR) in acidic electrolyte as compared with hydrothermally synthesized Ru/C [124]. The sample composed by tetrahedrons shows an average size of 8.3 nm, Figure 17b,c, which are in agreement with the models of *fcc* tetrahedrons from each zone axes, explicitly indicating that these particles are *fcc* tetrahedral single crystals enclosed by (111) facets. High angular annular dark field-scanning transmission electron microscopy image and their corresponding contrast line profile are shown in Figure 17d,e.

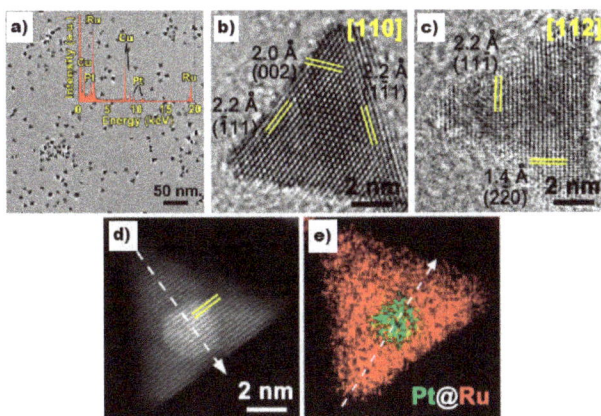

Figure 17. (a) $Pt_{10}@Ru_{90}$ nanotetrahedrons TEM image and its EDS (Energy dispersive spectroscopy) spectrum; (**b,c**) HRTEM (High resolution transmission electron microscopy) images, corresponding to FFT (Fast Fourier Transform) patterns, and models of *fcc* tetrahedrons from (110) and (112) zone axes, respectively; (**d**) HAADF-STEM (high-angle annular dark-field imaging-scanning transmission electron microscope) image; and (**e**) EDS mapping [124]. Reprinted with permission from [124]. Copyright American Chemical Society, 2015.

The HOR polarization curves obtained on *fcc* Pt@Ru, Ru/C (as-prepared), and commercial Pt/C loaded on glassy carbon RDE with a rotation speed of 2500 rpm in H_2 saturated 0.1 M $HClO_4$ solution are presented in Figure 18. A significant HOR activity was achieved on *fcc* Pt@Ru as compared to Ru/C. $Pt_{10}@Ru_{90}$/C showed the highest activity among the core-shell samples, i.e., 0.19 A·mg^{-1} and 0.30 mA·cm^{-2}. The mass, and area activities are higher, by one order of magnitude, than the as-prepared Ru/C (0.016 A·mg^{-1} and 0.017 mA·cm^{-2}) and comparable with those of commercial platinum (0.23 A·mg^{-1} and 0.27 mA·cm^{-2}).

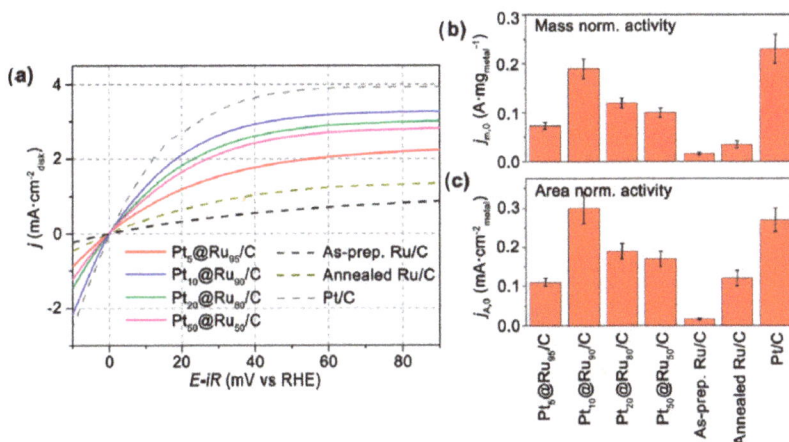

Figure 18. (**a**) Hydrogen oxidation reaction (HOR) Current-potential characteristics for Pt@Ru/C, Ru/C (as-prepared), Ru/C (annealed), and Pt/C (commercial); total metal mass: 3 μg. Test conditions: $\upsilon = 5$ mV·s^{-1}; at a rotation speed of 2500 rpm in a 0.1 M $HClO_4$ solution at 25 °C. (**b**) Mass and (**c**) ECSA (electrochemical surface area) normalized exchange current densities [124]. Reprinted with permission from [124]. Copyright American Chemical Society, 2015.

In search of new materials for HOR, palladium phosphide electrocatalyst with a small particle size was synthesized. Its improved stability, in acid medium, and the increased electrocatalytic activity towards the HOR with respect to commercial Pd/C was obtained [125]. Transition metals alloyed with phosphorus produced metal-phosphide, resistant to corrosion of the underlying transition metal, presenting metallic conductivity, or low electrical resistivity as low-band gap semiconductors [126]. Furthermore, they can be synthesized following a variety of routes and presenting high stability in basic and acidic media in comparison to the unalloyed metals [127–129]. The catalytic activity of transition metal phosphides and metal-phosphide materials namely, Ni [130–133], Co [134], Mo [135] and W [136] towards the HER has been studied. From these experiments it has been determined that the HOR performance of Pd-phosphide is comparable to that of palladium supported on carbon (Figure 19). The materials show a similar curve for the HOR, although there is a reduction in the limiting current value for PdP_2/C. Materials containing phosphide showed a weak wave before the limiting current which it is associated with the hydrogen oxidation on the phosphide lattice. This indicates that H_2 diffusion could be higher in the particles than in the palladium precursor. However, a faster onset for the HOR than the Pd/C precursor was visualized on the phosphorus material by dividing the current density by the corresponding limiting current.

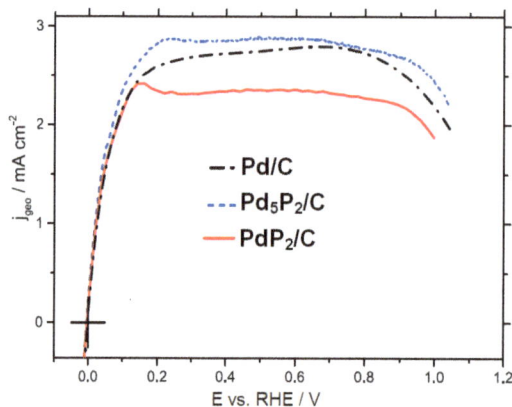

Figure 19. HOR on Pd/C, Pd_5P_2/C and PdP_2/C electrocatalysts in 0.1 M $HClO_4$ at 298 K [125].

Regarding the HOR in alkaline conditions, a pioneering study investigating non-PGM catalysts for HOR in AEMFCs, was carried out by Lu et al. [137]. In this work, Ni nanoparticles were decorated with Cr. This system generated an initial maximum power density of ca. 70 $mW \cdot cm^{-2}$, operating at 60 °C. Another recent report has described nickel-based anode catalysts in alkaline media to oxidize hydrazine [138].

In a recent contribution Ni was combined with Mo and Co in ternary alloys: NiMoCo, which were tested for HOR in alkaline conditions [119]. This latter showed an improved HOR activity versus Ni alone; however, the recorded high HOR activity could be sustained only for anodic potentials below of 0.1 V, apparently because of the excessive surface oxidation above this potential. The HOR on both CoNiMo and Pt increases with applied overpotential below ~0.1 V, Figure 20. While the HOR increases beyond 0.1 V on Pt, on CoNiMo this reaction decreases certainly due to the passivation of Ni on the catalyst surface as suggested [119,139].

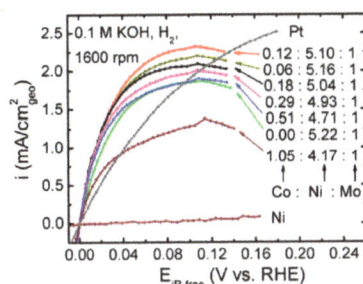

Figure 20. Linear scan voltammograms for the HOR on electroplated Ni, NiMo, CoNiMo. Atomic ratios between Co, Ni and Mo in the plating solution. Reprinted with permission from [119]. Copyright Royal Society of Chemistry, 2014.

Recent papers have proposed the use of various nanoscale electrocatalysts, such as Ni films partially covered by a palladium layer which provide one of the most active materials for the HOR [140]. These Pd/Ni electrodes were synthesized by an electrolytic deposition of palladium over a polycrystalline Ni surface varying the composition of the surface. The catalytic performance of the Pd/Ni electrodes towards the HOR was evaluated in alkaline medium, showing an improvement for this bimetallic catalyst. Limiting currents at 0.55 V in a H_2 saturated KOH solution, as well as the current exchange values are shown in Figure 21. Ni material did not present any catalytic activity in these conditions; this could be the result of a Ni surface covered with oxygenated species which block the surface and avoid the hydrogen adsorption. Nevertheless, a minimal palladium amount (0.5%) deposited onto Ni, activates the compound to perform the HOR. An increase of the coverage up to 17% in palladium shows two interesting phenomena; the kinetic current experiences a pseudo-linear increase from 1.5% to 17% palladium coverage. Contrary to the kinetic current increase, the onset potential presents a decrease as a function of the palladium loading. Both parameters did not present further improvement with a Pd deposition above 17%, Figure 21.

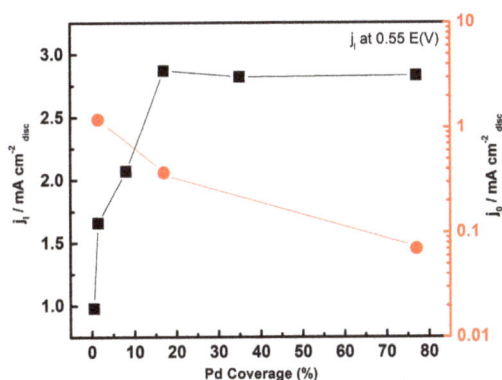

Figure 21. Limiting currents of Pd/Ni at 0.55 V as a function of Pd coverage percentage (Black squares) in H_2 saturated 0.1 M KOH, and comparison with the exchange current density from the Tafel plot (red dots). Data extracted from [140].

From this study the authors determined that the 17% Pd coverage corresponds to 7 atomic layers (approximately a thickness of 1.5 nm). It was also determined that Pd/C catalysts, under the same work conditions conveyed values below 0.1 mA/cm$^2_{Pd}$ at an overvoltage $\eta = 0.05$ V with a palladium loading of 5 µg/cm^2 [141], while Pd/Ni with 0.9 µg/cm^2 reaches 0.30 mA/cm$^2_{disc}$ and 0.57 mA/cm$^2_{Pd}$

at an overvoltage $\eta = 0.05$ V [120]. For Ni covered by a high amount of Pd, the exchange current density value reached a very similar one to that of Pd/C. This value < 0.1 mA/cm$^2_{Pd}$ is the usual value for palladium in alkaline medium, this was corroborated in a similar study where a gold surface was covered by palladium [142]. Coming back to the Pd/Ni study, the highest exchange current values were obtained with a low Pd coverage, even above 1.5% on Ni (Figure 21 (red dots)) [140].

To replace a portion of the expensive platinum metal without compromising the HOR electroactivity in alkaline medium, the synergistic ligand and strain effects were put into advantage by alloying Pt NWs (nanowires) to metals such as: Fe, Co, Ru, Cu, Au [143,144]. The synthesis of such materials was done by a solution-based technique [145]. Pt NWs were synthesized using a hexachloroplatinic acid hydrate solution combined with cetyltrimethylammonium bromide (CTAB) in chloroform. Subsequently, water was added to this mixture to be stirred for an additional 30 min at 1600 rpm. After 30 min reaction, a sodium borohydride solution was poured in, as the reducing agent, and the reduction process was confirmed by a color change from gray to black. The entire mixture was stirred for an additional 20 min at 1600 rpm, subsequently centrifuged and the supernatant was discarded. The binary NWs were analogously synthesized using commercial precursor. To generate a series of PtRu, PtFe, PtCu, PtCo, and PtAu ultrathin nanowires, ruthenium (III) chloride hydrate, iron(III) nitrate nonahydrate, cupric chloride, cobalt chloride and hydrogen tetrachloroaurate (III) hydrate were used, respectively. A cleaning protocol was used to electrochemically eliminate the CTAB surfactant and activate the electrode surfaces [146]. Figure 22a shows the performance of as-synthesized Pt NWs with that of binary alloy catalysts made of 70 atom % Pt and 30 atom % Fe, Co, and Ru, respectively. Pt$_7$Ru$_3$ NW catalyst was able to achieve a diffusion-limited current at lower potentials, indicative of an improved HOR kinetics. Pt alloyed with both Au and Cu (30 at. %) are depicted in Figure 22b. Specifically, the Pt$_7$Cu$_3$ NW catalyst attain the diffusion-limited current density at ~0.2 V, whereas the Pt$_7$Au$_3$ NW catalyst only attained this value at ~0.25 V. This observation would indicate relatively slower HOR kinetics for this particular Pt$_7$Au$_3$ composition, with respect to other binary catalysts. The experimental HOR exchange current densities obtained at 2500 rpm with respect to the calculated hydrogen binding energy (HBE) values are found in Figure 22c [147]. Summing up, the exchange current densities have been used to evaluate catalytic activity, close to the reversible potential, since this is the potential where HOR is kinetically limited [148].

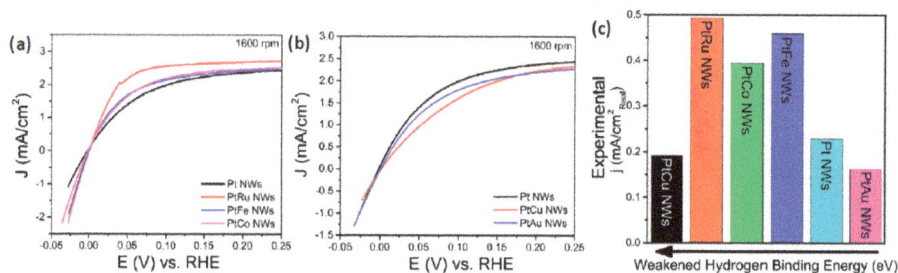

Figure 22. (a, b) HOR linear sweep voltammograms acquired in a hydrogen-saturated 0.1 M KOH solution at 1600 rpm (current normalized to the geometric surface area of the electrode). (c) HOR exchange current densities on the basis of calculated surface hydrogen binding energy (HBE) values [144]. Reprinted with permission from [144]. Copyright American Chemical Society, 2016.

There is no kinetic data in alkaline medium for rhodium electrodes. In this context, a study with nanostructured Rh electrodes with application in the HOR was carried out through the evaluation of the elementary kinetic parameters of the Tafel-Heyrovsky-Volmer (THV) mechanism [149]. By means of the sputtering of Rh on a glassy carbon substrate an electrode was prepared, this process was carried out in an inert atmosphere [150]. The analysis of the limiting current at an overpotential of $\eta = 0.3$ V as a function of the rotation rate, under the kinetic mechanism of THV, is shown in a Koutecky-Levich plot, Figure 23. At this overpotential, one can read a kinetic current density of 3.44 mA/cm².

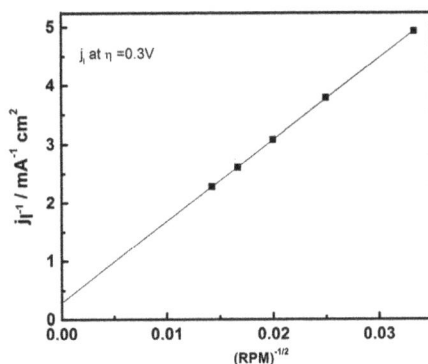

Figure 23. Koutecky-Levich analysis of the limiting current dependence with the rotation rate under the kinetic mechanism of Tafel-Heyrovsky-Volmer (THV) in 0.1 M NaOH, 298 K. Data extracted at η = 0.3 V from [149].

To establish the THV-mechanism, Equations (3), (6) and (7), certain constrains that imply the Koutecky-Levich expression were considered [151]. We can observe how in the potential range of interest, the HOR followed preferentially the Tafel-Volmer route [149]. The equilibrium reaction rate value obtained from the correlated experimental results was $j_0 = 1.18 \times 10^{-4}$ A·cm⁻², this value was significantly lower than that obtained for Rh in acid medium (6.77×10^{-4} A·cm⁻²) [150].

3.2. Oxygen Reduction Reaction—ORR, and Carbon Supports

In addition to the precious metals, required to improve the kinetics of the ORR, the oxygen reduction reaction is also affected by the ionic species contained in the acid or alkaline media, where the electrocatalysts are evaluated [152]. The kinetics, and the electroreduction mechanism for oxygen is carried out through an adsorption of the oxygen molecules on the catalysts surface. This process can be affected by blocking species composed by the anions of the electrolyte or oxygen containing molecules (OH-ads). The presence of these species affects the coordination and the adsorption energy of the molecular oxygen over the catalyst surface, thus influencing negatively the reaction ORR mechanism [153]. Previous studies demonstrated that this deficiency of the electrochemical exchange is due to the effect of the blocking species: OH⁻ and anions present in electrolyte solutions. Such chemical species induce even a higher blocking effect on Pt-based electrocatalysts [154].

Various studies [155,156] have demostrated that the decrease in the ORR activity in sulfuric acid solutions (H_2SO_4) is caused by a substantial adsorption of sulfate (SO_4^{2-}) or bisulfate (HSO_4^-) onto the Pt surface which cause an adsorption mixed potential between the anions $H_2SO_4^-/SO_4^{2-}$ and the molecular oxygen. Another kind of species which affect the oxygen adsorption onto the catalysts surface are the phosphate anions (PO_4^{3-}) present in the phosphoric acid solution (H_3PO_4). Thus, it was determined that prechloric acid ($HClO_4$) is an ideal electrolytic solution to perform the ORR in acid medium since the perchlorate anions (ClO_4^-) practically do not interfere in the oxygen adsorption process at the catalysts surface [157].

On the other hand, the ORR kinetics is carried out in a more efficient manner in alkaline medium than acid medium [158]. This allows the use of catalysts with a lower price than Pt, such as silver, palladium, copper, nickel, among others [159]. The main technical disadvantages presented by the ordinary alkaline fuel cells (AFC's) are the durability of the electrodes, in highly caustic medium, the adequate water supply, at the electrodes to avoid both the anodic flooding, and the properly cathode drying, and the progressive electrolyte carbonatation, which is caused by the CO_2 present in the atmosphere or produced from the oxidation of the fuel used [160]. The carbonates formation has a particular importance since these species precipitate over the electrodes, damaging and blocking the microporous structure of the catalysts and therefore the catalytic active sites reducing progressively the fuel cell performance [1]. To avoid the aforementioned issues, it was developed AFC's with flowing electrolytes streams, compared with the AFC's systems with stationary flow, these novel fuel cells present a better performance and lifetime [160–162].

Regarding catalysts supports, for Pt and non-precious metal based catalytic centers, carbon materials and oxides are most commonly used and investigated materials. Low and high graphitic carbon are, e.g., carbon black (CB), single-walled or multi-walled carbon nanotube (CNT), graphene oxide (GO) [55,163–182] and carbon nitride (CN) [183–186] among others. The semi-conducting oxides, such as TiO_2, doped-TiO_2 and doped-SnO_2, have been reported as support for catalytic centers [187–193]. The conductivity of oxides can be improved by synthesizing oxide/carbon composites [194]. The supports of this kind of composite materials are widely used for Pt or other catalytic centers [194–198]. The utilization of supporting materials can affect the nanoparticle (NPs) morphology, dispersion, and create a support-metal interaction (SMSI) [199–202]. These facts can favor the catalytic process, and hence, the support has a great impact on the design of composites based catalysts, aiming at improving the kinetics and stability for the electrocatalytic processes in low temperature fuel cells.

$$TiO_2 + 2H_2 + 3Pt \rightarrow TiPt_3 + 2H_2O \qquad (8)$$

The SMSI effect is directly related with the electrochemical activity of novel electrocatalysts. This interaction was firstly found in metal NPs supported by oxide [199,200]. Tauster et al. demonstrated the loss of H_2 and CO absorption capacity on TiO_2 supported group 8 metals (Ru, Rh, Pd, Os, Ir and Pt) [200]. This phenomenon revealed the formation of a chemical bond between the metal and TiO_2, cf. Equation (8) [200].

Later, similar effect was discovered in, such as La_2O_3 [203], and ZrO_2 [204] supported noble metals. In addition to oxides, noble metals supported onto carbon-based materials showed also SMSI, as revealed by the electron transfer at Pt-C interface [201]. For non-precious metals, it is well known that Co–N chemical bonds form from cobalt-oxides supported N-doped carbon materials [205–207]. The SMSI effect was also reported in Ni/Al_2O_3, [208], Ni/HZSM-5 (zeolite) [209] composites. Recently, Alonso-Vante et al. [195,197,198,210] reported SMSI in Pt and rare earth oxide interface, in these experimentations different from classical SMSI, the SMSI was identified as structural surface modification of Pt NPs. Therefore, from this point of view, the SMSI widely exist in metal-support composites, induced by the surface/structure modification of the catalytic surfaces. The catalytic performance can thus be improved via tailoring SMSI, via, the use of oxide, oxide-carbon composites, and highly graphitic carbon supports [163,198,211,212], as revealed through the use of various analytical tools of data collected from XRD (X-ray diffraction), XPS or/and electrochemical CO-stripping [163,198,211,212]. A way to probe the SMSI for Pt surface, and structure modification can be brought into light from XRD data, since the Pt diffraction peak broadening is affected by crystallite size, micro-strain and stacking fault [211], on the other hand, the Pt XPS spectra may reflect the charge transfer at the interface, induced by chemical bonding [163]. The SMSI on Pt NPs surface can be probed further via CO-stripping [213].

3.2.1. Low Graphitic Carbon

With respect to low graphitic carbon materials, commercially available carbon black from Vulcan or Ketjen is usually applied to support nanoparticles (NPs) [164,214–219]. They were reported as supports for Pt, alloyed Pt-M (M = 3d transition metals), Pd and alloyed Pd-M NPs. For instance, commercial Pt NPs supported onto carbon black, have been widely used as a benchmark for electrocatalytic process in half-cells or fuel cells [195,197,198,220,221]. Pt-M nanoalloys supported by carbon black were developed to investigate the dealloying effect of M atoms on Pt surface towards ORR activity in acid medium [215,219]. Recently, supported by theoretical DFT (density functional theory) calculation, Chorkendorff's group reported experimental facts showing that the Pt-RE (RE = rare earth elements) alloys were the most active and stable Pt-based electrocatalysts for ORR [12,222–224]. Therefore, the synthesis of highly and homogeneously dispersed Pt-RE nanoalloys is an interesting phenomenon in the domain of energy conversion. However, the synthesis is extremely difficult via a soft chemical route, since the reduction potential of metallic REs ca. < −2.0 V/RHE [197,198]. This is the reason why a layer of RE oxides usually cover the surface of the obtained Pt NPs, even after heat-treatment at 300–900 °C [195,197,198,210,225,226]; surprisingly, the ORR activity on RE oxides modified Pt surface was improved with respect to pure Pt surface. Therefore, it is very interesting to study the effect of RE oxides on Pt surface towards ORR, using carbon black as supports. As shown in Figure 24a–d, for Pt-M_2O_3 (M = Y, Gd and Sm) supported onto carbon black, the particle size and morphology were almost the same for both Pt and Pt-M_2O_3 NPs, whereas the ORR, Figure 24e, activity was enhanced on Pt-M_2O_3 NPs. Among oxide modified samples, the oxygen reduction activity on Pt-Y_2O_3/C overcame that on Pt-Gd_2O_3/C and Pt-Sm_2O_3/C samples.

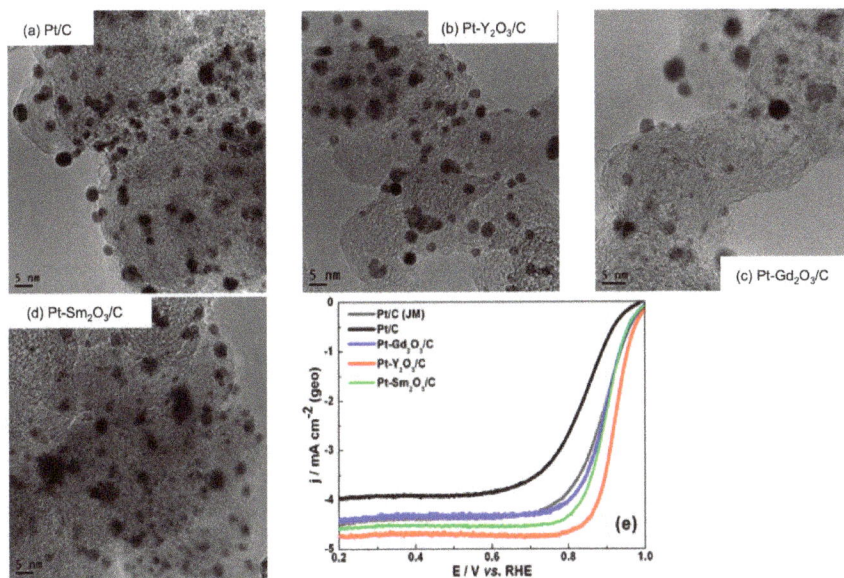

Figure 24. TEM images for (**a**) Pt/C and (**b**–**d**) Pt-M_2O_3/C (M = Y, Gd and Sm), respectively, and (**e**) the rotating disk electrode polarization curves (ORR) in oxygen saturated 0.1M $HClO_4$ at 900 rpm.

As indicated above, cf. Figure 6, different Pd NPs morphologies could be generated by adding different amounts of carbon support, showing the impact of carbon black support towards the nucleation of Pd NPs [57]. Moreover, the morphological impact of Pd NPs towards ORR and fuels (methanol, and formic acid) oxidation were identified [56,57]. In this regard, Co_3O_4 NPs, with different morphology deposited onto carbon black, showed different catalytic activity towards ORR [227]. The synthesis of Co_3O_4 NPs was done in two steps: (1) dissolving Co^{3+} precursor into a solution of water and DMF (dimethyformamide); and (2) adding ammonia into the solution to form $Co(OH)_3$ followed by a heat-treatment at 330 °C [227]. Adjusting the ratio of water and DFM to 1:0 (Co_3O_4-10), 3:1 (Co_3O_4-31), 1:1 (Co_3O_4-11) and 1:3 (Co_3O_4-13), a different morphology of Co_3O_4 NPs was, indeed, generated, Figure 25a–d [227]. This phenomenon indicates that nucleation of $Co(OH)_3$ onto carbon support was controlled by the ratio of aqueous and organic solvents during the synthesis [227]. The ORR Tafel plot, Figure 25e further shows the morphological impact of Co_3O_4 NPs towards catalytic activity [227].

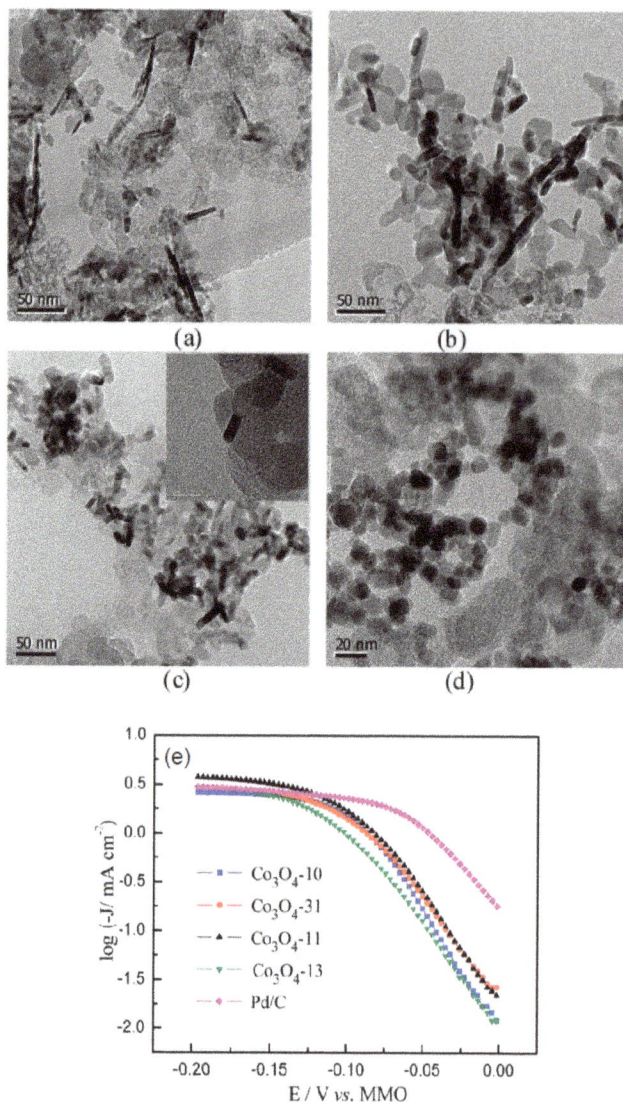

Figure 25. The TEM images for (**a**) Co_3O_4-10; (**b**) Co_3O_4-31; (**c**) Co_3O_4-11 and (**d**) Co_3O_4-13 supported by carbon black; (**e**) Tafel plot for ORR on Co_3O_4-10, Co_3O_4-31, Co_3O_4-11, Co_3O_4-13 and Pd/C (20 wt %). Reprinted with permission from [227]. Copyright the Royal Society of Chemistry, 2012.

Concerning the carbon nitride supports, Di Noto et al. [183] described the synthesis of a series of Pd-based catalysts with a general formula $K_n[Pd_xCo_yC_zN_lH_m]$ and a molar ratio $y/x > 1$ and $y/x < 1$. The most relevant results rely on the ORR activity presented by the Pd-based catalysts with $y/x > 1$ annealed at 700 °C and 900 °C which overpassed the activity presented by a commercial Pt/C electrocatalyst. With respect to HOR and methanol tolerance no-improvement against Pt/C was obtained. To explain this ORR catalytic improvement, the authors suggested that a high concentration of cobalt atoms provokes the Pd-rich domains coalescence improving the mass utilization efficiency of the active sites. Ongoing with the incorporation of active metals in a carbon nitride matrix, this group synthesized two Pd-based electrocatalysts [184]. The materials $PdNi-CN_h$ 900 and $PdCoAu-CN_h$ 600 were produced by a pyrolysis method. The samples were physical-chemically characterized and evaluated towards the ORR with and without the presence of methanol and HOR. The main physical-chemical results revealed the formation of a disordered graphitic carbon nitride domain in $PdNi-CN_h$ 900. On the other hand, $PdCoAu-CN_h$ 600 presented a poorer nitrogen atoms incorporation due to a less graphitic carbon nitride matrix. No enhancement towards in the ORR and the HOR against the commercial Pt reference was observed. However, both electrocatalysts showed a better methanol tolerance than commercial Pt. Di Noto el al. [186] also worked in the production of Pt-based-carbon nitride (CN) core-shell electrocatalysts, and their application as cathodes in a single fuel cell system. PtNi-CN and PtFe-CN were synthesized via pyrolysis process and decorating a hybrid organic-inorganic material. In general, both PtNi-CN and PtFe-CN improved the fuel cell performance in comparison with the commercial Pt assembly. PtNi-CN presented a higher selectivity and a four-electron pathway to perform the ORR than the PtFe-CN. This enhanced activity was the result of a higher density of neighboring platinum active sites in the PtNi-CN surface [186].

Continuing in the same context, carbon nanohorns (CNH) is another interesting low graphitic carbon material, showing horn shape carbon particles self-assembling to a dahlia-like morphology [228–230]. Our group reported recently N-doped (NCNH) supported $CoSe_2$, Figure 26a, for ORR in alkaline medium [229]. The ORR activity on $CoSe_2/NCNH$ was improved with respect to $CoSe_2/CNH$, $CoSe_2/C$ and NCNH [229]. The XPS results revealed a charge transfer from N atom to $CoSe_2$ nanocrystals in the $CoSe_2/NCNH$ sample [229], thus indicating a support catalyst center interaction in $CoSe_2/NCNH$, responsible for the ORR kinetics enhancement [229].

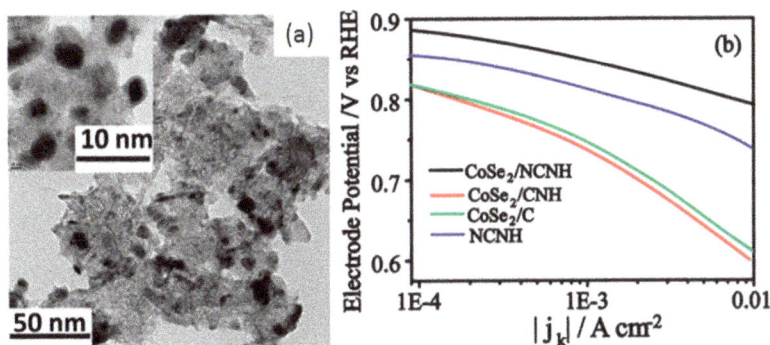

Figure 26. (a) TEM image for $CoSe_2/NCNH$ sample; (b) Tafel plot for ORR on $CoSe_2/NCNH$, $CoSe_2/CNH$, $CoSe_2/C$ and NCNH. Reprinted with permission from [229]. Copyright John Wiley & Sons, 2015.

3.2.2. High Graphitic Carbon

High graphitic carbon materials, such as carbon nanotubes (CNT) and graphene oxides (GO) are widely used as supports for both precious and non-precious catalytic centers [104,171,177,231–241] The high graphitic carbons are less corrosive than low graphitic carbon as shown in Figure 27a [164]. Taking CNT as an example, the Raman spectra show that the oxidized multi-walled CNT (ox.MWCNT) has dominant graphitic sp^2 domains, as compared to carbon (Vulcan XC-72) which shows highly disordered sp^3 domains, Figure 27b. The surface ratio $(I_D)/(I_G)$ domain was 2.1 and 1.2 for carbon Vulcan and ox.MWCNT, respectively.

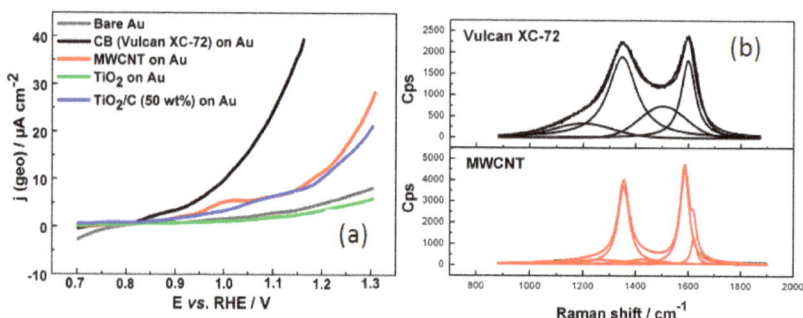

Figure 27. (**a**) The oxidation curves for carbon- and oxide-based supports in 0.1 M HClO₄. Reprinted with permission from [164], Copyright Elsevier, 2015. (**b**) The Raman spectra for carbon Vulcan XC-72 and oxidized multi-walled CNT (MWCNT).

When a catalytic center, e.g., Pt, is chemically deposited onto highly graphitic carbon the interaction takes place with both sp³ and sp² domains, Figure 28a,b, shows the close Pt particle size and morphology [201]. However, the surface chemistry is totally different as depicted in Figure 28c. There is a sole CO-oxidation peak on the Pt/Vulcan sample, whereas there are two peaks on the Pt/MWCNT-m sample [201]. Based on Raman, XPS and DFT calculation results one can conclude that the peak I should be related to monoxide carbon oxidation on the platinum surface interacted with the sp² domain of the support, and peak II attributed to oxidation on Pt site interacted with the sp³ domain [201]. The ORR stability, Figure 28d,e, in acid medium was improved on CNTs support with respect to carbon, and attributed to the interaction between Pt NPs and sp² domains of CNT.

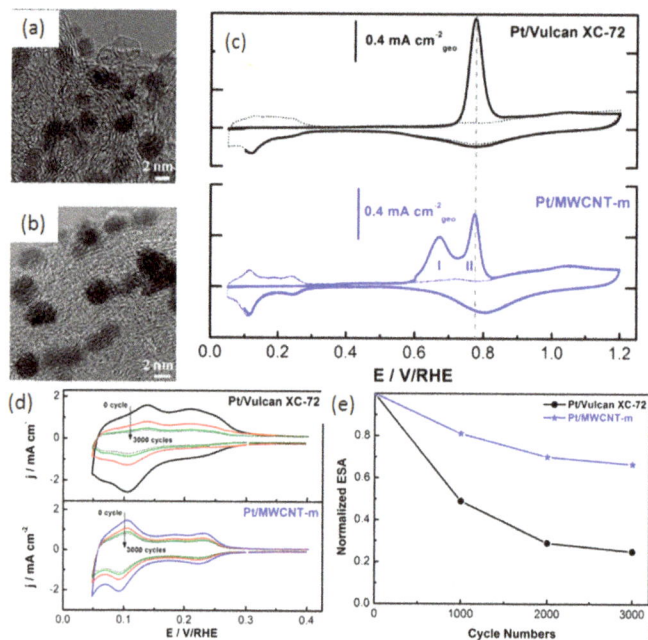

Figure 28. (**a**) Vulcan carbon XC-72 and (**b**) homemade MWCNTs (marked as MWCNT-m); (**c**) CO-stripping curves for Pt/C and Pt/MWCNT-m; (**d**) Cyclic voltammograms on Pt/Vulcan and Pt/MWCNT-m and (**e**) evolution of Pt active surface during cycling. Reprinted with permission from [201]. Copyright American Chemical Society, 2013.

Graphene, a highly graphitic carbon material, has been used for catalytic centers support. An increased ORR activity of Pt NPs deposited onto N-doped reduced graphene oxide (NRGO) was reported [242]. In this work, Pt NPs supported by GO, graphite, carbon black (Vulcan XC-72), reduced graphene oxide (RGO) and NRGO were synthesized and studied for ORR [242]. According to XPS results, Figure 29a,b, the electron transfer from Pt to N moieties was identified, and responsible for modifying the electronic state of Pt atoms, supported by DFT calculation [242], and the increased ORR activity, see Figure 29d. The finger print, onto the various supports, of Pt NPs are contrasted in Figure 29c.

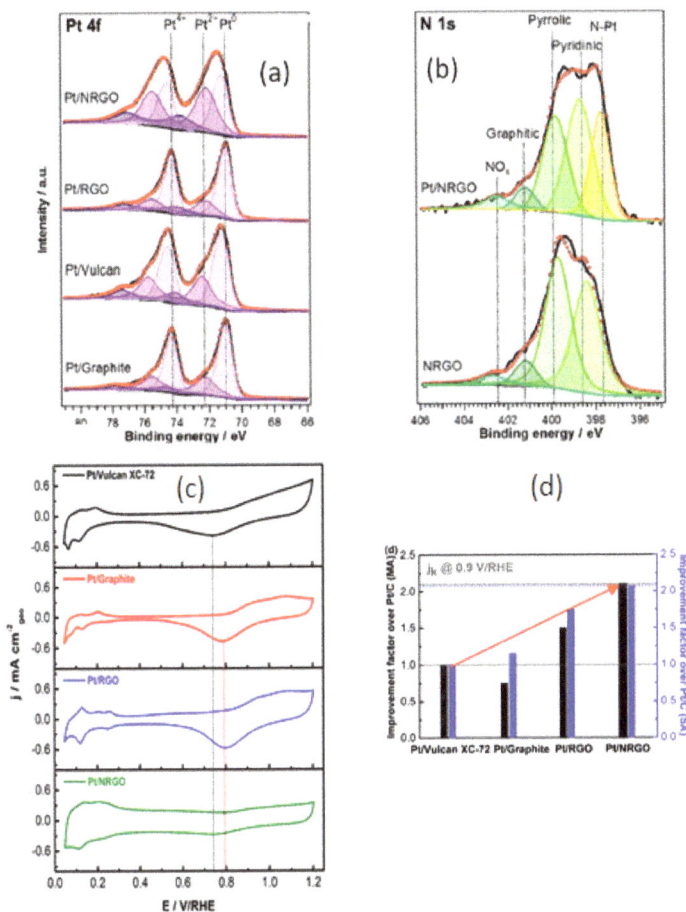

Figure 29. XPS spectra of (**a**) Pt 4f and (**b**) N 1s; (**c**) cyclic voltammograms and (**d**) normalized ORR specific and mass activity for Pt/Vulcan, Pt/graphite, Pt/RGO and Pt/NRGO. Reprinted with permission from [242]. Copyright the Royal Society of Chemistry, 2015.

The support also interacts with non-precious metal centers. This phenomenon is clearly shown for $CoSe_2$ in Figure 30a,b, where the ORR activity in alkaline medium was improved when the chalcogenide was supported on functionalized CNT (f-CNT) as compared to that on carbon black [243]. In acid medium, the f-CNT supported $CoSe_2$ was the most active catalyst for ORR, as compared to carbon black, TiO_2-C composite and NCNH supported $CoSe_2$ [243]. In alkaline medium, however, $CoSe_2$/NCNH showed the highest ORR activity, followed by $CoSe_2$/f-CNT [243]. One should note that the NCNH support was very active towards ORR in alkaline medium [229]. Therefore, it is not surprising that the oxygen reduction activity was further increased on $CoSe_2/NCNH$.

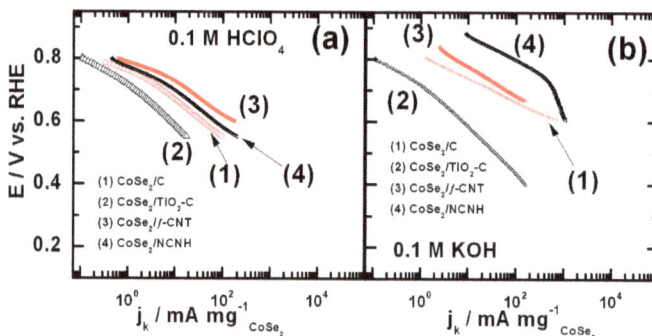

Figure 30. Tafel plot for ORR on $CoSe_2/C$, $CoSe_2/TiO_2$-C, $CoSe_2/f$-CNT and $CoSe_2/NCNH$ in (**a**) acid and (**b**) alkaline medium.

An important contribution in the metal-support interaction domain was reported by Lee et al. [244]. They developed three novel Pd-Pt electrocatalysts supported onto three different carbon supports, namely, defective graphene nanosheets (GNSs), herringbone graphite nanofibers (GNFs), and single-walled carbon nanotubes (SWCNTs). The synthesis of the electrocatalyst Pd_3Pt_1 supported onto such different supports was carried out by chemical reduction of Pd and Pt precursor deposited on the various carbon supports. Among other characterization techniques, Raman spectrum was one of the most interesting tools to determine that GNS presented the highest disordered carbon structure with a D/G ratio of 1.23 over the values of 1.12 and 0.61 for Pd_3Pt/GNF and $Pd_3Pt/SWCNT$, respectively. Pd_3Pt/GNS presented the highest activity toward ORR in acid medium producing the highest percentage of water (70.35%). Because the use of the same active metal supported onto various carbon supports, the authors attribute the enhanced performance of Pd_3Pt/GNs to a metal-support interaction. Two years later this group reported the synthesis of a Pd-based electrocatalyst supported onto reduced graphene nanosheets (RGN) which were produced by an electrostatic method (RGNSECM) but also via a chemical reduction method (RGNCM) [245]. They deposited cubic Pd nanoparticles (10 nm) onto these two RGN substrates by an electrostatic deposition process. XPS analysis reveals one of the main characteristics to understand the metal-support interaction effect; the carbon contents in RGNSECM and RGNCM were 95.9% and 77.2%, respectively. Additionally, the oxygen content for these supports were 4.1% for RGNSECM and 22.8% for RGNCM. This analysis confirms the higher number of physical defects on RGNSECM resulting in more oxygen-containing functional groups which improve the supports vacancies to perform a better oxygen reduction. The electrochemical evaluations carried out in alkaline medium pointed out a higher mass activity for Pd cubes/RGNSECM but also a best catalytic response in the ORR overcoming the electrocatalytic response of RGNCM and a commercial Pd/C electrode. Furthermore, RGNSECM produced the less amount of hydrogen peroxide showing an oxygen reduction tendency of four-electrons but also a higher stability. These excellent properties of RGNSECM were related to the substrate defects, which induce a strong binding trap for Pd, a higher charge transfer, and a downshift in the d-band of the Pd clusters.

3.2.3. Oxide and Oxide-Carbon Composites

Oxides, such as semiconducting TiO_2 have been used as support materials for Pt-based catalytic centers, since much more corrosion-resistant than carbon materials, cf. Figure 27a. To favor the charge transfer between the oxide support and catalytic center, the conductivity of Titania must be improved via doping. Recently, Mo- and Ru-doped TiO_2 ($Ti_{0.7}M_{0.3}O_2$, M = Mo and Ru) were reported as support for Pt NPs [193,246]. In both cases, the enhanced electron transfer between oxide and Pt was identified, leading to different Pt surface chemistry, catalytic activity and stability [193,246]. As depicted in Figure 31a, the X-ray absorption spectra show that the white-line intensity in X-ray Absorption Near Edge Structure (XANES) zone of $Pt/Ti_{0.7}M_{0.3}O_2$ was lower than Pt/C and PtCo/C [193]. In Figure 31b, the $Pt/Ti_{0.7}M_{0.3}O_2$ sample showed the lowest number of unfilled d-states, followed by PtCo/C, Pt/C and Pt foil, indicating that the electron transfer is carried out from the support to platinum, known as SMSI effect, induces the d-band vacancy of platinum. The effect is more significant on $Ti_{0.7}M_{0.3}O_2$ than on Pt. The ORR kinetics, in acid medium, shown in Figure 31c, is enhanced on oxide supported Pt with respect to Pt/C and PtCo/C. These results show that increased SMSI favors catalytic activity on Pt surface towards ORR.

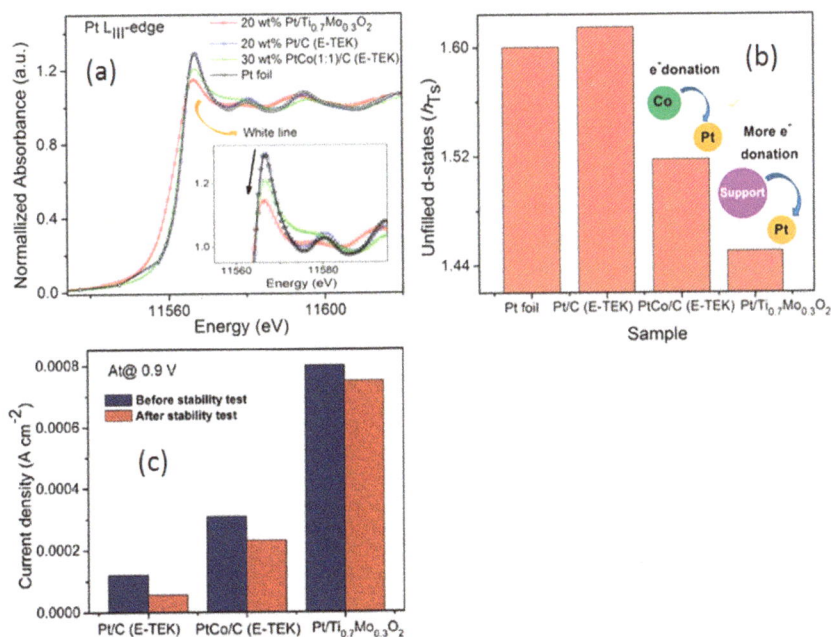

Figure 31. (a) X-ray Absorption Near Edge Structure (XANES) spectra of Pt L_{III}-edge; (b) values of unfilled d-state and (c) ORR kinetic current before/after stability test for Pt/C, PtCo/C and $Pt/Ti_{0.7}M_{0.3}O_2$. Reprinted with permission from [193]. Copyright American Chemical Society, 2011.

In addition, oxide-carbon composites are good candidates as catalytic center supports [194,196–198,210,247]. Our group reported Pt NPs supported by TiO_2-C, and Y doped TiO_2-C [194]. It was found that TiO_2 became more conducting after doping with Yttrium (Y:TiO_2-C). This material was as conducting as the carbon black, ensuring, in this way, the efficiency of electron transfer between the catalyst center and the support. The SMSI effect between Pt NPs and support was improved on photo-deposited Pt NPs onto oxides with respect to Pt/TiO_2-C prepared via a soft chemical route [194]. The SMSI effect can be easily visualized in the CO-stripping process, see Figure 32a, where one can observe that the photo-deposition (PD) favors the interaction between Pt and sp^2 domain of carbon support, whereas the carbonyl chemical route (CCR) favors the Pt-sp^3 interaction. Such an observation demonstrated a random dispersion of Pt NPs via CCR, while PD is more selective to deposit Pt NPs onto sp^2 domains of carbon. On TiO_2-C and Y:TiO_2-C, cf. Figure 32a, the Pt NPs interact with the sp^2 domain of carbon, and oxide. Concerning the ORR activity, in acid medium, there was a slight enhancement on Pt/Y:TiO_2-C, with respect to platinum samples without

Y, Figure 32b. The oxygen reduction stability on Pt/Y:TiO$_2$-C, studied via an accelerated stability test (AST), from 0.6 to 1.0 V/RHE, is close to that of Y-free samples. However, when cycling from 0.6 to 1.2 V/RHE, the stability of Pt NPs was clearly put in evidence on the oxide, Figure 32c–f. Such an observation indicated that the SMSI of Pt-Y:TiO$_2$ stabilized the Pt NPs on the support.

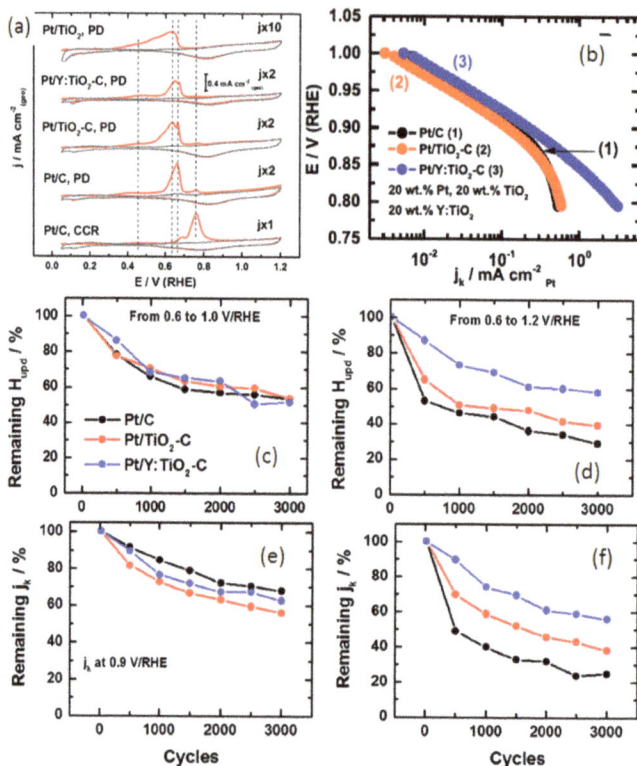

Figure 32. (a) CO-stripping for Pt/C (CCR), Pt/C (PD), Pt/TiO$_2$ (PD), Pt/TiO$_2$-C (PD) and Pt/Y:TiO$_2$-C (PD). (b) Tafel plot and (c–f) stability results towards ORR for photo-deposited Pt/C, Pt/TiO$_2$-C and Pt/Y:TiO$_2$-C. Reprinted with permission from [194]. Copyright Elsevier, 2015.

4. Micro-Fuel Cell Platform

Micro-Laminar Flow Fuel Cells

In the power generation field, as novel energy sources for residential and industrial applications, fuel cells have become essential devices. The fast advancement in the technology demands multi-functional, smaller and lighter electronic devices. This has stimulated manufacturers to look for alternative solutions [248]. Portable electronics, mobile phones, and computing gadgets are among the largest and fastest growing devices nowadays [249]. The statistics reveal that worldwide the number of mobile phone users increased from about 500 million in the year 2000 to about 5000 million in the year 2011 [250]. Recent data show a quasi-exponential growth of these values. The International Telecommunication Union's (ITU) latest report estimates that more than 7 billion mobile phone subscriptions were globally conducted by the end of 2014, with a population penetration rate of 97% [250]. Portable electronics currently rely on batteries as a power source, but the need for constant replacement and recharging, interrupted operation, disposal issues, and increasing power requirements has prompted the development of new power sources [251]. Thereby, micro fuel cells have become promising devices to satisfy the global power requirements [252] thanks to their easy operation, competitive efficiency and environmental friendly performance. Even nowadays, the fabrication of such devices confronts many fabrication challenges, namely, design fabrication, the gas diffusion layer (GDL), electrocatalysts improvement and the exchange membrane assembly (MEA) [253]. The development of such devices is documented in the reference [254]. A recent modification of micro fuel cell systems are the silicon-glass-bonding systems by Mex et al. [255]. Silicon-glass-bonding systems retain the Nafion® membranes disadvantage, methanol crossover, therefore, extra barrier layers have to be employed. A new thin MEA consisted of sputtered catalysts deposited onto porous silicon supports, ceramics or foils. One of the main characteristic of those films is the low permeation rate and chemical and thermal stability.

On the other hand, researchers from the Mechanical Technology Incorporated (MTI) MicroFuel Cells institute carried out the development of a simple direct methanol fuel cell design that can be scaled to applications for a wide range of portable devices. This novel device design does not require the use of pumps and water collectors; these modifications reduce the manufacture cost but also allows the use of concentrated fuels [256]. On the same line, people from the Lawrence Livermore National Laboratory developed the prototype of a miniature fuel cell [257]. This device incorporated a thin-film and a microfluidic fuel processor into a common package. Methanol from replaceable cartridges was used as fuel, the power generation device provided three times longer operating time than rechargeable batteries. A design configuration in a planar array [258,259] a "flip-flop" arrangement which allows a fully continuous electrolyte [259] is shown in Figure 33. The series combination performance of the two configurations is shown in Figure 34.

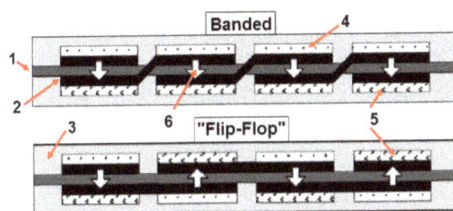

Figure 33. Banded and "Flip-flop" planar series interconnections. 1.—Membrane, 2.—Electrode, 3.—Flow Structure, 4.—Fuel, 5.—Oxidant, and 6.—Ion flux. The flip-flop configuration presents interconnects that avoid the membrane plane crossing, extracted from [258].

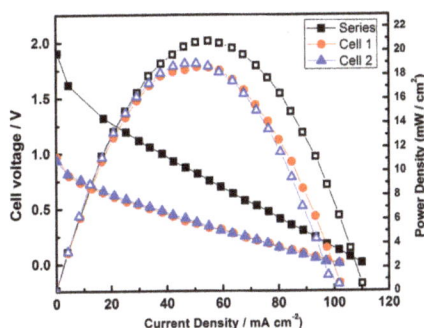

Figure 34. Cell voltage, and power density curves of two-cell assembly. Data extracted from [258].

The fabrication and evaluation of silicon wafers miniature fuel cells was reported by Yu et al. [260]. These devices consisted of a MEA between two silicon substrates. The design included feed holes and fuel distributors (channels in the silicon wafers). Pt/C electrocatalysts were used as electrodes (Pt loading = 4 mg/cm^2). To form the MEA, two electrodes with an area of 5 cm^2 were hot-pressed to a Nafion 112 piece. The results showed that the fuel cell performance was increased as a function of the composite thickness (silicon wafer layers). The best performance was presented by the miniature fuel cell with smaller size channels, these results are presented in Figure 35 [260].

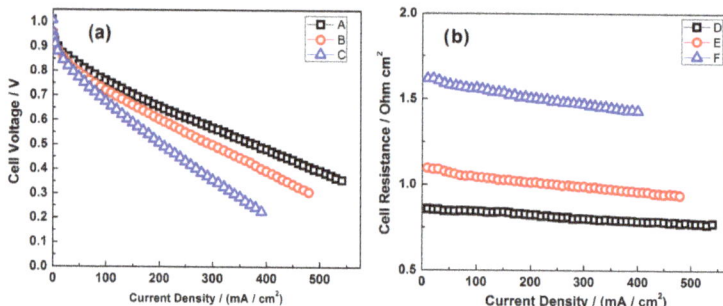

Figure 35. (**a**) Polarization of cells (A = cell 3, B = cell 2, C = cell 1); and (**b**) resistance of cells (D = cell 3, E = cell 2, F = cell 1). The cell temperature was 25 °C, dry H_2/O_2, the gas pressure was 0.10/0.10 MPa and the gas flow rate of H_2/O_2 at 50/50 mL/min. Data extracted from [260].

Until 2004, researchers were focused on tailoring, and miniaturizing, polymer exchange membrane fuel cells—PEM related issues [261]. It was not until 2005 that Choban et al. [262] worked on the first membraneless micro FC. This novel design eliminates the fuel crossover and cathode flooding issues presented in fuel cells. In this research, the membraneless micro FC performance was studied in acid, alkaline and also mixed-reactant media. In this last condition the cathode is in acid medium, while the anode is in alkaline medium or vice versa. In these flow conditions, the aqueous streams are merged into a microchannel and they continue flowing in a laminar flow through the channels configuration (Figure 36a).

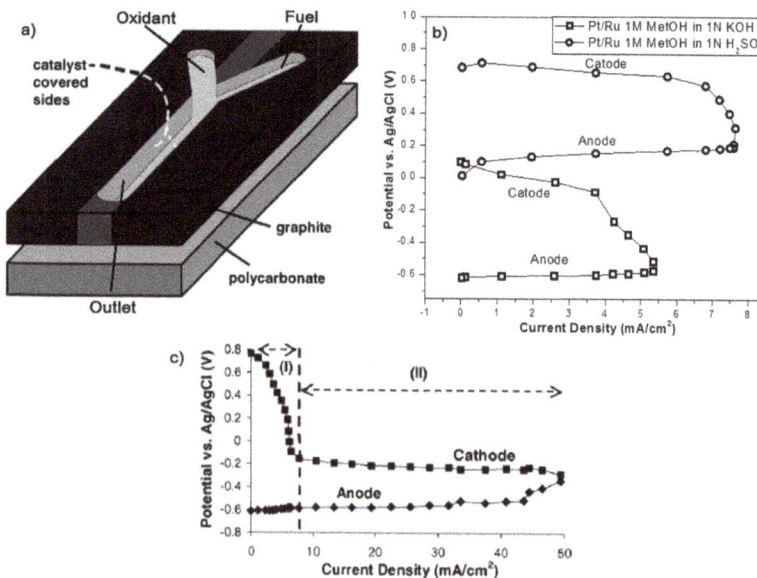

Figure 36. (**a**) Laminar flow fuel cell scheme. (**b**) Current-potential curves of individual electrode performances with an oxidant stream of oxygen-saturated. (**c**) Individual electrode performances (mixed-media configuration). Fuel: 1.0 M methanol in 1 N KOH; Oxidant: O_2-saturated 1 N H_2SO_4. Data extracted from [262].

In this experiment, unsupported Pt/Ru nanoparticles and Pt black were used as anode and cathode, respectively [262]. The individual electrodes polarization curves in different media are shown in Figure 36b, the results presented in this work are in agreement with the literature [263,264]. After several days, without operational interruption, any performance drop was appreciated. Also, any carbonate formation issue was found in this LFFC. In the mixed-reactant mode (anode-alkaline, cathode-acid), Figure 36c region II, we can observe how at constant cathodic potential the current potential value increased until the anodic value became the limiting factor.

The limiting behavior of mass-transfer is observed for the oxidation of methanol in these experimental conditions [265]. A higher overall cell potential, OCP and current density was obtained for the mixed-reactant configuration than those registered for all acidic and alkaline conditions.

The fuel cell operation in alkaline medium presents positive advantages in the kinetic of the anodic and cathodic reactions. However, the called "mixed-reactant" conditions allows the opportunity to increase the maximum OCP achievable in simple alkaline reactant conditions. Additionally, the design simplicity for this condition allows experiments to be carried out consecutively in a single fuel cell without modifying the system and changing only the pH of both electrolytic streams [262]. It is well known that the flow channel configuration [266] can induce the forced water removal inside the fuel cell [267]. According to this, Hsieh et al. [268] studied the flow fields configuration and effects under fixed operating conditions. Four flow fields were studied herein; the so-called interdigitated flow channel presented the most uniform transient current in comparison with the rest of channels designs: Parallel, mesh and serpentine. Finally, in this study it was developed a correlation of the water accumulated in the flow channels according to the time [268].

The development of a miniature PEM fuel cell stack with carbon bipolar plates was described in a previous contribution [269]. In Figure 37 is shown a representation of the PEM fuel cell stack. The stack is assembled in the following order: end plate, rubber gasket, bipolar end plate, rubber gasket, MEA, rubber gasket, bipolar plate, rubber gasket and end plate as depicted in the final fuel cell stack shown also in Figure 37. The way to keep together all those components is using screws which align the cell components and also allow the compression pressure control.

Figure 37. Three-cell fuel cell stack configuration. Scheme extracted from [270].

In this way, the stack was tested at different temperatures, Figure 38. The current-potential curves showed how the voltage increased as a function of the temperature increment. The optimal temperature conditions for this design was 60 to 80 °C and employing external hydration, because the dehydration of the membrane a cell drop performance occurred above 80 °C [270].

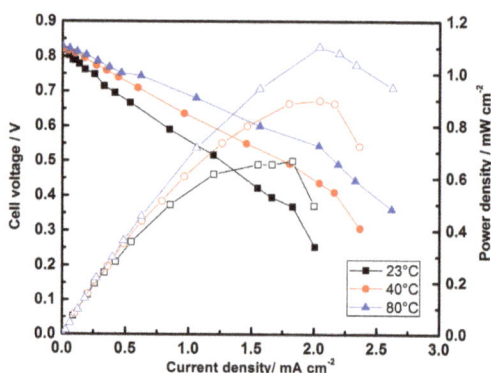

Figure 38. Temperature effect on the fuel cell polarization and the power density curves. Data extracted from [270].

Regarding the methanol crossover issue, and the methanol tolerant electrocatalysts development for the ORR, Whipple et al. [271], worked on the miniaturization of fuel cells tailoring the design and simplifying the fuel cell operation, Figure 39.

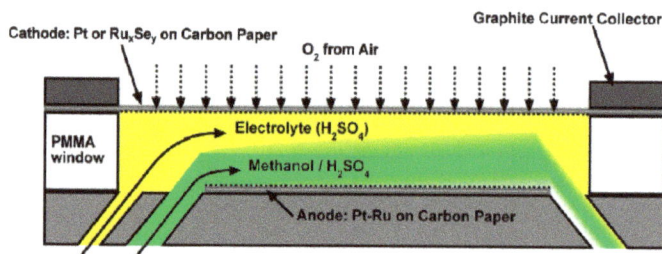

Figure 39. Scheme of the air-breathing laminar flow fuel cell (LFFC) [271]. Reprinted with permission from [271], Copyright Elsevier, 2009.

For this experiment two different cathodic streams without methanol and containing the same amount of methanol at anode were used in the operation of the laminar flow fuel cell. In the same way, an acid solution (0.5 M sulfuric acid) containing 0.1, 1, 2, 5, 7, 10, or 15 M CH_3OH was used as anolite. Figure 40 shows the individual cathodic and anodic polarization curve. These curves show the high depolarization of the Pt cathode in the presence of a stream containing methanol, Figure 40a2, this effect does not affect the same Pt electrode in the free-methanol stream, Figure 40a1. As we can observe, this depolarization effect does not occur when a Ru_xSe_y electrode is used as cathode; on the contrary, we can observe a slightly enhancement. This result proves that Ru_xSe_y electrocatalyst presents high tolerance to CH_3OH and does not present a depolarization effect caused by a mixed potential as occurs with Pt [82,272–274].

The Ru_xSe_y performance increase in presence of methanol can be explained as a result of a change in the oxygen diffusivity and solubility (Figure 40b2). Itoe et al. [275] showed that these two last parameters change in function of the CH_3OH–H_2SO_4 concentration ratio. This research group found that upon addition of CH_3OH to 0.5 M H_2SO_4, the initial oxygen diffusivity jumps by a factor of 2.5 to 3; after this, this value decreases until the initial value.

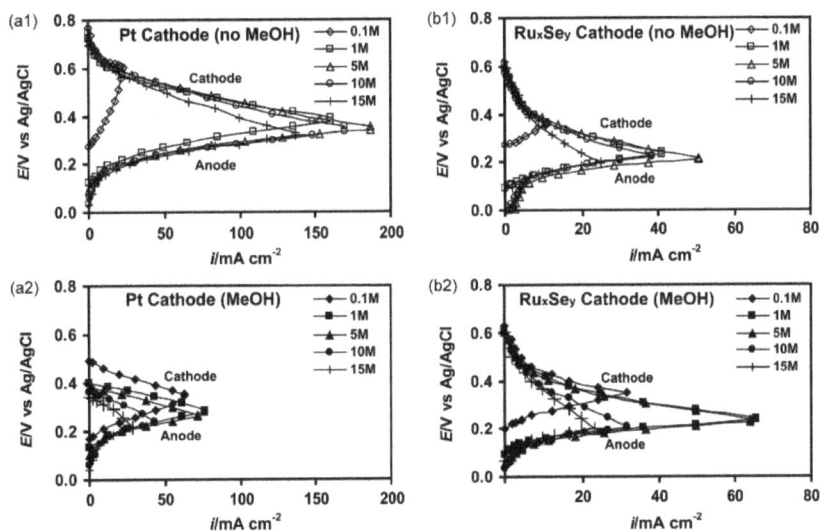

Figure 40. Current-potential curves of individual electrode performances for (a) Pt and (b) Ru_xSe_y cathodes working in a stream comprised of (1) 0.5 M H_2SO_4 and (2) 0.5 M H_2SO_4 + the methanol concentration present in the anode [271]. Reprinted with permission from [271], Copyright Elsevier, 2009.

This study demonstrated that the activity of the ruthenium chalcogenide (Ru_xSe_y) cathode increases in presence of CH_3OH concentration by 30 to 60% in the concentration range of 1 to 7 M [271].

The flow fields design of micro PEMFCs was reported by Lu et al. [253]. In this study, the authors evaluated the effect of different flow field topologies in the μ-PEMFCs performance, Figure 41. The electrochemical tests were performed at reactant flow rates of 15, 30 and 50 mL·min^{-1}. As a result of this experiment it was found that the μ-PEMFCs with different flow patterns present a similar tendency at flow rates of 50 mL·min^{-1}. On the contrary, the μ-PEMFC performance is rapidly deteriorated at low flow rates, this decrease of performance is caused by the micro channels flooding. The design with mixed and long multichannel yielded the best performance [253].

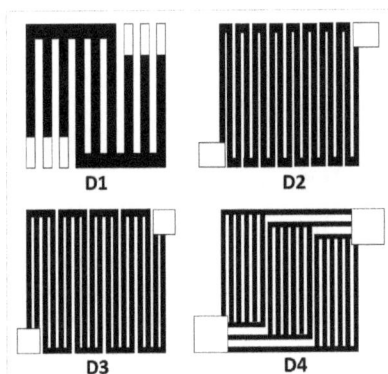

Figure 41. Flow design variations (D1–D4) for the μ-PEMFCs (micro proton exchange membrane fuel cell) reported in [253].

In Figure 42 we can observe the maximum power density value reached by each flow field (D1-D4) at a current density value of 175 mA·cm^{-2}. At a flow rate of 15 mL·min^{-1}, the power density value is in the range from 76 mW·cm^{-2} to 105 mW·cm^{-2} (38% discrepancy), 21% for the flow rate of 30 mL·min^{-1}, and finally 13% for the flow rates of 50 mL·min^{-1}.

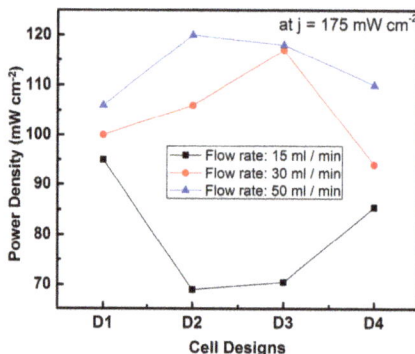

Figure 42. Power density curves of the different μ-PEMFCs designs with different flow rates at a density current value of 175 mW·cm^{-2}. Data extracted from [253].

Since we know that cluster-like compounds such as the ruthenium chalcogenide Ru$_x$Se$_y$ presents a high methanol tolerance, and selectivity to perform the ORR in presence of methanol [82,271–274], Gago et al. reported on the electrochemical measurements of 20 wt % Ru$_x$Se$_y$/C as cathode for the oxygen reduction in the presence of formic acid in a microfluidic formic acid fuel cell (μFAFC) [276]. A commercial Pt/C electrode was used also for comparison purposes. It was found that at formic acid concentration of 0.5 M, Ru$_x$Se$_y$/C and Pt/C reached maximum current density values of 11.44 mA·cm^{-2} and 4.44 mA·cm^{-2}, respectively. Meanwhile the power density values were similar for both materials, ca. 1.9 mW·cm^{-2}. On the other hand, at a concentration of 5 M formic acid the power density value for Ru$_x$Se$_y$ outperformed 9 times the power density value obtained by Pt/C, Figure 43, attesting that the Ru$_x$Se$_y$ electrocatalyst is more suitable than Pt/C to be used as cathode in a μFAFC [276].

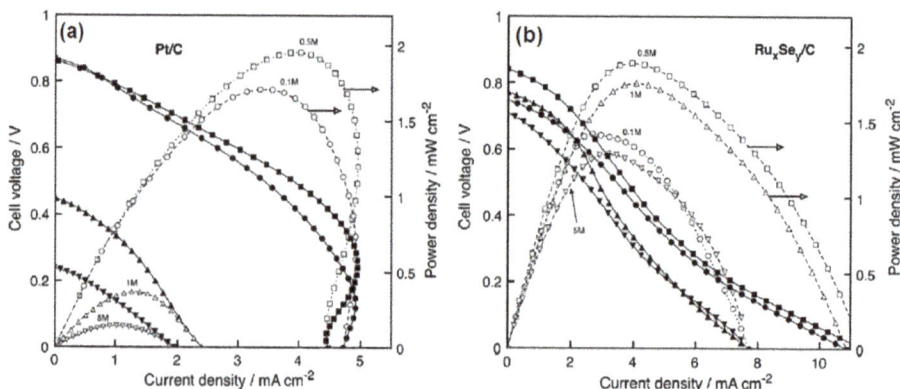

Figure 43. Polarization and power density curves for (**a**) 30 wt % Pt/C E-TEK and (**b**) Ru$_x$Se$_y$/C; in the μFAFC. 0.1 (circles), 0.5 (squares), 1 (triangles) and 5 M (inverted triangle), T = 25 °C, Flow rate = 1.2 mL·min^{-1}. 20 wt % Pd/C E-TEK was used as anode catalyst [276]. Reprinted with permission from [276]. Copyright Elsevier, 2011.

In 2012, Brushett et al. [160] performed a very interesting experiment on the alkaline microfluidic H_2/O_2 fuel cells field. They used a micro fuel cell to evaluate the electrochemical activity of Pt/C-based electrodes in alkaline medium. Their studies demonstrated that the behavior of such alkaline fuel cells (AFCs) is restricted by carbonated species presented in the electrolyte. These species affected only the performance of the anode, whereas the cathode performance remains practically unchanged. The carbonate formation (CO_3^{2-}/HCO_3^-) is one of the major technical issues in AFCs, this species formation reduces the concentration of OH^- in the electrolyte, hence it reduces the solution conductivity and the kinetic of the electrode reaction. One strategy to mitigate the water management and carbonate issues in alkaline FCs is with a recirculating electrolyte, the flow of the electrolyte helps the water management and removes carbonates resulting in a carbon dioxide tolerance up to ~100 ppm [160]. In this work, experiments to study the tolerance to CO_2 poisoning were performed as described as follows; in the first condition, CO_2 was injected in the anodic side of the µLFFC in stationary electrolyte conditions (3 M KOH). In order to determine a baseline, the first evaluation of the cell performance was carried out without the presence of CO_2. After this, a flow of 15 sccm of CO_2 was injected to the anode for 10 min meanwhile a nitrogen flow of 15 sccm was flowed in the cathode. After the gas exposure, the device was electrochemically evaluated to determine any change in performance. Afterwards, a second CO_2 exposure (10 min) followed by the electrochemical evaluation were performed. Subsequently, a rinse of the chamber was done by flowing KOH solution (10 min). Finally, the fuel cell was tested in a stationary mode in order to determine the KOH rinse effect. To perform a second study, the research group followed the same conditions of the first tests but introducing the CO_2 at the anode with a flowing of 3M KOH solution at a continuous flowing rate (0.3 mL/min) [277].

Figure 44a shows the electrochemical evaluation of the fuel cell operating with a stationary electrolyte. As a result of the CO_2 exposure the fuel cell suffers a power density decrease after the first exposure stage, and this power density reduction drops even further after the second CO_2 exposure. Nevertheless, the maximum power density value of 103 mW·cm^{-2} was regained after the KOH rinse, Figure 44a. The reduction of current density registered was attributed to precipitants which can be eliminated by the effective KOH rinse. On the other hand, in Figure 44b we can observe the polarization and power density curves for the flowing electrolyte conditions. From these curves we can observe how CO_2 does not affect the fuel cell performance because any carbonate species formed inside the fuel cell is immediately flushed. This study confirms that the fuel cell electrodes in flowing electrolyte conditions are not affected by the CO_2 exposure [277].

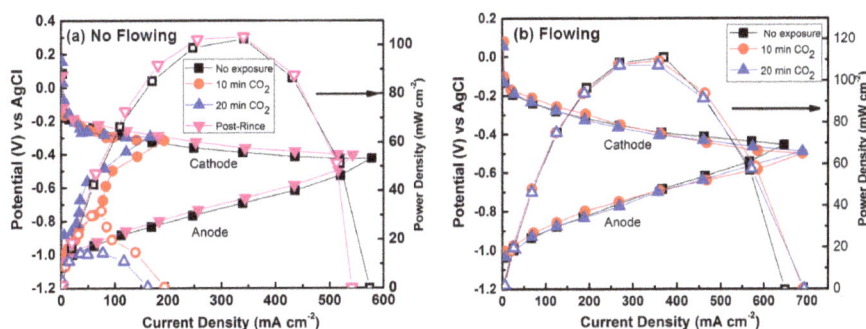

Figure 44. Anodic-cathodic polarization curves and power density curves for the alkaline microfluidic H_2/O_2 fuel cell exposed to CO_2 (**a**) stationary electrolyte, and (**b**) flowing a KOH solution. In both studies the H_2/O_2 flow rate was 50 sccm, the electrodes loading was 2 mg·Pt/C/cm^2, and experiments were performed at room temperature. Data extracted from [277].

Coming back to the transition metal chalcogenides and the membraneless fuel cells, two novel catalysts; carbon-supported Pt_xS_y and $CoSe_2$ were tested in a novel µLFFC [278]. A schematic image of the fabrication and assembly of the micro fuel cell components is shown in the Figure 45, detailed information about the components and operation can be consulted in reference [278].

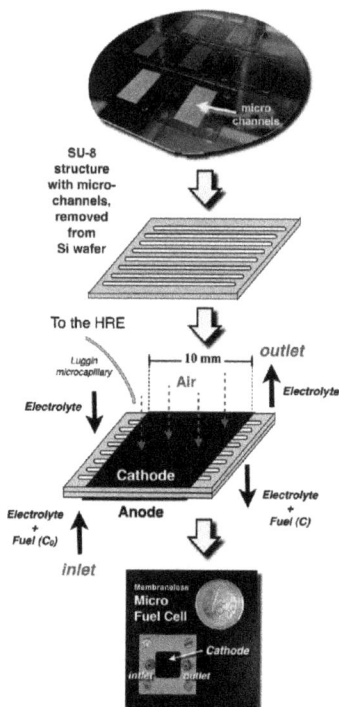

Figure 45. Fabrication and operating scheme of the µLFFC (micro-laminar flow fuel cell) [278]. Reprinted with permission from [278]. Copyright Elsevier, 2012.

In this study, the authors reported how such chalcogenide materials can be employed for developing cheaper, simpler, and smaller micro devices [278]. In Figure 46, we can observe how at a fuel concentration of 5 M CH_3OH the power density curves reached maximum values of 6.5 mW·cm^{-2} for Pt, 4 mW·cm^{-2} for Pt_xS_y and 0.23 mW·cm^{-2} for $CoSe_2$ when these material were used as cathodes in the micro fuel cell assembly. Moreover, when the micro fuel cell was operated in mixed-reactant mode, the maximum value for the Pt power density decreased by 80%; and 35% for Pt_xS_y; nevertheless, $CoSe_2$ did not present any change in performance [278].

These results suggested further that a µLFFC with selective electrodes can operate in mixed-reactant mode (one electrolyte + fuel stream). This configuration makes easier the fuel delivery and the pumping system reducing the manufacturing costs. Additionally, the crossover issue is no longer a major problem on this kind of micro-devices [278,279].

The use of methanol tolerant electrocatalysts such as Pt_xTi_y and Pt_xSe_y in an air-breathing methanol microfluidic fuel cell was further reported [280]. This study presented the improvement of a mixed-reactant micro fuel cell (MRFC) performance in comparison to a laminar flow fuel configuration (LFFC). We can observe that for the LFFC configuration, the anolyte and the catholyte contain 0.5 M H_2SO_4, Figure 47a. The 5M CH_3OH fuel is contained only in the anodic side. In the case of MRFC configuration, Figure 47b, the fuel (5M CH_3OH) and the supporting electrolyte (0.5 M H_2SO_4) are contained in only one reservoir [280].

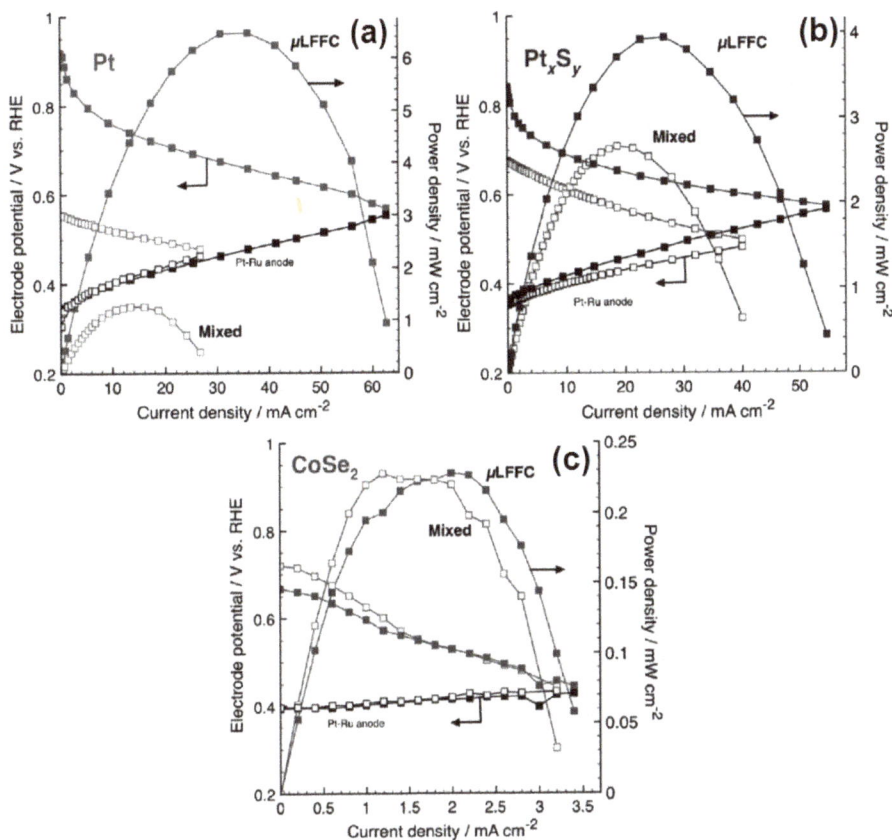

Figure 46. Anodic and cathodic polarization and power density curves of the μLFFCs with (**a**) Pt; (**b**) Pt$_x$S$_y$; and (**c**) CoSe$_2$ cathode catalysts. 5 M CH$_3$OH was used as fuel in the anodic stream (solid symbols) and in both streams (empty symbols) [278]. Reprinted with permission from [278]. Copyright Elsevier, 2012.

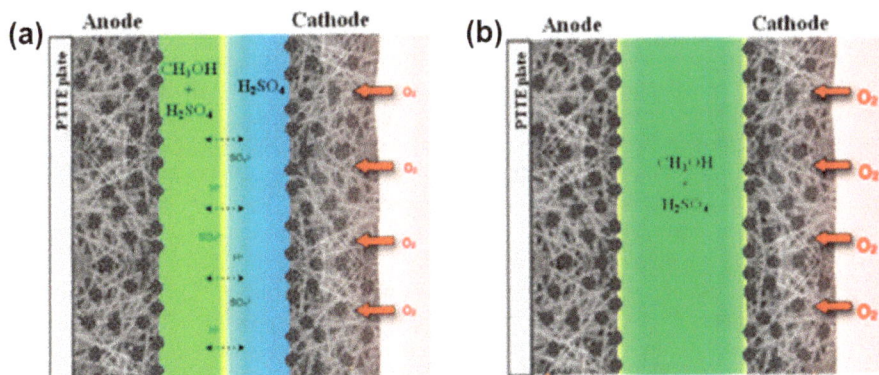

Figure 47. Laminar flow fuel cell configurations; (**a**) Insolated fuel cell and (**b**) mixed-reactant fuel cell [280]. Reprinted with permission from [280]. Copyright the Royal Society of Chemistry, 2014.

The individual anodic and cathodic polarizations for the Pt/C, Pt_xTi_y/C and Pt_xSe_y/C electrocatalysts are presented in Figure 48a–c. In the mixed-reactant condition a mixed potential between the oxygen reduction and the methanol oxidation is formed at the surface of the non-tolerant cathode. Because an inhibition effect caused by the Ti atoms on the Pt surface, the Pt_xTi_y/C overcame the potential value obtained for Pt/C [63]. Interestingly, the Pt_xSe_y/C presented a similar but enhanced effect, this electrocatalyst presented a selectivity to reduces de oxygen instead of oxidizing the methanol contained in the electrolytic solution. Thus, it allows the minimization of the methanol poisoning by the use of Se coordinated atoms onto the Pt surface [280].

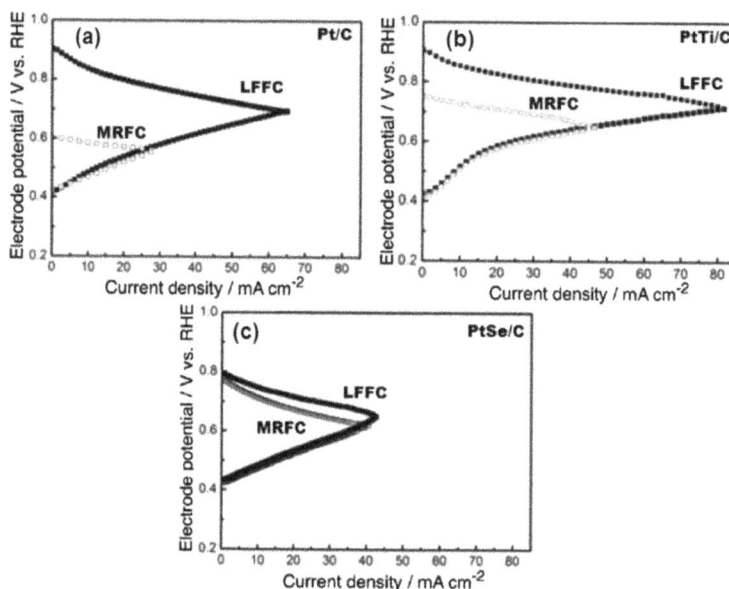

Figure 48. Current-potential polarizations for air-breathing methanol laminar flow fuel cell and mixed reactant flow cell (LFFC and MRFC, respectively) systems for (a) Pt/C; (b) Pt_xTi_y/C and (c) Pt_xSe_y/C cathode catalysts [280]. Reprinted with permission from [280]. Copyright the Royal Society of Chemistry, 2014.

Furthermore, recent studies showed that is possible to improve the electrocatalytic activity of non-precious $CoSe_2$ electrocatalysts evaluated in alkaline medium, with different carbon supports [229]. From this study, it was determined that a $CoSe_2$/NCNH electrocatalyst showed a lower onset overpotential than $CoSe_2$ supported on carbon, and carbon nanohorns ($CoSe_2$/C and $CoSe_2$/CNH, respectively). To validate the half-cell results, a self-breathing µLFFC was constructed and the $CoSe_2$/NCNH was used as a methanol tolerant cathodic electrocatalyst. The highest maximum power density value of 10.05 mW·cm^{-2} was achieved with $CoSe_2$/NCNH (5 M methanol as the fuel), Figure 49. This enhancement presented by $CoSe_2$/NCNH suggests a modification of the $CoSe_2$ active center by an interaction with the NCNH support [229].

Nanostructured palladium electrocatalysts supported on Vulcan carbon XC-72 were synthesized via the carbonyl chemical route and studied in micro-laminar flow fuel cells [56]. When these materials were evaluated to perform the oxidation of organic molecules, they presented a relationship between the electrocatalysts morphology and their electrochemical activity. A 20 wt % metal loading Pd/C electrocatalyst was selected to be used as anode in a direct formic acid µLFFC. As we can observe in Figure 50a, the power density in the direct formic acid µLFFC reached 14.7 mW·cm^{-2} with Pd/C-2, while its commercial counterpart Pd/C presented a power density value of 6.01 mW·cm^{-2}.

The individual polarization curves for the anode and cathode shown in Figure 50b, compares the performances of the anodes. While the cathodic polarization curves do not present variation. It turns out that the enhanced μLFFC performance is the result of a better oxidation reaction on the Pd/C-2 anode.

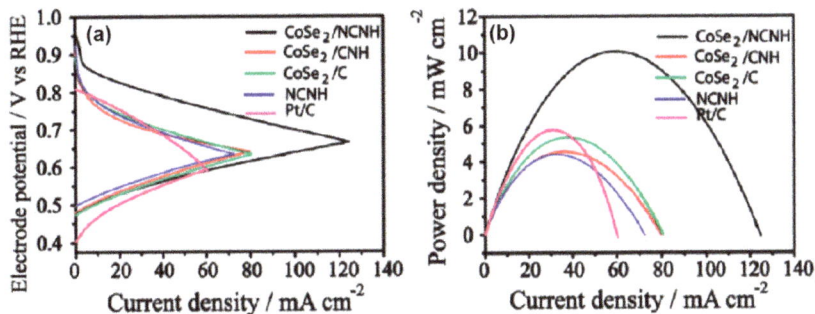

Figure 49. (**a**) Anodic and cathodic polarizations; and (**b**) power density curves of an air-breathing μLFFC with $CoSe_2$/NCNH, $CoSe_2$/CNH, $CoSe_2$/C, NCNHs and Pt/C as cathodes. A Pt-Ru/C electrode was used as the anode. A methanol solution 5 M was used as fuel in the alkaline solution [229]. Reprinted with permission from [229]. Copyright John Wiley and Sons, 2012.

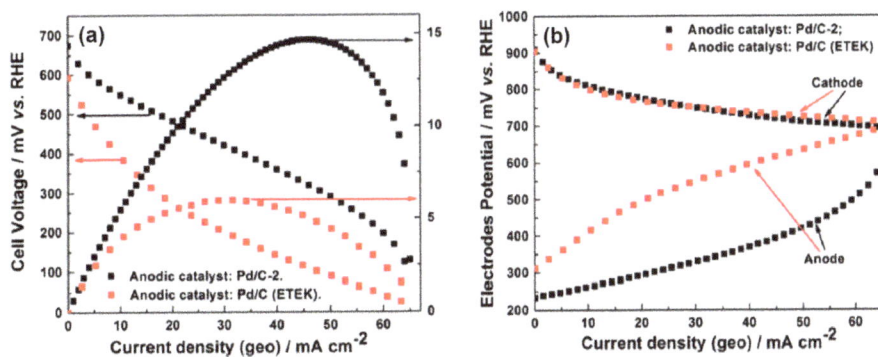

Figure 50. Formic Acid μLFFC. (**a**) cell polarization and power density curves; and (**b**) Anodic and cathodic polarization curves [56]. Reprinted with permission from [56]. Copyright Elsevier, 2016.

Pd/C-2 was also selected to perform the methanol oxidation reaction in a direct methanol μLFFC. Similarly, to the previous formic acid oxidation test, a commercial Pd/C electrode was used as reference material. From Figure 51a we can observe how the micro fuel cell configuration with Pd/C-2 as anode, presented a maximum power density value of 1.3 mW·cm^{-2} whereas the commercial counterpart reached only 0.45 mW·cm^{-2}. In Figure 51b we can observe that the power density difference is the result of the anodic polarizations [56].

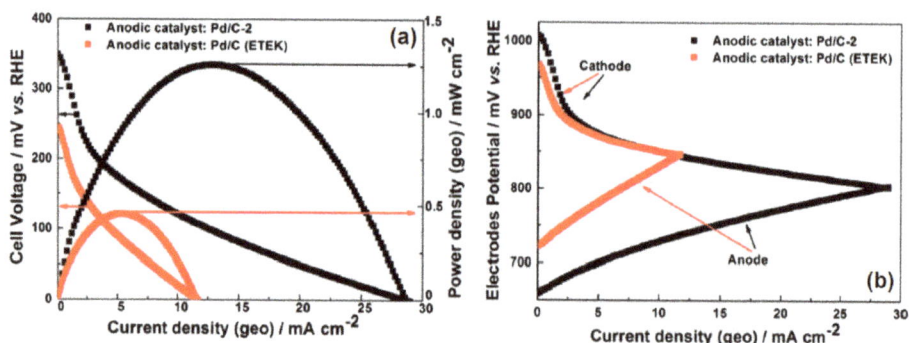

Figure 51. Direct Methanol µLFFC. (**a**) cell polarization and power density curves; and (**b**) electrode polarization curves [56]. Reprinted with permission from [56]. Copyright Elsevier, 2016.

5. Concluding Remarks

An attempt regarding the development of electrocatalytic materials has been outlined in this review. The activity and stability towards the ORR on Pt and Pd nanoparticles can be attained through ligand and support effects. The former focuses on synthesis of Pt or Pd nanoalloyed with a transition metals (ex. Co, Ni, etc.). The latter is tailoring the SMSI phenomenon, to attain the electronic, surface and structural modification of the catalytic center. The main trend in this direction is the application of highly graphitic carbon, carbon-oxide composites, and conducting/semiconducting oxide as supports. In this sense, significant progress in catalytic process on the catalytic center has been so far achieved. As pointed out herein, low temperature synthetic routes, e.g., carbonyl chemical or photo-deposition, was successfully employed.

The photo-deposition is particularly interesting for Pt-based NPs anchored onto high graphitic carbon, semiconducting oxide, and oxide-carbon composites. Strong metal support interaction (SMSI) between platinum and support could be reinforced, increasing the stability of catalytic center for ORR process. In addition, by means of Ti, Se or/and Cr as ligands, the tolerance of Pt was enhanced, favoring their application in a fuel cell using small organics, such as methanol as a fuel. Non-platinum metal chalcogenides, such as Ru_xSe_y and $CoSe_2$, are methanol tolerant electrocatalysts for ORR. Compared with Pt or Pd-based catalysts, the non-precious metal ones showed less catalytic activity, but an enhanced tolerance to methanol leading to the use of mixed-reactant systems, since cross-over effect in direct methanol fuel cell can be reduced, as described through the micro-fuel cells platform. Pt- or Pd-based electrocatalysts are still the most active materials for the hydrogen oxidation reaction (HOR). Nevertheless, non-precious metal nanoalloy, such as CoNiMo, was identified as active for HOR. Research on such kind of electrocatalyst has still a long way to go.

Concerning the oxidation process of organic fuels, e.g., methanol, and formic acid, Pd-based catalysts showed increased activity with respect to Pt ones. Chalcogenide materials containing Pt and Pd were measured for ORR and various fuel oxidation in a micro-fluid fuel cell. The non-precious metal catalysts become more and more attractive. The mechanism for the catalytic activity of such centers is still to be developed. It seems that the morphological impact plays an important role on the electrocatalytic activity. Thus, efforts should be further devoted to this issue.

Acknowledgments: The authors thank all the co-workers, cited in the literature list, and the University of Poitiers, France.

Author Contributions: N. A.-V. conceived the manuscript and warranted the scientific scope. The data collection and drafting was made by J.M. M.-H. in close collaboration with Y.L. All authors fully contributed to the revision of the manuscript.

Conflicts of Interest: The authors declare no conflict of interest.

References

1. Perry, M.L.; Fuller, T.F. A historical perspective of fuel cell technology in the 20th century. *J. Electrochem. Soc.* **2002**, *149*, S59–S67. [CrossRef]
2. Haiping, X.; Li, K.; Xuhui, W. Fuel cell power system and high power DC-DC converter. *IEEE Trans. Power Electr.* **2004**, *19*, 1250–1255.
3. Jayasayee, K.; Veen, J.A.R.V.; Manivasagam, T.G.; Celebi, S.; Hensen, E.J.M.; de Bruijn, F.A. Oxygen reduction reaction (ORR) activity and durability of carbon supported PtM (Co, Ni, Cu) alloys: Influence of particle size and non-noble metals. *Appl. Catal. B: Environ.* **2012**, *111–112*, 515–526. [CrossRef]
4. Gasteiger, H.A.; Kocha, S.S.; Sompalli, B.; Wagner, F.T. Activity benchmarks and requirements for Pt, Pt-alloy, and non-Pt oxygen reduction catalysts for PEMFCs. *Appl. Catal. B: Environ.* **2005**, *56*, 9–35. [CrossRef]
5. Ezeta, A.; Arce, E.M.; Solorza, O.; González, R.G.; Dorantes, H. Effect of the leaching of Ru-Se-Fe and Ru-Mo-Fe obtained by mechanical alloying on electrocatalytical behavior for the oxygen reduction reaction. *J. Alloys Compd.* **2009**, *483*, 429–431. [CrossRef]
6. Suryanarayana, C. Mechanical alloying and milling. *Prog. Mater. Sci.* **2001**, *46*, 1–184. [CrossRef]
7. Mukerjee, S.; Srinivasan, S.; Soriaga, M.P.; McBreen, J. Role of structural and electronic properties of Pt and Pt alloys on electrocatalysis of oxygen reduction: An in situ XANES and EXAFS investigation. *J. Electrochem. Soc.* **1995**, *142*, 1409–1422. [CrossRef]
8. Chen, S.; Ferreira, P.J.; Sheng, W.; Yabuuchi, N.; Allard, L.F.; Shao-Horn, Y. Enhanced activity for oxygen reduction reaction on "Pt3Co" nanoparticles: Direct evidence of percolated and sandwich-segregation structures. *J. Am. Chem. Soc.* **2008**, *130*, 13818–13819. [CrossRef] [PubMed]
9. Min, M.-K.; Cho, J.; Cho, K.; Kim, H. Particle size and alloying effects of Pt-based alloy catalysts for fuel cell applications. *Electrochim. Acta* **2000**, *45*, 4211–4217. [CrossRef]
10. Guo, S.; Li, D.; Zhu, H.; Zhang, S.; Markovic, N.M.; Stamenkovic, V.R.; Sun, S. FePt and CoPt nanowires as efficient catalysts for the oxygen reduction reaction. *Angew. Chem. Int. Ed.* **2013**, *52*, 3465–3468. [CrossRef] [PubMed]
11. Shao, M.; Chang, Q.; Dodelet, J.-P.; Chenitz, R. Recent advances in electrocatalysts for oxygen reduction reaction. *Chem. Rev.* **2016**, *116*, 3594–3657. [CrossRef] [PubMed]
12. Greeley, J.; Stephens, I.E.L.; Bondarenko, A.S.; Johansson, T.P.; Hansen, H.A.; Jaramillo, T.F.; Rossmeisl, J.; Chorkendorff, I.; Nørskov, J.K. Alloys of platinum and early transition metals as oxygen reduction electrocatalysts. *Nat. Chem.* **2009**, *1*, 552–556. [CrossRef] [PubMed]
13. Pedersen, C.M.; Escudero-Escribano, M.; Velázquez-Palenzuela, A.; Christensen, L.H.; Chorkendorff, I.; Stephens, I.E.L. Benchmarking Pt-based electrocatalysts for low temperature fuel cell reactions with the rotating disk electrode: Oxygen reduction and hydrogen oxidation in the presence of CO. *Electrochim. Acta* **2015**, *179*, 647–657. [CrossRef]
14. Alonso-Vante, N.; Jaegermann, W.; Tributsch, H.; Hoenle, W.; Yvon, K. Electrocatalysis of oxygen reduction by chalcogenides containing mixed transition metal clusters. *J. Am. Chem. Soc.* **1987**, *109*, 3251–3257. [CrossRef]
15. Alonso-Vante, N. Chevrel phase and cluster-like chalcogenide materials. In *Handbook of Fuel Cells*; Vielstich, W., Lamm, A., Gasteiger, H., Eds.; John Wiley & Sons, Ltd.: Chichester, UK, 2003; Volume 2, pp. 534–543.
16. Friedrich, K.A.; Geyzers, K.P.; Linke, U.; Stimming, U.; Stumper, J. Co adsorption and oxidation on a Pt(111) electrode modified by ruthenium deposition: An ir spectroscopic study. *J. Electroanal. Chem.* **1996**, *402*, 123–128. [CrossRef]
17. Piazza, S.; Kühne, H.M.; Tributsch, H. Photoelectrochemical and electrocatalytic behaviour of p-type ruthenium disulphide electrodes. *J. Electroanal. Chem.* **1985**, *196*, 53–67. [CrossRef]
18. Bewick, A.; Gutiérrez, C.; Larramona, G. An international journal devoted to all aspects of electrode kinetics, interfacial structure, properties of electrolytes, colloid and biological electrochemistryin-situ IR spectroscopy study of the ruthenium electrode in acid and alkaline solutions. *J. Electroanal. Chem.* **1992**, *332*, 155–167. [CrossRef]
19. Alonso-Vante, N. Spectral sensitization of titanium dioxide electrodes via Ru-based chalcogenides thin layers. *Sol. Energy Mater. Sol. C* **1994**, *31*, 509–524. [CrossRef]

20. Chang, C.C.; Wen, T.C. Kinetics of oxygen reduction at RuO_2-coated titanium electrode in alkaline solution. *J. Appl. Electrochem.* **1997**, *27*, 355–363. [CrossRef]
21. González-Cruz, R.; Solorza-Feria, O. Oxygen reduction in acid media by a $Ru_xFe_ySe_z(CO)_n$ cluster catalyst dispersed on a glassy carbon-supported nafion film. *J. Sol. State Electrochem.* **2003**, *7*, 289–295. [CrossRef]
22. Cheng, H.; Yuan, W.; Scott, K. The influence of a new fabrication procedure on the catalytic activity of ruthenium–selenium catalysts. *Electrochim. Acta* **2006**, *52*, 466–473. [CrossRef]
23. Nagabhushana, K.S.; Dinjus, E.; Bönnemann, H.; Zaikovskii, V.; Hartnig, C.; Zehl, G.; Dorbandt, I.; Fiechter, S.; Bogdanoff, P. Reductive annealing for generating Se doped 20 wt % Ru/C cathode catalysts for the oxygen reduction reaction. *J. Appl. Electrochem.* **2007**, *37*, 1515–1522. [CrossRef]
24. Montiel, M.; García-Rodríguez, S.; Hernández-Fernández, P.; Díaz, R.; Rojas, S.; Fierro, J.L.G.; Fatás, E.; Ocón, P. Relevance of the synthesis route of Se-modified Ru/C as methanol tolerant electrocatalysts for the oxygen reduction reaction. *J. Power Sources* **2010**, *195*, 2478–2487. [CrossRef]
25. Suárez-Alcántara, K.; Solorza-Feria, O. Kinetics and PEMFC performance of $Ru_xMo_ySe_z$ nanoparticles as a cathode catalyst. *Electrochim. Acta* **2008**, *53*, 4981–4989. [CrossRef]
26. Koper, M.T.M. Electrocatalysis on bimetallic and alloy surfaces. *Surf. Sci.* **2004**, *548*, 1–3. [CrossRef]
27. He, T.; Kreidler, E.; Xiong, L.; Ding, E. Combinatorial screening and nano-synthesis of platinum binary alloys for oxygen electroreduction. *J. Power Sources* **2007**, *165*, 87–91. [CrossRef]
28. Anastasijević, N.A.; Vesović, V.; Adžić, R.R. Determination of the kinetic parameters of the oxygen reduction reaction using the rotating ring-disk electrode. *J. Electroanal. Chem. Interfacial Electrochem.* **1987**, *229*, 305–316. [CrossRef]
29. Kalevaru, V.N.; Benhmid, A.; Radnik, J.; Pohl, M.M.; Bentrup, U.; Martin, A. Marked influence of support on the catalytic performance of PdSb acetoxylation catalysts: Effects of Pd particle size, valence states, and acidity characteristics. *J. Catal.* **2007**, *246*, 399–412. [CrossRef]
30. Bagotzky, V.S.; Shumilova, N.A.; Samoilov, G.P.; Khrushcheva, E.I. Electrochemical oxygen reduction on nickel electrodes in alkaline solutions—II. *Electrochim. Acta* **1972**, *17*, 1625–1635. [CrossRef]
31. Ju, W.; Favaro, M.; Durante, C.; Perini, L.; Agnoli, S.; Schneider, O.; Stimming, U.; Granozzi, G. Pd nanoparticles deposited on nitrogen-doped HOPG: New insights into the Pd-catalyzed oxygen reduction reaction. *Electrochim. Acta* **2014**, *141*, 89–101. [CrossRef]
32. Chen, J.; Takanabe, K.; Ohnishi, R.; Lu, D.; Okada, S.; Hatasawa, H.; Morioka, H.; Antonietti, M.; Kubota, J.; Domen, K. Nano-sized tin on carbon black as an efficient electrocatalyst for the oxygen reduction reaction prepared using an $MPG-C_3N_4$ template. *Chem. Commun.* **2010**, *46*, 7492–7494. [CrossRef] [PubMed]
33. Wu, J.; Gross, A.; Yang, H. Shape and composition-controlled platinum alloy nanocrystals using carbon monoxide as reducing agent. *Nano Lett.* **2011**, *11*, 798–802. [CrossRef] [PubMed]
34. Vukmirovic, M.B.; Zhang, J.; Sasaki, K.; Nilekar, A.U.; Uribe, F.; Mavrikakis, M.; Adzic, R.R. Platinum monolayer electrocatalysts for oxygen reduction. *Electrochim. Acta* **2007**, *52*, 2257–2263. [CrossRef]
35. Salvador-Pascual, J.J.; Citalán-Cigarroa, S.; Solorza-Feria, O. Kinetics of oxygen reduction reaction on nanosized Pd electrocatalyst in acid media. *J. Power Sources* **2007**, *172*, 229–234. [CrossRef]
36. Sánchez-Sánchez, C.M.; Bard, A.J. Hydrogen peroxide production in the oxygen reduction reaction at different electrocatalysts as quantified by scanning electrochemical microscopy. *Anal. Chem.* **2009**, *81*, 8094–8100. [CrossRef] [PubMed]
37. Shao, M.-H.; Sasaki, K.; Adzic, R.R. Pd–Fe nanoparticles as electrocatalysts for oxygen reduction. *J. Am. Chem. Soc.* **2006**, *128*, 3526–3527. [CrossRef] [PubMed]
38. Shao, M.; Liu, P.; Zhang, J.; Adzic, R. Origin of enhanced activity in palladium alloy electrocatalysts for oxygen reduction reaction. *J. Phys. Chem. B* **2007**, *111*, 6772–6775. [CrossRef] [PubMed]
39. Liu, H.; Manthiram, A. Tuning the electrocatalytic activity and durability of low cost $Pd_{70}Co_{30}$ nanoalloy for oxygen reduction reaction in fuel cells. *Electrochem. Commun.* **2008**, *10*, 740–744. [CrossRef]
40. Kirubakaran, A.; Jain, S.; Nema, R.K. A review on fuel cell technologies and power electronic interface. *Renew. Sust. Energy Rev.* **2009**, *13*, 2430–2440. [CrossRef]
41. Mora-Hernández, J.M.; Ezeta-Mejía, A.; Reza-San Germán, C.; Citalán-Cigarroa, S.; Arce-Estrada, E.M. Electrochemical activity towards ORR of mechanically alloyed PdCo supported on vulcan carbon and carbon nanospheres. *J. Appl. Electrochem.* **2014**, *44*, 1307–1315. [CrossRef]

42. Lee, Y.-W.; Ko, A.R.; Kim, D.-Y.; Han, S.-B.; Park, K.-W. Octahedral Pt-Pd alloy catalysts with enhanced oxygen reduction activity and stability in proton exchange membrane fuel cells. *RSC Adv.* **2012**, *2*, 1119–1125. [CrossRef]
43. Antolini, E. Palladium in fuel cell catalysis. *Energy Environ. Sci.* **2009**, *2*, 915–931. [CrossRef]
44. Calvo, S.R.; Balbuena, P.B. Density functional theory analysis of reactivity of Pt_xPd_y alloy clusters. *Surf. Sci.* **2007**, *601*, 165–171. [CrossRef]
45. Shen, P.K.; Xu, C. Alcohol oxidation on nanocrystalline oxide Pd/C promoted electrocatalysts. *Electrochem. Commun.* **2006**, *8*, 184–188. [CrossRef]
46. Zhang, S.; Shao, Y.; Yin, G.; Lin, Y. Electrostatic self-assembly of a Pt-around-Au nanocomposite with high activity towards formic acid oxidation. *Angew. Chem. Int. Ed.* **2010**, *49*, 2211–2214. [CrossRef] [PubMed]
47. Ha, S.; Larsen, R.; Masel, R.I. Performance characterization of Pd/C nanocatalyst for direct formic acid fuel cells. *J. Power Sources* **2005**, *144*, 28–34. [CrossRef]
48. Zhu, Y.; Khan, Z.; Masel, R.I. The behavior of palladium catalysts in direct formic acid fuel cells. *J. Power Sources* **2005**, *139*, 15–20. [CrossRef]
49. Alonso-Vante, N. Tailoring of metal cluster-like materials for the molecular oxygen reduction reaction. *Pure Appl. Chem.* **2008**, *80*, 2103–2114. [CrossRef]
50. Gago Aldo, S.; Habrioux, A.; Alonso-Vante, N. Tailoring nanostructured catalysts for electrochemical energy conversion systems. *Nanotechnol. Rev.* **2012**, *1*, 427–453. [CrossRef]
51. Alonso-Vante, N. Carbonyl tailored electrocatalysts. *Fuel Cells* **2006**, *6*, 182–189. [CrossRef]
52. Le Rhun, V.; Garnier, E.; Pronier, S.; Alonso-Vante, N. Electrocatalysis on nanoscale ruthenium-based material manufactured by carbonyl decomposition. *Electrochem. Commun.* **2000**, *2*, 475–479. [CrossRef]
53. Yang, H.; Alonso-Vante, N.; Lamy, C.; Akins, D.L. High methanol tolerance of carbon-supported Pt-Cr alloy nanoparticle electrocatalysts for oxygen reduction. *J. Electrochem. Soc.* **2005**, *152*, A704–A709. [CrossRef]
54. Yang, H.; Alonso-Vante, N.; Léger, J.-M.; Lamy, C. Tailoring, structure, and activity of carbon-supported nanosized Pt–Cr alloy electrocatalysts for oxygen reduction in pure and methanol-containing electrolytes. *J. Phys. Chem. B* **2004**, *108*, 1938–1947. [CrossRef]
55. Yang, H.; Vogel, W.; Lamy, C.; Alonso-Vante, N. Structure and electrocatalytic activity of carbon-supported Pt–Ni alloy nanoparticles toward the oxygen reduction reaction. *J. Phys. Chem. B* **2004**, *108*, 11024–11034. [CrossRef]
56. Luo, Y.; Mora-Hernández, J.M.; Alonso-Vante, N. Morphological impact onto organic fuel oxidation of nanostructured palladium synthesized via carbonyl chemical route. *J. Electroanal. Chem.* **2016**, *765*, 79–86. [CrossRef]
57. Luo, Y.; Mora-Hernández, J.M.; Estudillo-Wong, L.A.; Arce-Estrada, E.M.; Alonso-Vante, N. Nanostructured palladium tailored via carbonyl chemical route towards oxygen reduction reaction. *Electrochim. Acta* **2015**, *173*, 771–778. [CrossRef]
58. Longoni, G.; Chini, P. Synthesis and chemical characterization of platinum carbonyl dianions $[Pt_3(CO)_6]_n{}^{2-}$ (*n* = .apprx.10,6,5,4,3,2,1). A new series of inorganic oligomers. *J. Am. Chem. Soc.* **1976**, *98*, 7225–7231. [CrossRef]
59. Manzo-Robledo, A.; Boucher, A.C.; Pastor, E.; Alonso-Vante, N. Electro-oxidation of carbon monoxide and methanol on carbon-supported Pt–Sn nanoparticles: A dems study. *Fuel Cells* **2002**, *2*, 109–116. [CrossRef]
60. Yang, H.; Coutanceau, C.; Léger, J.-M.; Alonso-Vante, N.; Lamy, C. Methanol tolerant oxygen reduction on carbon-supported Pt–Ni alloy nanoparticles. *J. Electroanal. Chem.* **2005**, *576*, 305–313. [CrossRef]
61. Ma, J.; Gago, A.S.; Alonso-Vante, N. Performance study of platinum nanoparticles supported onto MWCNT in a formic acid microfluidic fuel cell system. *J. Electrochem. Soc.* **2013**, *160*, F859–F866. [CrossRef]
62. Kondo, T.; Iwasaki, Y.; Honma, Y.; Takagi, Y.; Okada, S.; Nakamura, J. Formation of nonbonding π electronic states of graphite due to Pt-C hybridization. *Phys. Rev. B* **2009**, *80*, 233408. [CrossRef]
63. Ma, J.; Habrioux, A.; Miyao, T.; Kakinuma, K.; Inukai, J.; Watanabe, M.; Alonso-Vante, N. Correlation between surface chemical composition with catalytic activity and selectivity of organic-solvent synthesized Pt-Ti nanoparticles. *J. Mater. Chem. A* **2013**, *1*, 8798–8804. [CrossRef]
64. Alonso-Vante, N.; Schubert, B.; Tributsch, H. Transition metal cluster materials for multi-electron transfer catalysis. *Mater. Chem. Phys.* **1989**, *22*, 281–307. [CrossRef]
65. Chevrel, R.; Sergent, M.; Prigent, J. Sur de nouvelles phases sulfurées ternaires du molybdène. *J. Solid State Chem.* **1971**, *3*, 515–519. (In French) [CrossRef]

66. Perrin, A.; Chevrel, R.; Sergent, M.; Fischer, Ø. Synthesis and electrical properties of new chalcogenide compounds containing mixed (Mo, Me)$_6$ octahedral clusters (Me = Ru or Rh). *J. Sol. State Chem.* **1980**, *33*, 43–47. [CrossRef]

67. Perrin, A.; Sergent, M.; Fischer, O. New compounds of the type Mo$_2$Re$_4$X$_8$ (M = S, Se) containing octahedral Mo$_2$Re$_4$ clusters. *Mater. Res. Bull.* **1978**, *13*, 259–264. [CrossRef]

68. Alonso-Vante, N.; Tributsch, H. Energy conversion catalysis using semiconducting transition metal cluster compounds. *Nature* **1986**, *323*, 431–432. [CrossRef]

69. Alonso-Vante, N.; Tributsch, H. what. *J. Electroanal. Chem. Interfacial Electrochem.* **1987**, *229*, 223–237.

70. Kühne, H.M.; Tributsch, H. Oxygen and chlorine evolution on ruthenium-iron-disulphide mediated by low energy photons. *Ber. Bunsenges. Physikalische Chem.* **1984**, *88*, 10–16. [CrossRef]

71. Kühne, H.M.; Tributsch, H. Energetics and dynamics of the interface of RuS$_2$ and implications for photoelectrolysis of water. *J. Electroanal. Chem. Interfacial Electrochem.* **1986**, *201*, 263–282. [CrossRef]

72. Vante, N.A.; Schubert, B.; Tributsch, H.; Perrin, A. Influence of d-state density and chemistry of transition metal cluster selenides on electrocatalysis. *J. Catal.* **1988**, *112*, 384–391. [CrossRef]

73. Hughbanks, T.; Hoffmann, R. Molybdenum chalcogenides: Clusters, chains, and extended solids. The approach to bonding in three dimensions. *J. Am. Chem. Soc.* **1983**, *105*, 1150–1162. [CrossRef]

74. Jaegermann, W.; Pettenkofer, C.; Alonso Vante, N.; Schwarzlose, T.; Tributsch, H. Chevrel phase type compounds: Electronic, chemical and structural factors in oxygen reduction electrocatalysis. *Ber. Bunsenges. Phys. Chem.* **1990**, *94*, 513–520. [CrossRef]

75. Solorza-Feria, O.; Ellmer, K.; Giersig, M.; Alonso-Vante, N. Novel low-temperature synthesis of semiconducting transition metal chalcogenide electrocatalyst for multielectron charge transfer: Molecular oxygen reduction. *Electrochim. Acta* **1994**, *39*, 1647–1653. [CrossRef]

76. Kochubey, D.I.; Nikitenko, S.G.; Parmon, V.N.; Gruzdkov, Yu.A.; Tributsch, H.; Alonso-Vante, N. In situ EXAFS-electrochemical study of reduction of molecular oxygen on Mo☐Ru☐Se thin layers electrodes in acidic media. *Phys. B Condens. Matter* **1995**, *208–209*, 694–696. [CrossRef]

77. Schmidt, T.J.; Paulus, U.A.; Gasteiger, H.A.; Alonso-Vante, N.; Behm, R.J. Oxygen reduction on Ru$_{1.92}$Mo$_{0.08}$SeO$_4$, Ru/carbon, and Pt/carbon in pure and methanol-containing electrolytes. *J. Electrochem. Soc.* **2000**, *147*, 2620–2624. [CrossRef]

78. Dassenoy, F.; Vogel, W.; Alonso-Vante, N. Structural studies and stability of cluster-like Ru$_x$Se$_y$ electrocatalysts. *J. Phys. Chem. B* **2002**, *106*, 12152–12157. [CrossRef]

79. Alonso-Vante, N.; Borthen, P.; Fieber-Erdmann, M.; Strehblow, H.H.; Holub-Krappe, E. An in situ grazing incidence X-ray absorption study of ultra thin ruxsey cluster-like electrocatalyst layers. *Electrochim. Acta* **2000**, *45*, 4227–4236. [CrossRef]

80. Malakhov, I.V.; Nikitenko, S.G.; Savinova, E.R.; Kochubey, D.I.; Alonso-Vante, N. In situ EXAFS study to probe active centers of Ru chalcogenide electrocatalysts during oxygen reduction reaction. *J. Phys. Chem. B* **2002**, *106*, 1670–1676. [CrossRef]

81. Babu, P.K.; Lewera, A.; Jong, H.C.; Hunger, R.; Jaegermann, W.; Alonso-Vante, N.; Wieckowski, A.; Oldfield, E. Selenium becomes metallic in Ru-Se fuel cell catalysts: An EC-NMR and XPS investigation. *J. Am. Chem. Soc.* **2007**, *129*, 15140–15141. [CrossRef] [PubMed]

82. Cao, D.; Wieckowski, A.; Inukai, J.; Alonso-Vante, N. Oxygen reduction reaction on ruthenium and rhodium nanoparticles modified with selenium and sulfur. *J. Electrochem. Soc.* **2006**, *153*, A869–A874. [CrossRef]

83. Ma, J.; Gago, A.S.; Vogel, W.; Alonso-Vante, N. Tailoring and tuning the tolerance of a Pt chalcogenide cathode electrocatalyst to methanol. *ChemCatChem* **2013**, *5*, 701–705. [CrossRef]

84. Nie, Y.; Li, L.; Wei, Z. Recent advancements in Pt and Pt-free catalysts for oxygen reduction reaction. *Chem. Soc. Rev.* **2015**, *44*, 2168–2201. [CrossRef] [PubMed]

85. Borup, R.; Meyers, J.; Pivovar, B.; Kim, Y.S.; Mukundan, R.; Garland, N.; Myers, D.; Wilson, M.; Garzon, F.; Wood, D.; et al. Scientific aspects of polymer electrolyte fuel cell durability and degradation. *Chem. Rev.* **2007**, *107*, 3904–3951. [CrossRef] [PubMed]

86. Su, D.S.; Sun, G. Nonprecious-metal catalysts for low-cost fuel cells. *Angew. Chem. Int. Ed.* **2011**, *50*, 11570–11572. [CrossRef] [PubMed]

87. Behret, H.; Binder, H.; Sandstede, G. Electrocatalytic oxygen reduction with thiospinels and other sulphides of transition metals. *Electrochim. Acta* **1975**, *20*, 111–117. [CrossRef]

88. Feng, Y.; Alonso-Vante, N. Nonprecious metal catalysts for the molecular oxygen-reduction reaction. *Phys. Status Solidi* **2008**, *245*, 1792–1806. [CrossRef]

89. Puntes, V.F.; Krishnan, K.M.; Alivisatos, A.P. Colloidal nanocrystal shape and size control: The case of cobalt. *Science* **2001**, *291*, 2115–2117. [CrossRef] [PubMed]

90. Yang, H.T.; Shen, C.M.; Wang, Y.G.; Su, Y.K.; Yang, T.Z.; Gao, H.J. Stable cobalt nanoparticles passivated with oleic acid and triphenylphosphine. *Nanotechnology* **2004**, *15*, 70. [CrossRef]

91. Yin, Y.; Rioux, R.M.; Erdonmez, C.K.; Hughes, S.; Somorjai, G.A.; Alivisatos, A.P. Formation of hollow nanocrystals through the nanoscale kirkendall effect. *Science* **2004**, *304*, 711–714. [CrossRef] [PubMed]

92. Sidik, R.A.; Anderson, A.B. Co_9S_8 as a catalyst for electroreduction of O_2: Quantum chemistry predictions. *J. Phys. Chem. B* **2006**, *110*, 936–941. [CrossRef] [PubMed]

93. Vayner, E.; Sidik, R.A.; Anderson, A.B.; Popov, B.N. Experimental and theoretical study of cobalt selenide as a catalyst for O_2 electroreduction. *J. Phys. Chem. C* **2007**, *111*, 10508–10513. [CrossRef]

94. Feng, Y.; He, T.; Alonso-Vante, N. In situ free-surfactant synthesis and ORR- electrochemistry of carbon-supported Co_3S_4 and $CoSe_2$ nanoparticles. *Chem. Mater.* **2008**, *20*, 26–28. [CrossRef]

95. Feng, Y.; He, T.; Alonso-Vante, N. Oxygen reduction reaction on carbon-supported $CoSe_2$ nanoparticles in an acidic medium. *Electrochim. Acta* **2009**, *54*, 5252–5256. [CrossRef]

96. Susac, D.; Sode, A.; Zhu, L.; Wong, P.C.; Teo, M.; Bizzotto, D.; Mitchell, K.A.R.; Parsons, R.R.; Campbell, S.A. A methodology for investigating new nonprecious metal catalysts for PEM fuel cells. *J. Phys. Chem. B* **2006**, *110*, 10762–10770. [CrossRef] [PubMed]

97. Susac, D.; Zhu, L.; Teo, M.; Sode, A.; Wong, K.C.; Wong, P.C.; Parsons, R.R.; Bizzotto, D.; Mitchell, K.A.R.; Campbell, S.A. Characterization of FeS_2-based thin films as model catalysts for the oxygen reduction reaction. *J. Phys. Chem. C* **2007**, *111*, 18715–18723. [CrossRef]

98. Zhu, L.; Susac, D.; Teo, M.; Wong, K.C.; Wong, P.C.; Parsons, R.R.; Bizzotto, D.; Mitchell, K.A.R.; Campbell, S.A. Investigation of CoS_2-based thin films as model catalysts for the oxygen reduction reaction. *J. Catal.* **2008**, *258*, 235–242. [CrossRef]

99. Ahlberg, E.; Elfström Broo, A. Oxygen reduction at sulphide minerals. 1. A rotating ring disc electrode (RRDE) study at galena and pyrite. *Int. J. Miner. Process.* **1996**, *46*, 73–89. [CrossRef]

100. Bonakdarpour, A.; Delacote, C.; Yang, R.; Wieckowski, A.; Dahn, J.R. Loading of Se/Ru/C electrocatalyst on a rotating ring-disk electrode and the loading impact on a H_2O_2 release during oxygen reduction reaction. *Electrochem. Commun.* **2008**, *10*, 611–615. [CrossRef]

101. Alonso-Vante, N. Platinum and non-platinum nanomaterials for the molecular oxygen reduction reaction. *ChemPhysChem* **2010**, *11*, 2732–2744. [CrossRef] [PubMed]

102. Feng, Y.; Alonso-Vante, N. Carbon-supported cubic $CoSe_2$ catalysts for oxygen reduction reaction in alkaline medium. *Electrochim. Acta* **2012**, *72*, 129–133. [CrossRef]

103. Feng, Y.J.; He, T.; Alonso-Vante, N. Carbon-supported $CoSe_2$ nanoparticles for oxygen reduction reaction in acid medium. *Fuel Cells* **2010**, *10*, 77–83.

104. Liang, Y.; Li, Y.; Wang, H.; Zhou, J.; Wang, J.; Regier, T.; Dai, H. Co_3O_4 nanocrystals on graphene as a synergistic catalyst for oxygen reduction reaction. *Nat. Mater.* **2011**, *10*, 780–786. [CrossRef] [PubMed]

105. Sabatier, P. Hydrogénations et déshydrogénations par catalyse. *Ber. Dtsch. Chem. Ges.* **1911**, *44*, 1984–2001. [CrossRef]

106. Yang, B.; Burch, R.; Hardacre, C.; Headdock, G.; Hu, P. Understanding the optimal adsorption energies for catalyst screening in heterogeneous catalysis. *ACS Catal.* **2014**, *4*, 182–186. (In French) [CrossRef]

107. Maruyama, J.; Abe, I. Formation of platinum-free fuel cell cathode catalyst with highly developed nanospace by carbonizing catalase. *Chem. Mater.* **2005**, *17*, 4660–4667. [CrossRef]

108. Zhang, L.; Zhang, J.; Wilkinson, D.P.; Wang, H. Progress in preparation of non-noble electrocatalysts for PEM fuel cell reactions. *J. Power Sources* **2006**, *156*, 171–182. [CrossRef]

109. Wood, T.E.; Tan, Z.; Schmoeckel, A.K.; O'Neill, D.; Atanasoski, R. Non-precious metal oxygen reduction catalyst for PEM fuel cells based on nitroaniline precursor. *J. Power Sources* **2008**, *178*, 510–516. [CrossRef]

110. Rand, D.A.J.; Woods, R. Cyclic voltammetric studies on iridium electrodes in sulphuric acid solutions. *J. Electroanal. Chem. Interfacial Electrochem.* **1974**, *55*, 375–381. [CrossRef]

111. Croissant, M.J.; Napporn, T.; Léger, J.M.; Lamy, C. Electrocatalytic oxidation of hydrogen at platinum-modified polyaniline electrodes. *Electrochim. Acta* **1998**, *43*, 2447–2457. [CrossRef]

112. Cao, L.; Sun, G.; Li, H.; Xin, Q. Carbon-supported IrSn catalysts for direct ethanol fuel cell. *Fuel Cells Bull.* **2007**, *2007*, 12–16. [CrossRef]

113. Lee, K.; Zhang, L.; Zhang, J. Ir$_x$CO$_{1-x}$ (x = 0.3–1.0) alloy electrocatalysts, catalytic activities, and methanol tolerance in oxygen reduction reaction. *J. Power Sources* **2007**, *170*, 291–296. [CrossRef]

114. Santos, E.; Schmickler, W. Electrocatalysis of hydrogen oxidation—Theoretical foundations. *Angew. Chem. Int. Ed.* **2007**, *46*, 8262–8265. [CrossRef] [PubMed]

115. Mokrane, S.; Makhloufi, L.; Alonso-Vante, N. Electrochemical behaviour of platinum nanoparticles supported on polypyrrole (PPY)/C composite. *ECS Trans.* **2008**, *6*, 93–103.

116. Shao, M. Palladium-based electrocatalysts for hydrogen oxidation and oxygen reduction reactions. *J. Power Sources* **2011**, *196*, 2433–2444. [CrossRef]

117. Kwon, K.; Jin, S.-A.; Lee, K.H.; You, D.J.; Pak, C. Performance enhancement of Pd-based hydrogen oxidation catalysts using tungsten oxide. *Catal. Today* **2014**, *232*, 175–178. [CrossRef]

118. Zhang, H.; Shen, P.K. Recent development of polymer electrolyte membranes for fuel cells. *Chem. Rev.* **2012**, *112*, 2780–2832. [CrossRef] [PubMed]

119. Sheng, W.; Bivens, A.P.; Myint, M.; Zhuang, Z.; Forest, R.V.; Fang, Q.; Chen, J.G.; Yan, Y. Non-precious metal electrocatalysts with high activity for hydrogen oxidation reaction in alkaline electrolytes. *Energy Environ. Sci.* **2014**, *7*, 1719–1724. [CrossRef]

120. Spendelow, J.S.; Wieckowski, A. Electrocatalysis of oxygen reduction and small alcohol oxidation in alkaline media. *PhysChemChemPhys* **2007**, *9*, 2654–2675. [CrossRef] [PubMed]

121. Neyerlin, K.C.; Gu, W.; Jorne, J.; Gasteiger, H.A. Study of the exchange current density for the hydrogen oxidation and evolution reactions. *J. Electrochem. Soc.* **2007**, *154*, B631–B635. [CrossRef]

122. Durst, J.; Siebel, A.; Simon, C.; Hasche, F.; Herranz, J.; Gasteiger, H.A. New insights into the electrochemical hydrogen oxidation and evolution reaction mechanism. *Energy Environ. Sci.* **2014**, *7*, 2255–2260. [CrossRef]

123. Greeley, J.; Jaramillo, T.F.; Bonde, J.; Chorkendorff, I.; Norskov, J.K. Computational high-throughput screening of electrocatalytic materials for hydrogen evolution. *Nat. Mater.* **2006**, *5*, 909–913. [CrossRef] [PubMed]

124. Gu, J.; Guo, Y.; Jiang, Y.-Y.; Zhu, W.; Xu, Y.-S.; Zhao, Z.-Q.; Liu, J.-X.; Li, W.-X.; Jin, C.-H.; Yan, C.-H.; et al. Robust phase control through hetero-seeded epitaxial growth for face-centered cubic Pt@Ru nanotetrahedrons with superior hydrogen electro-oxidation activity. *J. Phys. Chem. C* **2015**, *119*, 17697–17706. [CrossRef]

125. Kucernak, A.R.J.; Fahy, K.F.; Sundaram, V.N.N. Facile synthesis of palladium phosphide electrocatalysts and their activity for the hydrogen oxidation, hydrogen evolutions, oxygen reduction and formic acid oxidation reactions. *Catal. Today* **2016**, *262*, 48–56. [CrossRef]

126. Carenco, S.; Portehault, D.; Boissière, C.; Mézailles, N.; Sanchez, C. Nanoscaled metal borides and phosphides: Recent developments and perspectives. *Chem. Rev.* **2013**, *113*, 7981–8065. [CrossRef] [PubMed]

127. Henkes, A.E.; Vasquez, Y.; Schaak, R.E. Converting metals into phosphides: A general strategy for the synthesis of metal phosphide nanocrystals. *J. Am. Chem. Soc.* **2007**, *129*, 1896–1897. [CrossRef] [PubMed]

128. Barry, B.M.; Gillan, E.G. Low-temperature solvothermal synthesis of phosphorus-rich transition-metal phosphides. *Chem. Mater.* **2008**, *20*, 2618–2620. [CrossRef]

129. Alexander, A.-M.; Hargreaves, J.S.J. Alternative catalytic materials: Carbides, nitrides, phosphides and amorphous boron alloys. *Chem. Soc. Rev.* **2010**, *39*, 4388–4401. [CrossRef] [PubMed]

130. Shervedani, R.K.; Lasia, A. Studies of the hydrogen evolution reaction on Ni-P electrodes. *J. Electrochem. Soc.* **1997**, *144*, 511–519. [CrossRef]

131. Hu, C.C.; Bai, A. Optimization of hydrogen evolving activity on nickel–phosphorus deposits using experimental strategies. *J. Appl. Electrochem.* **2001**, *31*, 565–572. [CrossRef]

132. Popczun, E.J.; McKone, J.R.; Read, C.G.; Biacchi, A.J.; Wiltrout, A.M.; Lewis, N.S.; Schaak, R.E. Nanostructured nickel phosphide as an electrocatalyst for the hydrogen evolution reaction. *J. Am. Chem. Soc.* **2013**, *135*, 9267–9270. [CrossRef] [PubMed]

133. Kucernak, A.R.J.; Naranammalpuram Sundaram, V.N. Nickel phosphide: The effect of phosphorus content on hydrogen evolution activity and corrosion resistance in acidic medium. *J. Mater. Chem. A* **2014**, *2*, 17435–17445. [CrossRef]

134. Popczun, E.J.; Read, C.G.; Roske, C.W.; Lewis, N.S.; Schaak, R.E. Highly active electrocatalysis of the hydrogen evolution reaction by cobalt phosphide nanoparticles. *Angew. Chem. Int. Ed.* **2014**, *53*, 5427–5430. [CrossRef] [PubMed]

135. Xiao, P.; Sk, M.A.; Thia, L.; Ge, X.; Lim, R.J.; Wang, J.-Y.; Lim, K.H.; Wang, X. Molybdenum phosphide as an efficient electrocatalyst for the hydrogen evolution reaction. *Energy Environ. Sci.* **2014**, *7*, 2624–2629. [CrossRef]

136. McEnaney, J.M.; Chance Crompton, J.; Callejas, J.F.; Popczun, E.J.; Read, C.G.; Lewis, N.S.; Schaak, R.E. Electrocatalytic hydrogen evolution using amorphous tungsten phosphide nanoparticles. *Chem. Commun.* **2014**, *50*, 11026–11028. [CrossRef] [PubMed]

137. Lu, S.; Pan, J.; Huang, A.; Zhuang, L.; Lu, J. Alkaline polymer electrolyte fuel cells completely free from noble metal catalysts. *Proc. Natl. Acad. Sci. USA* **2008**, *105*, 20611–20614. [CrossRef]

138. Sakamoto, T.; Asazawa, K.; Sanabria-Chinchilla, J.; Martinez, U.; Halevi, B.; Atanassov, P.; Strasser, P.; Tanaka, H. Combinatorial discovery of Ni-based binary and ternary catalysts for hydrazine electrooxidation for use in anion exchange membrane fuel cells. *J. Power Sources* **2014**, *247*, 605–611. [CrossRef]

139. Brown, D.E.; Mahmood, M.N.; Man, M.C.M.; Turner, A.K. Preparation and characterization of low overvoltage transition metal alloy electrocatalysts for hydrogen evolution in alkaline solutions. *Electrochim. Acta* **1984**, *29*, 1551–1556. [CrossRef]

140. Bakos, I.; Paszternák, A.; Zitoun, D. Pd/Ni synergestic activity for hydrogen oxidation reaction in alkaline conditions. *Electrochim. Acta* **2015**, *176*, 1074–1082. [CrossRef]

141. Subbaraman, R.; Danilovic, N.; Lopes, P.P.; Tripkovic, D.; Strmcnik, D.; Stamenkovic, V.R.; Markovic, N.M. Origin of anomalous activities for electrocatalysts in alkaline electrolytes. *J. Phys. Chem. C* **2012**, *116*, 22231–22237. [CrossRef]

142. Henning, S.; Herranz, J.; Gasteiger, H.A. Bulk-palladium and palladium-on-gold electrocatalysts for the oxidation of hydrogen in alkaline electrolyte. *J. Electrochem. Soc.* **2015**, *162*, F178–F189. [CrossRef]

143. Subbaraman, R.; Tripkovic, D.; Strmcnik, D.; Chang, K.-C.; Uchimura, M.; Paulikas, A.P.; Stamenkovic, V.; Markovic, N.M. Enhancing hydrogen evolution activity in water splitting by tailoring Li$^+$-Ni(OH)$_2$-Pt interfaces. *Science* **2011**, *334*, 1256–1260. [CrossRef] [PubMed]

144. Scofield, M.E.; Zhou, Y.; Yue, S.; Wang, L.; Su, D.; Tong, X.; Vukmirovic, M.B.; Adzic, R.R.; Wong, S.S. Role of chemical composition in the enhanced catalytic activity of Pt-based alloyed ultrathin nanowires for the hydrogen oxidation reaction under alkaline conditions. *ACS Catal.* **2016**, *6*, 3895–3908. [CrossRef]

145. Scofield, M.E.; Koenigsmann, C.; Wang, L.; Liu, H.; Wong, S.S. Tailoring the composition of ultrathin, ternary alloy ptrufe nanowires for the methanol oxidation reaction and formic acid oxidation reaction. *Energy Environ. Sci.* **2015**, *8*, 350–363. [CrossRef]

146. Alia, S.M.; Larsen, B.A.; Pylypenko, S.; Cullen, D.A.; Diercks, D.R.; Neyerlin, K.C.; Kocha, S.S.; Pivovar, B.S. Platinum-coated nickel nanowires as oxygen-reducing electrocatalysts. *ACS Catal.* **2014**, *4*, 1114–1119. [CrossRef]

147. Kandoi, S.; Ferrin, P.A.; Mavrikakis, M. Hydrogen on and in selected overlayer near-surface alloys and the effect of subsurface hydrogen on the reactivity of alloy surfaces. *Top. Catal.* **2010**, *53*, 384–392. [CrossRef]

148. Rheinländer, P.J.; Herranz, J.; Durst, J.; Gasteiger, H.A. Kinetics of the hydrogen oxidation/evolution reaction on polycrystalline platinum in alkaline electrolyte reaction order with respect to hydrogen pressure. *J. Electrochem. Soc.* **2014**, *161*, F1448–F1457. [CrossRef]

149. Montero, M.A.; Gennero de Chialvo, M.R.; Chialvo, A.C. Kinetics of the hydrogen oxidation reaction on nanostructured rhodium electrodes in alkaline solution. *J. Power Sources* **2015**, *283*, 181–186. [CrossRef]

150. Montero, M.A.; Fernández, J.L.; Gennero de Chialvo, M.R.; Chialvo, A.C. Characterization and kinetic study of a nanostructured rhodium electrode for the hydrogen oxidation reaction. *J. Power Sources* **2014**, *254*, 218–223. [CrossRef]

151. Montero, M.A.; Fernández, J.L.; Gennero de Chialvo, M.R.; Chialvo, A.C. Kinetic study of the hydrogen oxidation reaction on nanostructured iridium electrodes in acid solutions. *J. Phys. Chem. C* **2013**, *117*, 25269–25275. [CrossRef]

152. Pasti, I.A.; Gavrilov, N.M.; Mentus, S.V. Potentiodynamic investigation of oxygen reduction reaction on polycrystalline platinum surface in acidic solutions: The effect of the polarization rate on the kinetic parameters. *Int. J. Electrochem. Sci.* **2012**, *7*, 11076–11090.

153. Lopes, P.P.; Strmcnik, D.; Jirkovsky, J.S.; Connell, J.G.; Stamenkovic, V.; Markovic, N. Double layer effects in electrocatalysis: The oxygen reduction reaction and ethanol oxidation reaction on Au(111), Pt(111) and Ir(111) in alkaline media containing Na and Li cations. *Catal. Today* **2016**, *262*, 41–47. [CrossRef]

154. Wang, J.X.; Markovic, N.M.; Adzic, R.R. Kinetic analysis of oxygen reduction on Pt(111) in acid solutions: Intrinsic kinetic parameters and anion adsorption effects. *J. Phys. Chem. B* **2004**, *108*, 4127–4133. [CrossRef]
155. Markovic, N.; Gasteiger, H.; Ross, P.N. Kinetics of oxygen reduction on Pt(hkl) electrodes: Implications for the crystallite size effect with supported Pt electrocatalysts. *J. Electrochem. Soc.* **1997**, *144*, 1591–1597. [CrossRef]
156. Paulus, U.A.; Schmidt, T.J.; Gasteiger, H.A.; Behm, R.J. Oxygen reduction on a high-surface area Pt/vulcan carbon catalyst: A thin-film rotating ring-disk electrode study. *J. Electroanal. Chem.* **2001**, *495*, 134–145. [CrossRef]
157. Strmcnik, D.; Escudero-Escribano, M.; Kodama, K.; StamenkovicVojislav, R.; Cuesta, A.; Marković, N.M. Enhanced electrocatalysis of the oxygen reduction reaction based on patterning of platinum surfaces with cyanide. *Nat. Chem.* **2010**, *2*, 880–885. [CrossRef] [PubMed]
158. Antolini, E.; Gonzalez, E.R. Alkaline direct alcohol fuel cells. *J. Power Sources* **2010**, *195*, 3431–3450. [CrossRef]
159. Blizanac, B.B.; Ross, P.N.; Markovic, N.M. Oxygen electroreduction on Ag(111): The Ph effect. *Electrochim. Acta* **2007**, *52*, 2264–2271. [CrossRef]
160. Cifrain, M.; Kordesch, K.V. Advances, aging mechanism and lifetime in AFCs with circulating electrolytes. *J. Power Sources* **2004**, *127*, 234–242. [CrossRef]
161. Koscher, G.A.; Kordesch, K. Alkaline methanol–air system. *J. Sol. State Electrochem.* **2003**, *7*, 632–636. [CrossRef]
162. Lin, B.Y.S.; Kirk, D.W.; Thorpe, S.J. Performance of alkaline fuel cells: A possible future energy system? *J. Power Sources* **2006**, *161*, 474–483. [CrossRef]
163. Lewera, A.; Timperman, L.; Roguska, A.; Alonso-Vante, N. Metal–support interactions between nanosized Pt and metal oxides (WO$_3$ and TiO$_2$) studied using X-ray photoelectron spectroscopy. *J. Phys. Chem. C* **2011**, *115*, 20153–20159. [CrossRef]
164. Luo, Y.; Alonso-Vante, N. The effect of support on advanced Pt-based cathodes towards the oxygen reduction reaction. State of the art. *Electrochim. Acta* **2015**, *179*, 108–118. [CrossRef]
165. Yang, W.; Wang, X.; Yang, F.; Yang, C.; Yang, X. Carbon nanotubes decorated with Pt nanocubes by a noncovalent functionalization method and their role in oxygen reduction. *Adv. Mater.* **2008**, *20*, 2579–2587. [CrossRef]
166. Orfanidi, A.; Daletou, M.K.; Neophytides, S.G. Preparation and characterization of Pt on modified multi-wall carbon nanotubes to be used as electrocatalysts for high temperature fuel cell applications. *Appl. Catal. B: Environ.* **2011**, *106*, 379–389. [CrossRef]
167. Li, W.; Liang, C.; Zhou, W.; Qiu, J.; Zhou, Z.; Sun, G.; Xin, Q. Preparation and characterization of multiwalled carbon nanotube-supported platinum for cathode catalysts of direct methanol fuel cells. *J. Phys. Chem. B* **2003**, *107*, 6292–6299. [CrossRef]
168. Chen, Y.; Wang, J.; Liu, H.; Banis, M.N.; Li, R.; Sun, X.; Sham, T.-K.; Ye, S.; Knights, S. Nitrogen doping effects on carbon nanotubes and the origin of the enhanced electrocatalytic activity of supported Pt for proton-exchange membrane fuel cells. *J. Phys. Chem. C* **2011**, *115*, 3769–3776. [CrossRef]
169. Wang, S.; Jiang, S.P.; White, T.J.; Guo, J.; Wang, X. Electrocatalytic activity and interconnectivity of Pt nanoparticles on multiwalled carbon nanotubes for fuel cells. *J. Phys. Chem. C* **2009**, *113*, 18935–18945. [CrossRef]
170. Hijazi, I.; Bourgeteau, T.; Cornut, R.; Morozan, A.; Filoramo, A.; Leroy, J.; Derycke, V.; Jousselme, B.; Campidelli, S. Carbon nanotube-templated synthesis of covalent porphyrin network for oxygen reduction reaction. *J. Am. Chem. Soc.* **2014**, *136*, 6348–6354. [CrossRef] [PubMed]
171. Hasche, F.; Oezaslan, M.; Strasser, P. Activity, stability and degradation of multi walled carbon nanotube (MXCNT) supported Pt fuel cell electrocatalysts. *PCCP* **2010**, *12*, 15251–15258. [CrossRef] [PubMed]
172. Kongkanand, A.; Vinodgopal, K.; Kuwabata, S.; Kamat, P.V. Highly dispersed Pt catalysts on single-walled carbon nanotubes and their role in methanol oxidation. *J. Phys. Chem. B* **2006**, *110*, 16185–16188. [CrossRef] [PubMed]
173. Wang, J.J.; Yin, G.P.; Zhang, J.; Wang, Z.B.; Gao, Y.Z. High utilization platinum deposition on single-walled carbon nanotubes as catalysts for direct methanol fuel cell. *Electrochim. Acta* **2007**, *52*, 7042–7050. [CrossRef]
174. Su, L.; Jia, W.; Zhang, L.; Beacham, C.; Zhang, H.; Lei, Y. Facile synthesis of a platinum nanoflower monolayer on a single-walled carbon nanotube membrane and its application in glucose detection. *J. Phys. Chem. C* **2010**, *114*, 18121–18125. [CrossRef]

175. Seo, M.H.; Choi, S.M.; Seo, J.K.; Noh, S.H.; Kim, W.B.; Han, B. The graphene-supported palladium and palladium–yttrium nanoparticles for the oxygen reduction and ethanol oxidation reactions: Experimental measurement and computational validation. *Appl. Catal. B: Environ.* **2013**, *129*, 163–171. [CrossRef]

176. Rao, C.V.; Reddy, A.L.M.; Ishikawa, Y.; Ajayan, P.M. Synthesis and electrocatalytic oxygen reduction activity of graphene-supported Pt_3Co and Pt_3Cr alloy nanoparticles. *Carbon* **2011**, *49*, 931–936. [CrossRef]

177. Kou, R.; Shao, Y.; Wang, D.; Engelhard, M.H.; Kwak, J.H.; Wang, J.; Viswanathan, V.V.; Wang, C.; Lin, Y.; Wang, Y.; et al. Enhanced activity and stability of Pt catalysts on functionalized graphene sheets for electrocatalytic oxygen reduction. *Electrochem. Commun.* **2009**, *11*, 954–957. [CrossRef]

178. Li, Y.; Tang, L.; Li, J. Preparation and electrochemical performance for methanol oxidation of Pt/graphene nanocomposites. *Electrochem. Commun.* **2009**, *11*, 846–849. [CrossRef]

179. Moldovan, M.S.; Bulou, H.; Dappe, Y.J.; Janowska, I.; Bégin, D.; Pham-Huu, C.; Ersen, O. On the evolution of Pt nanoparticles on few-layer graphene supports in the high-temperature range. *J. Phys. Chem. C* **2012**, *116*, 9274–9282. [CrossRef]

180. Shao, Y.; Zhang, S.; Wang, C.; Nie, Z.; Liu, J.; Wang, Y.; Lin, Y. Highly durable graphene nanoplatelets supported Pt nanocatalysts for oxygen reduction. *J. Power Sources* **2010**, *195*, 4600–4605. [CrossRef]

181. Jahan, M.; Bao, Q.; Loh, K.P. Electrocatalytically active graphene–porphyrin MOF composite for oxygen reduction reaction. *J. Am. Chem. Soc.* **2012**, *134*, 6707–6713. [CrossRef] [PubMed]

182. Jin, T.; Guo, S.; Zuo, J.-L.; Sun, S. Synthesis and assembly of Pd nanoparticles on graphene for enhanced electrooxidation of formic acid. *Nanoscale* **2013**, *5*, 160–163. [CrossRef] [PubMed]

183. Di Noto, V.; Negro, E.; Lavina, S.; Gross, S.; Pace, G. Pd-Co carbon-nitride electrocatalysts for polymer electrolyte fuel cells. *Electrochim. Acta* **2007**, *53*, 1604–1617. [CrossRef]

184. Di Noto, V.; Negro, E.; Polizzi, S.; Riello, P.; Atanassov, P. Preparation, characterization and single-cell performance of a new class of Pd-carbon nitride electrocatalysts for oxygen reduction reaction in PEMFCs. *Appl. Catal. B: Environ.* **2012**, *111–112*, 185–199. [CrossRef]

185. Di Noto, V.; Negro, E.; Polizzi, S.; Vezzù, K.; Toniolo, L.; Cavinato, G. Synthesis, studies and fuel cell performance of "core–shell" electrocatalysts for oxygen reduction reaction based on a $PtNi_x$ carbon nitride "shell" and a pyrolyzed polyketone nanoball "core". *Int. J. Hydrogen Energy* **2014**, *39*, 2812–2827. [CrossRef]

186. Di Noto, V.; Negro, E.; Polizzi, S.; Agresti, F.; Giffin, G.A. Synthesis–structure–morphology interplay of bimetallic "core–shell" carbon nitride nano-electrocatalysts. *ChemSusChem* **2012**, *5*, 2451–2459. [CrossRef] [PubMed]

187. Datye, A.K.; Kalakkad, D.S.; Yao, M.H.; Smith, D.J. Comparison of metal-support interactions in Pt/TiO_2 and Pt/CeO_2. *J. Catal.* **1995**, *155*, 148–153. [CrossRef]

188. Koudelka, M.; Monnier, A.; Sanchez, J.; Augustynski, J. Correlation between the surface composition of Pt/TiO_2 catalysts and their adsorption behaviour in aqueous solutions. *J. Mol. Catal.* **1984**, *25*, 295–305. [CrossRef]

189. Jiang, D.-e.; Overbury, S.H.; Dai, S. Structures and energetics of Pt clusters on TiO_2: Interplay between metal–metal bonds and metal–oxygen bonds. *J. Phys. Chem. C* **2012**, *116*, 21880–21885. [CrossRef]

190. Bauer, A.; Lee, K.; Song, C.; Xie, Y.; Zhang, J.; Hui, R. Pt nanoparticles deposited on TiO_2 based nanofibers: Electrochemical stability and oxygen reduction activity. *J. Power Sources* **2010**, *195*, 3105–3110. [CrossRef]

191. Huang, S.-Y.; Ganesan, P.; Popov, B.N. Electrocatalytic activity and stability of niobium-doped titanium oxide supported platinum catalyst for polymer electrolyte membrane fuel cells. *Appl. Catal. B: Environ.* **2010**, *96*, 224–231. [CrossRef]

192. Huang, S.-Y.; Ganesan, P.; Popov, B.N. Titania supported platinum catalyst with high electrocatalytic activity and stability for polymer electrolyte membrane fuel cell. *Appl. Catal. B: Environ.* **2011**, *102*, 71–77. [CrossRef]

193. Ho, V.T.T.; Pan, C.-J.; Rick, J.; Su, W.-N.; Hwang, B.-J. Nanostructured $Ti_{0.7}Mo_{0.3}O_2$ support enhances electron transfer to Pt: High-performance catalyst for oxygen reduction reaction. *J. Am. Chem. Soc.* **2011**, *133*, 11716–11724. [CrossRef] [PubMed]

194. Estudillo-Wong, L.A.; Luo, Y.; Diaz-Real, J.A.; Alonso-Vante, N. Enhanced oxygen reduction reaction stability on platinum nanoparticles photo-deposited onto oxide-carbon composites. *Appl. Catal. B: Environ.* **2015**, *187*, 291–300. [CrossRef]

195. Luo, Y.; Calvillo, L.; Daiguebonne, C.; Daletou, M.K.; Granozzi, G.; Alonso-Vante, N. A highly efficient and stable oxygen reduction reaction on Pt/CeO$_x$/C electrocatalyst obtained via a sacrificial precursor based on a metal-organic framework. *Appl. Catal. B: Environ.* **2016**, *189*, 39–50. [CrossRef]

196. Ruiz Camacho, B.; Morais, C.; Valenzuela, M.A.; Alonso-Vante, N. Enhancing oxygen reduction reaction activity and stability of platinum via oxide-carbon composites. *Catal. Today* **2013**, *202*, 36–43. [CrossRef]

197. Luo, Y.; Habrioux, A.; Calvillo, L.; Granozzi, G.; Alonso-Vante, N. Thermally induced strains on the catalytic activity and stability of Pt–M$_2$O$_3$/C (M = Y or Gd) catalysts towards oxygen reduction reaction. *ChemCatChem* **2015**, *7*, 1573–1582. [CrossRef]

198. Luo, Y.; Habrioux, A.; Calvillo, L.; Granozzi, G.; Alonso-Vante, N. Yttrium oxide/gadolinium oxide-modified platinum nanoparticles as cathodes for the oxygen reduction reaction. *ChemPhysChem* **2014**, *15*, 2136–2144. [CrossRef] [PubMed]

199. Tauster, S.J. Strong metal-support interactions. *Acc. Chem. Res.* **1987**, *20*, 389–394. [CrossRef]

200. Tauster, S.J.; Fung, S.C.; Garten, R.L. Strong metal-support interactions. Group 8 noble metals supported on titanium dioxide. *J. Am. Chem. Soc.* **1978**, *100*, 170–175. [CrossRef]

201. Ma, J.; Habrioux, A.; Morais, C.; Lewera, A.; Vogel, W.; Verde-Gómez, Y.; Ramos-Sanchez, G.; Balbuena, P.B.; Alonso-Vante, N. Spectroelectrochemical probing of the strong interaction between platinum nanoparticles and graphitic domains of carbon. *ACS Catal.* **2013**, *3*, 1940–1950. [CrossRef]

202. Ma, J.; Habrioux, A.; Pisarek, M.; Lewera, A.; Alonso-Vante, N. Induced electronic modification of Pt nanoparticles deposited onto graphitic domains of carbon materials by uv irradiation. *Electrochem. Commun.* **2013**, *29*, 12–16. [CrossRef]

203. Fleisch, T.H.; Hicks, R.F.; Bell, A.T. An XPS study of metal-support interactions on PdSiO$_2$ and PdLa$_2$O$_3$. *J. Catal.* **1984**, *87*, 398–413. [CrossRef]

204. Dall'Agnol, C.; Gervasini, A.; Morazzoni, F.; Pinna, F.; Strukul, G.; Zanderighi, L. Hydrogenation of carbon monoxide: Evidence of a strong metal-support interaction in RhZrO$_2$ catalysts. *J. Catal.* **1985**, *96*, 106–114. [CrossRef]

205. Wu, G.; Zelenay, P. Nanostructured nonprecious metal catalysts for oxygen reduction reaction. *Acc. Chem. Res.* **2013**, *46*, 1878–1889. [CrossRef] [PubMed]

206. Wu, G.; Mack, N.H.; Gao, W.; Ma, S.; Zhong, R.; Han, J.; Baldwin, J.K.; Zelenay, P. Nitrogen-doped graphene-rich catalysts derived from heteroatom polymers for oxygen reduction in nonaqueous lithium–O$_2$ battery cathodes. *ACS Nano* **2012**, *6*, 9764–9776. [CrossRef] [PubMed]

207. Wu, G.; More, K.L.; Johnston, C.M.; Zelenay, P. High-performance electrocatalysts for oxygen reduction derived from polyaniline, iron, and cobalt. *Science* **2011**, *332*, 443–447. [CrossRef] [PubMed]

208. Fu, T.; Wang, M.; Cai, W.; Cui, Y.; Gao, F.; Peng, L.; Chen, W.; Ding, W. Acid-resistant catalysis without use of noble metals: Carbon nitride with underlying nickel. *ACS Catal.* **2014**, *4*, 2536–2543. [CrossRef]

209. Jiang, T.; Zhang, Q.; Wang, T.-J.; Zhang, Q.; Ma, L.-L. High yield of pentane production by aqueous-phase reforming of xylitol over Ni/HZSM-5 and Ni/MCM22 catalysts. *Energy Convers. Manag.* **2012**, *59*, 58–65. [CrossRef]

210. Luo, Y.; Estudillo-Wong, L.A.; Alonso-Vante, N. Carbon supported Pt-Y$_2$O$_3$ and Pt-Gd$_2$O$_3$ nanoparticles prepared via carbonyl chemical route towards oxygen reduction reaction: Kinetics and stability. *Int. J. Hydrogen Energy* **2016**, in press. [CrossRef]

211. Vogel, W.; Timperman, L.; Alonso-Vante, N. Probing metal substrate interaction of Pt nanoparticles: Structural XRD analysis and oxygen reduction reaction. *Appl. Catal. A* **2010**, *377*, 167–173. [CrossRef]

212. Ma, J.; Habrioux, A.; Gago, A.; Alonso-Vante, N. Towards understanding the essential role played by the platinum-support interaction on electrocatalytic activity. *ECS Trans.* **2013**, *45*, 25–33. [CrossRef]

213. Ma, J.; Habrioux, A.; Guignard, N.; Alonso-Vante, N. Functionalizing effect of increasingly graphitic carbon supports on carbon-supported and TiO$_2$–carbon composite-supported Pt nanoparticles. *J. Phys. Chem. C* **2012**, *116*, 21788–21794. [CrossRef]

214. Matter, P.H.; Zhang, L.; Ozkan, U.S. The role of nanostructure in nitrogen-containing carbon catalysts for the oxygen reduction reaction. *J. Catal.* **2006**, *239*, 83–96. [CrossRef]

215. Wang, D.; Xin, H.L.; Hovden, R.; Wang, H.; Yu, Y.; Muller, D.A.; DiSalvo, F.J.; Abruña, H.D. Structurally ordered intermetallic platinum–cobalt core–shell nanoparticles with enhanced activity and stability as oxygen reduction electrocatalysts. *Nat. Mater.* **2013**, *12*, 81–87. [CrossRef] [PubMed]

216. Wang, D.; Xin, H.L.; Wang, H.; Yu, Y.; Rus, E.; Muller, D.A.; DiSalvo, F.J.; Abruña, H.D. Facile synthesis of carbon-supported Pd–Co core–shell nanoparticles as oxygen reduction electrocatalysts and their enhanced activity and stability with monolayer pt decoration. *Chem. Mater.* **2012**, *24*, 2274–2281. [CrossRef]

217. Wang, D.; Xin, H.L.; Yu, Y.; Wang, H.; Rus, E.; Muller, D.A.; Abruña, H.D. Pt-decorated PdCo@Pd/C core–shell nanoparticles with enhanced stability and electrocatalytic activity for the oxygen reduction reaction. *J. Am. Chem. Soc.* **2010**, *132*, 17664–17666. [CrossRef] [PubMed]

218. Lee, J.-S.; Park, G.S.; Lee, H.I.; Kim, S.T.; Cao, R.; Liu, M.; Cho, J. Ketjenblack carbon supported amorphous manganese oxides nanowires as highly efficient electrocatalyst for oxygen reduction reaction in alkaline solutions. *Nano Lett.* **2011**, *11*, 5362–5366. [CrossRef] [PubMed]

219. Cui, C.; Gan, L.; Heggen, M.; Rudi, S.; Strasser, P. Compositional segregation in shaped Pt alloy nanoparticles and their structural behaviour during electrocatalysis. *Nat. Mater.* **2013**, *12*, 765–771. [CrossRef] [PubMed]

220. Chhina, H.; Campbell, S.; Kesler, O. An oxidation-resistant indium tin oxide catalyst support for proton exchange membrane fuel cells. *J. Power Sources* **2006**, *161*, 893–900. [CrossRef]

221. Millington, B.; Whipple, V.; Pollet, B.G. A novel method for preparing proton exchange membrane fuel cell electrodes by the ultrasonic-spray technique. *J. Power Sources* **2011**, *196*, 8500–8508. [CrossRef]

222. Malacrida, P.; Escudero-Escribano, M.; Verdaguer-Casadevall, A.; Stephens, I.E.L.; Chorkendorff, I. Enhanced activity and stability of Pt-La and Pt-Ce alloys for oxygen electroreduction: The elucidation of the active surface phase. *J. Mater. Chem. A* **2014**, *2*, 4234–4243. [CrossRef]

223. Escudero-Escribano, M.; Verdaguer-Casadevall, A.; Malacrida, P.; Grønbjerg, U.; Knudsen, B.P.; Jepsen, A.K.; Rossmeisl, J.; Stephens, I.E.L.; Chorkendorff, I. Pt$_5$Gd as a highly active and stable catalyst for oxygen electroreduction. *J. Am. Chem. Soc.* **2012**, *134*, 16476–16479. [CrossRef] [PubMed]

224. Hernandez-Fernandez, P.; Masini, F.; McCarthy, D.N.; Strebel, C.E.; Friebel, D.; Deiana, D.; Malacrida, P.; Nierhoff, A.; Bodin, A.; Wise, A.M.; et al. Mass-selected nanoparticles of Pt$_x$Y as model catalysts for oxygen electroreduction. *Nat. Chem.* **2014**, *6*, 732–738. [CrossRef] [PubMed]

225. Jeon, M.K.; McGinn, P.J. Carbon supported Pt–Y electrocatalysts for the oxygen reduction reaction. *J. Power Sources* **2011**, *196*, 1127–1131. [CrossRef]

226. Nishanth, K.G.; Sridhar, P.; Pitchumani, S. Enhanced oxygen reduction reaction activity through spillover effect by Pt–Y(OH)$_3$/C catalyst in direct methanol fuel cells. *Electrochem. Commun.* **2011**, *13*, 1465–1468. [CrossRef]

227. Xu, J.; Gao, P.; Zhao, T.S. Non-precious Co$_3$O$_4$ nano-rod electrocatalyst for oxygen reduction reaction in anion-exchange membrane fuel cells. *Energy Environ. Sci.* **2012**, *5*, 5333–5339. [CrossRef]

228. Unni, S.M.; Mora-Hernandez, J.M.; Kurungot, S.; Alonso-Vante, N. CoSe$_2$ supported on nitrogen-doped carbon nanohorns as a methanol-tolerant cathode for air-breathing microlaminar flow fuel cells. *ChemElectroChem* **2015**, *2*, 1339–1345. [CrossRef]

229. Zhu, S.; Xu, G. Single-walled carbon nanohorns and their applications. *Nanoscale* **2010**, *2*, 2538–2549. [CrossRef] [PubMed]

230. Kasuya, D.; Yudasaka, M.; Takahashi, K.; Kokai, F.; Iijima, S. Selective production of single-wall carbon nanohorn aggregates and their formation mechanism. *J. Phys. Chem. B* **2002**, *106*, 4947–4951. [CrossRef]

231. Guo, D.-J.; Jing, Z.-H. A novel co-precipitation method for preparation of Pt-CeO$_2$ composites on multi-walled carbon nanotubes for direct methanol fuel cells. *J. Power Sources* **2010**, *195*, 3802–3805. [CrossRef]

232. Wang, X.; Li, W.; Chen, Z.; Waje, M.; Yan, Y. Durability investigation of carbon nanotube as catalyst support for proton exchange membrane fuel cell. *J. Power Sources* **2006**, *158*, 154–159. [CrossRef]

233. Wu, G.; More, K.L.; Xu, P.; Wang, H.-L.; Ferrandon, M.; Kropf, A.J.; Myers, D.J.; Ma, S.; Johnston, C.M.; Zelenay, P. A carbon-nanotube-supported graphene-rich non-precious metal oxygen reduction catalyst with enhanced performance durability. *Chem. Commun.* **2013**, *49*, 3291–3293. [CrossRef] [PubMed]

234. Jaouen, F.; Proietti, E.; Lefevre, M.; Chenitz, R.; Dodelet, J.-P.; Wu, G.; Chung, H.T.; Johnston, C.M.; Zelenay, P. Recent advances in non-precious metal catalysis for oxygen-reduction reaction in polymer electrolyte fuel cells. *Energy Environ. Sci.* **2011**, *4*, 114–130. [CrossRef]

235. Li, Q.; Xu, P.; Gao, W.; Ma, S.; Zhang, G.; Cao, R.; Cho, J.; Wang, H.-L.; Wu, G. Graphene/graphene-tube nanocomposites templated from cage-containing metal-organic frameworks for oxygen reduction in Li–O$_2$ batteries. *Adv. Mater.* **2014**, *26*, 1378–1386. [CrossRef] [PubMed]

236. Estudillo-Wong, L.A.; Vargas-Gómez, A.M.; Arce-Estrada, E.M.; Manzo-Robledo, A. TiO$_2$/C composite as a support for Pd-nanoparticles toward the electrocatalytic oxidation of methanol in alkaline media. *Electrochim. Acta* **2013**, *112*, 164–170. [CrossRef]

237. Xiong, L.; Manthiram, A. Synthesis and characterization of methanol tolerant Pt/TiO$_x$/C nanocomposites for oxygen reduction in direct methanol fuel cells. *Electrochim. Acta* **2004**, *49*, 4163–4170. [CrossRef]

238. Andrew Lin, K.-Y.; Hsu, F.-K.; Lee, W.-D. Magnetic cobalt-graphene nanocomposite derived from self-assembly of MOFs with graphene oxide as an activator for peroxymonosulfate. *J. Mater. Chem. A* **2015**, *3*, 9480–9490. [CrossRef]

239. Guo, D.-J.; Cui, S.-K.; Sun, H. Preparation of Pt–CeO$_2$/MWNT nano-composites by reverse micellar method for methanol oxidation. *J. Nanoparticle Res.* **2009**, *11*, 707–712. [CrossRef]

240. Von Kraemer, S.; Wikander, K.; Lindbergh, G.; Lundblad, A.; Palmqvist, A.E.C. Evaluation of TiO$_2$ as catalyst support in Pt-TiO$_2$/C composite cathodes for the proton exchange membrane fuel cell. *J. Power Sources* **2008**, *180*, 185–190. [CrossRef]

241. Sun, Z.; Wang, X.; Liu, Z.; Zhang, H.; Yu, P.; Mao, L. Pt–Ru/CeO$_2$/carbon nanotube nanocomposites: An efficient electrocatalyst for direct methanol fuel cells. *Langmuir* **2010**, *26*, 12383–12389. [CrossRef] [PubMed]

242. Ma, J.; Habrioux, A.; Luo, Y.; Ramos-Sanchez, G.; Calvillo, L.; Granozzi, G.; Balbuena, P.B.; Alonso-Vante, N. Electronic interaction between platinum nanoparticles and nitrogen-doped reduced graphene oxide: Effect on the oxygen reduction reaction. *J. Mater. Chem. A* **2015**, *3*, 11891–11904. [CrossRef]

243. Sreekuttan, M.U.; Campos-Roldan, C.A.; Mora-Hernandez, J.M.; Luo, Y.; Estudillo-Wong, L.A.; Alonso-Vante, N. The effect of carbon-based substrates onto non-precious and precious electrocatalytic centers. *ECS Trans.* **2015**, *69*, 35–42. [CrossRef]

244. Lee, C.-L.; Huang, C.-H.; Huang, K.-L.; Tsai, Y.-L.; Yang, C.-C. A comparison of three carbon nanoforms as catalyst supports for the oxygen reduction reaction. *Carbon* **2013**, *60*, 392–400. [CrossRef]

245. Liu, Z.-T.; Huang, K.-L.; Wu, Y.-S.; Lyu, Y.-P.; Lee, C.-L. A comparison of physically and chemically defective graphene nanosheets as catalyst supports for cubic Pd nanoparticles in an alkaline oxygen reduction reaction. *Electrochim. Acta* **2015**, *186*, 552–561. [CrossRef]

246. Thanh Ho, V.T.; Pillai, K.C.; Chou, H.-L.; Pan, C.-J.; Rick, J.; Su, W.-N.; Hwang, B.-J.; Lee, J.-F.; Sheu, H.-S.; Chuang, W.-T. Robust non-carbon Ti$_{0.7}$Ru$_{0.3}$O$_2$ support with co-catalytic functionality for Pt: Enhances catalytic activity and durability for fuel cells. *Energy Environ. Sci.* **2011**, *4*, 4194–4200. [CrossRef]

247. Timperman, L.; Alonso-Vante, N. Oxide substrate effect toward electrocatalytic enhancement of platinum and ruthenium–selenium catalysts. *Electrocatalysis* **2011**, *2*, 181–191. [CrossRef]

248. Scotti, G.; Kanninen, P.; Matilainen, V.-P.; Salminen, A.; Kallio, T. Stainless steel micro fuel cells with enclosed channels by laser additive manufacturing. *Energy* **2016**, *106*, 475–481. [CrossRef]

249. Wang, R.; Xu, Z. Recycling of non-metallic fractions from waste electrical and electronic equipment (WEEE): A review. *Waste Manag.* **2014**, *34*, 1455–1469. [CrossRef] [PubMed]

250. Cadena, L.E.S.; Arroyo, Z.G.; Lara, M.A.G.; Quiroz, Q.D. Cell-phone recycling by solvolysis for recovery of metals. *J. Mater. Sci. Chem. Eng.* **2015**, *3*, 52–57. [CrossRef]

251. Xiao, Z.; Feng, C.; Chan, P.C.H.; Hsing, I.M. Monolithically integrated planar microfuel cell arrays. *Sensors Actuators B: Chem.* **2008**, *132*, 576–586. [CrossRef]

252. Cha, S.-W.; O'Hayre, R.; Lee, S.J.; Saito, Y.; Prinz, F.B. Geometric scale effect of flow channels on performance of fuel cells. *J. Electrochem. Soc.* **2004**, *151*, A1856–A1864. [CrossRef]

253. Lu, Y.; Reddy, R.G. Performance of micro-PEM fuel cells with different flow fields. *J. Power Sources* **2010**, *195*, 503–508. [CrossRef]

254. Hockaday, R.; Navas, C. Micro-fuel cells ^{TM} for portable electronics. *Fuel Cells Bull.* **1999**, *2*, 9–12. [CrossRef]

255. Mex, L.; Ponath, N.; Müller, J. Miniaturized fuel cells based on microsystem technologies. *Fuel Cells Bull.* **2001**, *4*, 9–12. [CrossRef]

256. Breakthrough in micro fuel cell design, architecture. *Fuel Cells Bull.* **2002**, 1. Available online: http://thirdworld.nl/breakthrough-in-micro-fuel-cell-design-architecture (accessed on 15 September 2016).

257. Micro fuel cell developed at llnl. *Fuel Cells Bull.* **2002**, 4. Available online: http://thirdworld.nl/breakthrough-in-micro-fuel-cell-design-architecture (accessed on 15 September 2016).

258. Lee, S.J.; Chang-Chien, A.; Cha, S.W.; O'Hayre, R.; Park, Y.I.; Saito, Y.; Prinz, F.B. Design and fabrication of a micro fuel cell array with "flip-flop" interconnection. *J. Power Sources* **2002**, *112*, 410–418. [CrossRef]

259. Blum, A.; Duvdevani, T.; Philosoph, M.; Rudoy, N.; Peled, E. Water-neutral micro direct-methanol fuel cell (DMFC) for portable applications. *J. Power Sources* **2003**, *117*, 22–25. [CrossRef]
260. Yu, J.; Cheng, P.; Ma, Z.; Yi, B. Fabrication of miniature silicon wafer fuel cells with improved performance. *J. Power Sources* **2003**, *124*, 40–46. [CrossRef]
261. Choban, E.R.; Markoski, L.J.; Wieckowski, A.; Kenis, P.J.A. Microfluidic fuel cell based on laminar flow. *J. Power Sources* **2004**, *128*, 54–60. [CrossRef]
262. Choban, E.R.; Spendelow, J.S.; Gancs, L.; Wieckowski, A.; Kenis, P.J.A. Membraneless laminar flow-based micro fuel cells operating in alkaline, acidic, and acidic/alkaline media. *Electrochim. Acta* **2005**, *50*, 5390–5398. [CrossRef]
263. McLean, G.F.; Niet, T.; Prince-Richard, S.; Djilali, N. An assessment of alkaline fuel cell technology. *Int. J. Hydrogen Energy* **2002**, *27*, 507–526. [CrossRef]
264. Tripković, A.V.; Popović, K.D.; Lović, J.D.; Jovanović, V.M.; Kowal, A. Methanol oxidation at platinum electrodes in alkaline solution: Comparison between supported catalysts and model systems. *J. Electroanal. Chem.* **2004**, *572*, 119–128. [CrossRef]
265. Yu, E.H.; Scott, K. Development of direct methanol alkaline fuel cells using anion exchange membranes. *J. Power Sources* **2004**, *137*, 248–256. [CrossRef]
266. Hsieh, S.-S.; Yang, S.-H.; Kuo, J.-K.; Huang, C.-F.; Tsai, H.-H. Study of operational parameters on the performance of micro PEMFCs with different flow fields. *Energy Convers. Manag.* **2006**, *47*, 1868–1878. [CrossRef]
267. Nakajima, H.; Konomi, T.; Kitahara, T. Direct water balance analysis on a polymer electrolyte fuel cell (PEFC): Effects of hydrophobic treatment and micro-porous layer addition to the gas diffusion layer of a PEFC on its performance during a simulated start-up operation. *J. Power Sources* **2007**, *171*, 457–463. [CrossRef]
268. Hsieh, S.-S.; Huang, Y.-J. Measurements of current and water distribution for a micro-pem fuel cell with different flow fields. *J. Power Sources* **2008**, *183*, 193–204. [CrossRef]
269. Park, B.Y.; Madou, M.J. Design, fabrication, and initial testing of a miniature pem fuel cell with micro-scale pyrolyzed carbon fluidic plates. *J. Power Sources* **2006**, *162*, 369–379. [CrossRef]
270. Lin, P.-C.; Park, B.Y.; Madou, M.J. Development and characterization of a miniature PEM fuel cell stack with carbon bipolar plates. *J. Power Sources* **2008**, *176*, 207–214. [CrossRef]
271. Whipple, D.T.; Jayashree, R.S.; Egas, D.; Alonso-Vante, N.; Kenis, P.J.A. Ruthenium cluster-like chalcogenide as a methanol tolerant cathode catalyst in air-breathing laminar flow fuel cells. *Electrochim. Acta* **2009**, *54*, 4384–4388. [CrossRef]
272. Cheng, H.; Yuan, W.; Scott, K. Influence of thermal treatment on RuSe cathode materials for direct methanol fuel cells. *Fuel Cells* **2007**, *7*, 16–20. [CrossRef]
273. Colmenares, L.; Jusys, Z.; Behm, R.J. Activity, selectivity, and methanol tolerance of Se-modified Ru/C cathode catalysts. *J. Phys. Chem. C* **2007**, *111*, 1273–1283. [CrossRef]
274. Cremers, C.; Scholz, M.; Seliger, W.; Racz, A.; Knechtel, W.; Rittmayr, J.; Grafwallner, F.; Peller, H.; Stimming, U. Developments for improved direct methanol fuel cell stacks for portable power. *Fuel Cells* **2007**, *7*, 21–31. [CrossRef]
275. Itoe, R.N.; Wesson, G.D.; Kalu, E.E. Evaluation of oxygen transport parameters in H_2SO_4-CH_3OH mixtures using electrochemical methods. *J. Electrochem. Soc.* **2000**, *147*, 2445–2450. [CrossRef]
276. Gago, A.S.; Morales-Acosta, D.; Arriaga, L.G.; Alonso-Vante, N. Carbon supported ruthenium chalcogenide as cathode catalyst in a microfluidic formic acid fuel cell. *J. Power Sources* **2011**, *196*, 1324–1328. [CrossRef]
277. Brushett, F.R.; Naughton, M.S.; Ng, J.W.D.; Yin, L.; Kenis, P.J.A. Analysis of Pt/C electrode performance in a flowing-electrolyte alkaline fuel cell. *Int. J. Hydrogen Energy* **2012**, *37*, 2559–2570. [CrossRef]
278. Gago, A.S.; Gochi-Ponce, Y.; Feng, Y.-J.; Esquivel, J.P.; Sabaté, N.; Santander, J.; Alonso-Vante, N. Tolerant chalcogenide cathodes of membraneless micro fuel cells. *ChemSusChem* **2012**, *5*, 1488–1494. [CrossRef] [PubMed]

279. Gago, A.S.; Esquivel, J.-P.; Sabaté, N.; Santander, J.; Alonso-Vante, N. Comprehensive characterization and understanding of micro-fuel cells operating at high methanol concentrations. *Beilstein J. Nanotechnol.* **2015**, *6*, 2000–2006. [CrossRef] [PubMed]

280. Ma, J.; Habrioux, A.; Morais, C.; Alonso-Vante, N. Electronic modification of Pt via Ti and Se as tolerant cathodes in air-breathing methanol microfluidic fuel cells. *PhysChemChemPhys* **2014**, *16*, 13820–13826. [CrossRef] [PubMed]

catalysts

MDPI

Review

Palladium-Based Catalysts as Electrodes for Direct Methanol Fuel Cells: A Last Ten Years Review

Juan Carlos Calderón Gómez, Rafael Moliner and Maria Jesus Lázaro *

Instituto de Carboquímica (CSIC), Miguel Luesma Castán 4, Zaragoza 50018, Spain;
jccalderon@icb.csic.es (J.C.C.G.); rmoliner@icb.csic.es (R.M.)
* Correspondence: mlazaro@icb.csic.es; Tel.: +34-976-733-977

Academic Editors: Vincenzo Baglio and David Sebastián
Received: 14 July 2016; Accepted: 22 August 2016; Published: 27 August 2016

Abstract: Platinum-based materials are accepted as the suitable electrocatalysts for anodes and cathodes in direct methanol fuel cells (DMFCs). Nonetheless, the increased demand and scarce world reserves of Pt, as well as some technical problems associated with its use, have motivated a wide research focused to design Pd-based catalysts, considering the similar properties between this metal and Pt. In this review, we present the most recent advancements about Pd-based catalysts, considering Pd, Pd alloys with different transition metals and non-carbon supported nanoparticles, as possible electrodes in DMFCs. In the case of the anode, different reported works have highlighted the capacity of these new materials for overcoming the CO poisoning and promote the oxidation of other intermediates generated during the methanol oxidation. Regarding the cathode, the studies have showed more positive onset potentials, as fundamental parameter for determining the mechanism of the oxygen reduction reaction (ORR) and thus, making them able for achieving high efficiencies, with less production of hydrogen peroxide as collateral product. This revision suggests that it is possible to replace the conventional Pt catalysts by Pd-based materials, although several efforts must be made in order to improve their performance in DMFCs.

Keywords: palladium catalysts; direct methanol fuel cells; methanol oxidation; oxygen reduction; non-Pt content

1. Introduction

Direct methanol fuel cells (DMFCs) are alternative energy sources being employed in portable and electronic devices, considering the evident increase of energy price during the last years. Methanol is the fuel used in these devices, which has been recognized due to its high energy density and its easy handling [1]. Therefore, many research articles have been devoted to the study of them, in order to promote their entrance into the market. However, the high price of DMFCs is the main factor avoiding its commercialization, in spite of the progress in prototypes and mass production design, which does not have a direct relationship with the materials costs [2]. The high cost of these devices comes from the use of platinum and platinum alloys for methanol oxidation and oxygen reduction, the electrochemical reactions performed on anode and cathode, respectively. Some possible alternatives focused in the use of low platinum contents have been studied [3], particularly in the case of the cathode, suggesting materials including heat treated transition metals macrocycles [4,5], ruthenium chalcogenides [6,7] and palladium alloys [7,8]. Particularly, replacement of platinum and its alloys in the anode require more investigation, being the best solution the use of non-platinum catalysts in alkaline electrolytes [9,10].

Within this context, palladium has been suggested for substituting the platinum in anodes and cathodes of DMFCs because of the similarities between these metals, as well as the major abundance and low cost of Pd [11]. Mining sources of palladium are more abundant than those for Pt, a fact that makes Pd cheaper ($654.1 per oz.) than Pt ($1796.9 per oz.) [12]. Furthermore, in the case of the anode, palladium has displayed high tolerance towards CO poisoning [13] and high catalytic activity for alcohols oxidation in alkaline medium [12,14,15]. Regarding the cathode, Pd-based catalysts have also arisen as an alternative to perform the oxygen reduction reaction (ORR), especially if this metal is combined with other transition metals, which induce changes in the Pd electronic structure, and thus, increasing its activity [16].

In this work, a review of the progress of Pd-based catalysts for both anodes and cathodes in DMFCs is presented, considering two categories: carbon-supported and non-carbon-supported catalysts. The effect of second metals and novel supports in the catalytic activity is discussed, in order to explain why these Pd catalysts could be suggested as electrodes for direct methanol fuel cells.

2. Methanol Oxidation on Pd-Based Catalysts

2.1. Carbon-Supported Pd-Alloys

The role of carbon materials as support for catalysts in all the categories of polymer electrolyte membrane fuel cells is well known, and is related with different factors: (1) enhancement of high electroactive area, caused by a better dispersion of nanoparticles [17]; (2) improved electroactive species diffusion through the porous structure of carbon support [17]; and (3) promotion of the electronic transfer, either for the presence of surface functional groups or the decrease in Fermi level of the catalysts [18]. Therefore, a way for increasing the activity of catalysts is upgrading the electrical and morphological properties of carbon supports, a fact applied for both Pt-based and non Pt-based electrocatalysts. In this sense, graphene has been suggested as a good alternative to support Pd nanoparticles in order to improve the performance of anodic catalysts during the methanol oxidation reaction (MOR). Zhang and co-workers have found enhanced electrocatalytic activity for methanol oxidation when these kinds of materials are employed as anode catalysts, which was attributed to the large electrochemically active surface area (ECSA) generated by the large amount of active edge sites present in the graphene. These sites are able to anchor the Pd nanoparticles and modify the electronic properties of them, obtaining high currents in the forward scan and thus improved efficiency toward the production of CO_2 as principal product during the methanol oxidation [19]. In other work [20], the synthesis of ultrafine Pd nanoparticles (NPs) supported on N- and S-modified graphenes was reported. In this case, the effects caused by the high surface area of graphene were represented in the increased electrocatalytic performance during the methanol electro-oxidation, as well as the catalytic stability in comparison with the corrosion resistance displayed by Pd nanoparticles supported on both, undoped graphene and Vulcan XC-72R carbon black. Another example of the remarkable benefits in the catalytic activity caused by graphenes can be seen in a recent work reported by Zhang et al. [21], which prepared hybrid composites between graphitic carbon nitride and reduced graphene oxides as support for Pd nanostructures. These materials displayed high forward peak current densities, which were explained from: (1) the large ECSA; (2) the presence of planar groups that modified interactions between the support and the nanoparticles; and (3) the long-term stability of the composites. In fact, the current densities overcame those observed for Pd nanoparticles supported on reduced graphenes and a commercial Pd catalyst supported on activated carbon. Table 1 displayed some electrochemical parameters determined for catalysts supported on graphenes.

Table 1. Electrochemical properties of Pd catalysts supported on graphenes, doped graphenes, RGO, reduced graphene oxide.

Catalyst	Electrochemical Parameter			Reference
	Onset Potential (V vs. RHE)	Anodic Peak Potential (V vs. RHE)	Anodic Peak Current (mA mg^{-1} Pd)	
Pd/CNNF-G	0.420	0.800	1780	
Pd/C$_3$N$_4$-RGO	0.450	0.800	1550	
Pd/RGO	0.570	0.810	860	[19]
Pd/CNT	0.570	0.850	700	
Pd/AC	0.580	0.870	550	
Pd/C$_3$N$_4$	0.780	0.870	80	
Catalyst	Onset Potential (V vs. RHE)	Anodic Peak Potential (V vs. Hg/HgO)	Anodic Peak Current (mA cm^{-2})	Reference
Pd/NS-G	0.472	0.880	12.5	
Pd/G	0.600	0.880	7	[20]
Pd/C	0.520	0.890	5.6	

Generally, Pd-based catalysts are supported on carbon blacks, bearing in mind its high conductivity and low cost. An example of these catalysts are the Pd-M alloys (M = Ag, Ni, Rh, and Au), which have displayed different activities towards the methanol oxidation. Incorporation of Ag into the crystalline structure of Pd is one of the most studied non-Pt alloys, because of its high performance. Yin et al. [22] reported upper activities for Pd-Ag catalysts supported on Vulcan XC-72 carbon black, especially at Pd:Ag atomic ratios close to 65:35 and 46:54. In other work, Wang et al. [23] supported Pd-Ag nanoparticles on carbon blacks, varying the Ag content and finding the Pd:Ag 1:1 atomic ratio as the most efficient composition in terms of electrocatalytic activity in alkaline medium. Ag induced a decrease in the methanol oxidation onset potential (0.436 V vs. RHE), in comparison with that observed for a Pd/C catalyst (0.536 V vs. RHE). Furthermore, the authors reported that addition of Ag promoted the easy removal of CO$_{ads}$, increasing the number of active sites able to adsorb and oxidize methanol. The causes associated with the activity enhancement by Ag presence in the alloy were explained from a displacement in the d-band center of Pd, affecting the electronic properties of this metal and the activation of water at lower potentials in comparison with those required for Pd, which participates in the oxidation of CO$_{ads}$. Although the authors demonstrated that presence of Ag in these carbon black-supported materials is profitable for the activity of Pd, the major increase in the activity of this noble metal was achieved for the Pd-Ag alloy supported on carbon nanotubes (CNTs). They demonstrated that catalysts supported on CNTs possessed bigger electroactive surface and better catalytic activity than Pd-Ag/C Vulcan XC-72R, possibly because of high specific surface area and big pore volume of CNTs. The Pd-Ag/CNTs exhibited more remarkable current densities during the electrochemical oxidation of methanol than Vulcan XC-72R-supported catalysts. Reduced graphene oxides have also been used as supports for Pd-Ag nanoparticles, displaying good performances, particularly for the oxidation of the intermediates generated during the MOR. It is well known that the novel properties of graphenes are principally related with the elevated conductivity caused by its graphite plane structure [24]. A work reported by Li and co-workers [25] showed an improved oxidation of intermediates on Pd-Ag/ reduced graphene oxide (RGO), by means of the oxidation peaks generated during the forward (I_f) and reverse scans (I_b), which appeared in the methanol oxidation cyclic voltammetries. The oxidation of these chemisorbed intermediates was efficiently performed in the backward scan, as they are not totally oxidized during the forward scan. The highest ratio between these forward and backward currents was displayed by the catalyst Pd-Ag (1:1.5)/RGO, indicating a high conversion of methanol to CO$_2$ with a low production of intermediates. However, the authors considered Pd-Ag (1:1)/RGO as the best catalyst, because of the highest current densities displayed during the methanol oxidation. As mentioned above, the improved performance of these graphene-supported materials was attributed to the presence of Ag and its synergistic effect with Pd for methanol electro-oxidation. In general, the role of Ag, with independence of the employed carbon support, is related to its oxophilic character, which promotes the OH_{ads} adsorption, making easy the methanol dehydrogenation and CO_{ads} oxidation steps.

$$CH_3OH \rightleftharpoons CH_3OH_{ads} \tag{1}$$

$$CH_3OH_{ads} \rightarrow CO_{ad} + 4H^+ + 4e^- \tag{2}$$

$$H_2O \rightarrow OH_{ad} + H^+ + e^- \tag{3}$$

$$CO_{ad} + OH_{ad} \rightarrow CO_2 + H^+ + e^- \tag{4}$$

In this mechanism, after adsorption of methanol (Equation (1)), its deprotonation is carried out (Equation (2)) forming CO_{ad} which is oxidized by OH_{ad} (produced from water dissociation in Equation (3)), generating CO_2 and liberating the catalytic surface (Equation (4)). As demonstrated in a recent differential electrochemical mass spectrometry (DEMS) study [26], these factors produce obtaining large CO_2 current efficiencies at low oxidation potentials and lower potentials for CO_{ads} oxidation, as well as a displacement of the methyl formate forming potential towards more positive potentials, guaranteeing the complete oxidation of methanol. The effects attributed to graphene are related to high surface area, high electrical conductivity, upgraded electronic transference and the obtaining of more small and stable metal nanoparticles [25]. Table 2 displays some electrochemical performances and parameters obtained during the electrochemical oxidation of methanol on carbon-supported Pd-Ag catalysts.

Table 2. Electrochemical properties of Pd-Ag catalysts supported on different carbon materials.

Catalyst	Electrochemical Parameter			Reference
	Onset Potential (V vs. RHE)	Anodic Peak Potential (V vs. RHE)	Anodic Peak Current (mA mg^{-1} Pd or Pt)	
Pd/C	0.545	0.915	210.5	
Pd$_{80}$Ag$_{20}$/C	0.475	0.904	691.6	
Pd$_{65}$Ag$_{35}$/C	0.435	0.865	629.6	[22]
Pd$_{46}$Ag$_{54}$/C	0.475	0.855	453.4	
Pt/C TKK	0.475	0.925	689.3	
Catalyst	**Onset Potential (V vs. RHE)**	**Anodic Peak Potential (V vs. RHE)**	**Anodic Peak Current (mA cm^{-2})**	**Reference**
Pd/C	0.536	0.928	0.557	
Pd-Ag(2:1)/C	0.446	0.886	0.635	
Pd-Ag(1:1)/C	0.436	0.856	0.678	[23]
Pd-Ag(1:1)/CNTs	0.436	0.886	0.950	
Pd-Ag(1:1.5)/C	0.446	0.856	0.707	
Catalyst	**Anodic Peak Potential (V vs. RHE)**	**Anodic Peak Current (mA mg^{-1} Pd)**	**Ratio Between Forward and Backward Anodic Currents**	**Reference**
Pd/C	0.915	311	1.41	
Pd-Ag(1:1)/GO	0.865	225	1.50	
Pd-Ag(1.5:1)/RGO	0.915	334	1.42	[25]
Pd-Ag(1:1)/RGO	0.875	630	3.15	
Pd-Ag(1:1.5)/RGO	0.895	585	6.55	
Pd-Ag(1:1)/RGO-SB	0.870	545	1.48	

Carbon black-supported Pd-Ni is another widely reported alloy useful for carrying out the methanol oxidation. Different studies have demonstrated that presence of NiO in Pd catalysts can reduce the onset potential for the MOR, as well as increase the oxidation current densities [27]. It seems that nickel produces a similar effect to that for ruthenium in terms of the capacity to form OH_{ads} at lower potentials than Pd. Furthermore, Ni can induce changes in the electronic properties of Pd, similarly to the silver effect cited above [28]. These facts are reflected in the elevated currents associated with the hydrogen adsorption/desorption process and the resistance towards the carbonaceous intermediates poisoning during the methanol oxidation [29]. In order to elucidate the origin of the improved activity in Pd-Ni catalysts supported on carbon blacks, Amin et al. determined the Ni surface coverage in these alloys, trying to correlate the presence of Ni oxides and hydroxides with the methanol oxidation current densities [30,31]. The authors found enhanced current densities at big Ni coverages, due to the Ni oxophilic character and its capacity to generate adsorbed Ni hydroxides at low potentials. Another conclusion from these works was the ability of the NiOH and NiOOH for oxidizing the CO_{ads} generated during the methanol oxidation, which occupies active Pd sites. Besides the incorporation of oxygenated species by Ni, it has also been suggested the appearance of some defects on palladium crystalline lattice, with a major activity towards the methanol oxidation [32]. Atomic ratio between Pd and Ni also played a key role in the activity of these carbon black-supported catalysts, as demonstrated by Calderon et al. in a recent work [33]. They synthesized Pd-Ni catalysts supported on chemically treated carbon blacks, which contained surface oxygen and nitrogen groups. The metal contents of these catalysts was close to 25 wt %, whereas the studied Pd:Ni atomic ratios were near 1:1 and 1:2. Although no evident effects related to the presence of O- and N-surface functional groups were observed, these catalysts exhibited a higher CO poisoning tolerance than that observed for a Pd/C catalyst, which was explained from the increased amount of OH_{ads} formed on surface Ni atoms, while the methanol oxidation produced higher current densities in the catalysts with Pd:Ni = 1:2, in comparison with current densities generated with the catalysts with the atomic ratio close to 1:1 and the catalysts without any content of Ni (catalyst Pd/C), being this fact a proof of the beneficial effects of Ni in this reaction. Regarding other novel carbon materials as supports for Pd-Ni catalysts, Singh et al. [34] synthesized PdNi alloys supported on multi-walled carbon nanotubes (MWCNTs), finding remarkable activities toward the methanol oxidation, which were explained from the increased electroactive area of these composites, promoted by the use of MWCNTs and the electronic properties of the alloys, as described above. All of the authors coincided in the change of these electronic properties, also promoted by Ni oxides and hydroxides, being crucial factors for enhancing the electrochemical activity of the catalysts.

Other authors have suggested the doping of MWCNTs, as a plausible alternative for improving the activity, using for instance MnO_2 to cover the surface of carbon materials [35]. The presence of MnO_2 upgraded the electrocatalytic activity of Pd/MWCNTs and the tolerance to carbonaceous species poisoning, considering the high current values in the forward anodic peak current density (I_f). This value indicated a more effective removal of the poisoning species from the catalyst surface, oxidizing them to carbon dioxide. The authors suggested that the better activity for the Pd-MnO_2/MWCNT catalyst could be explained from the synergetic effect between Pd and MnO_2, promoting the adsorption of OH_{ad} species able to convert the poisoning species to CO_2. Table 3 present some catalytic parameters determined for the MOR performed on Pd-Ni catalysts supported on carbon materials.

Rh is another metal recently proposed to be alloyed with Pd. This metal is known for applications as exhaust systems in automobiles [36] and CO-tolerant electrode for the oxidation of H_2 in high temperature-proton exchange membrane fuel cells if hydrogen comes from a gas reformate process [37]. Studies related with the ethanol oxidation on Pt-Rh alloys in acidic and alkaline media have also been reported, displaying good methanol conversions to CO_2 [38–41], and generating efficiencies explained from the high Rh oxophilicity. About the methanol oxidation, Jurzinsky and co-workers [42] prepared Pd-Rh catalysts supported on carbon blacks, observing lower onset potentials and higher mass current densities than those exhibited by a Pd/C catalyst. The principal properties conferring to this Pd-Rh a high catalytic activity are related with: (1) capacity to form OH_{ads} at low potential; and (2) high stability of these OH_{ads}. The authors suggested that addition of Rh increased the tolerance of Pd towards the poisoning caused by different intermediates and CO. By means of some DEMS experiments, they also demonstrated that Pd-Rh/C catalysts can increase the efficiency in the oxidation of methanol to CO_2. DMFCs single test demonstrated enhanced power densities when these catalysts were used as anodes, in comparison with the performances obtained with Pd/C and Pt/C as anodes. The behavior of Pd-Rh alloys supported on graphenes meant an enhancement of the methanol oxidation currents, as was reported by Hsieh et al. [43]. The enhanced catalytic activity was attributed to the well-dispersed nanoparticles supported on the graphenes, resulting in a high active surface area and more active sites than those formed when the alloys are supported on carbon blacks. Another important advantage for the use of graphenes as support could be related with the major conductivity of this material due to their graphite plane structure.

Table 3. Electrochemical properties of Pd-Ni catalysts supported on different carbon materials.

Catalyst	Electrochemical Parameter			Reference
	Onset Potential (V vs. RHE)	Anodic Peak Current (mA cm^{-2})	–	
Pd/C	0.555	14	–	
Pd-NiO(8:1)/C [a]	–	51	–	
Pd-NiO(6:1)/C [a]	–	61	–	[27]
Pd-NiO(4:1)/C [a]	–	74	–	
Pd-NiO(2:1)/C	0.535	63	–	
Pt/C	0.525	18	–	

Catalyst	Anodic Peak Potential (V vs. RHE)	Anodic Peak Current (mA cm^{-2})	–	Reference
Pd-Ni(1:1)/C	0.914	7,64	–	[30]
Pd-Ni(1–5wt %)/MWCNTs	0.969	341,68	–	[34]

Catalyst	Onset Potential (V vs. RHE)	Anodic Peak Potential (V vs. RHE)	Anodic Peak Current (mA cm^{-2})	Reference
Pd/C	0.611	1.006	1.41	
Pd-Ni(1:1)/C	0.421	0.941	1.50	[32]
Pt/C	0.441	1.006	1.48	

Catalyst	Onset Potential (V vs. RHE)	Anodic Peak Current (mA cm^{-2})	Reference
Pd-Ni/CB 1:1	0.452	0.912	
Pd-Ni/CBO 1:1	0.556	0.397	
Pd-Ni/CBN 1:1	0.511	0.536	
Pd-Ni/CB 1:2	0.498	1.100	[33]
Pd-Ni/CBO 1:2	0.458	1.126	
Pd-Ni/CBN 1:2	0.551	0.815	
Pd/C 1:1	0.600	0.310	

[a] Data not provided by the authors.

Cerium is another metal tested as a component of palladium alloys. A work reported by Alvi and co-workers [44] reported the synthesis of Pd-Ce nanoparticles supported on carbon nanofibers synthesized by electrospinning. They employed these materials as catalysts for the methanol oxidation, obtaining materials with crystalline structure and highly-dispersed Pd-Ce nanoparticles on the carbon nanofibers surface. Regarding the catalytic activity, these materials displayed an acceptable efficiency toward the methanol oxidation, with enhanced current densities depending of the increase in the methanol concentration. Regarding the use of Pd-Au catalysts, recently He and co-workers reported the deposition of Au@Pd core-shells on reduced graphene oxide, synthesized by chemical reduction in presence of cetyltrimethylammonium chloride and iodide ions [45]. This catalyst showed high catalytic activity and tolerance to carbonaceous species poisoning generated during the methanol oxidation, overcoming those observed for Pd nanoparticles supported on reduced graphene oxide and commercial Pd/C in alkaline media. This behavior was explained from the strong interactions between reduced graphenes and Au@Pd nanostructures. Some electrochemical properties reported by different authors, regarding to different palladium alloys are reported in Table 4.

Table 4. Electrochemical properties of Pd-alloys supported on different carbon materials.

Catalyst	Electrochemical Parameter		Reference
	Onset Potential (V vs. RHE)	Anodic Peak Current (mA mg^{-1} Pd)	
Pt/C (HiSPEC^{TM} 3000)	0.475	669.9	
Pd/C	0.585	543.8	
Rh/C	0.504	177.6	[42]
PdRh$_3$/C	0.500	369.2	
PdRh/C	0.445	933.9	
Pd$_3$Rh/C	0.497	955.7	
Catalyst	**Onset Potential (V vs. RHE)**	**Anodic Peak Current (mA mg^{-1} Pd)**	**Reference**
Pd/GO	0.260	73	
Pd$_{75}$Rh$_{25}$/GO	0.280	100	[43]
Pd$_{50}$Rh$_{50}$/GO	0.370	35	
Pd$_{25}$Rh$_{75}$/GO	0.310	31	
Catalyst	**Onset Potential (V vs. RHE)**	**Anodic Peak Current (mA mg^{-1} Pd)**	**Reference**
Pd/C	0.510	190	
Pd/MWCNTs	0.510	285	[35]
Pd-MnO$_2$/MWCNTs	0.460	420	
Catalyst	**Onset Potential (V vs. RHE)**	**Anodic Peak Current (mA cm^{-2})**	**Reference**
Au@Pd/RGO	0.500	28	
Pd/RGO	0.700	4	[45]
Pd/C	0.640	10	

In terms of the palladium amount employed in some alloys, the mentioned works indicate that Pd-Rh materials generate the highest performances in terms of the current densities produced during the methanol oxidation. Particularly, the work of Jurzinsky and co-workers [42] highlights the catalyst PdRh/C, supported on Vulcan carbon black XC-72R, with a metal content of 20 wt % and an atomic ratio Pd:Rh close to 1:1. This material exhibited the highest current density (close to 1 A mg^{-1} Pd), being one of the most promising catalysts because of its low Pd content. Silver seems to be other metal with a remarkable behavior due to the formation of OH$_{ads}$, which play a key role in the oxidation of methanol. According to the work of Yin et al. [22], the better performance was showed for the catalyst Pd$_{80}$Ag$_{20}$/C, also supported on Vulcan carbon black XC-72R, with current densities close to 700 mA mg^{-1} Pd, although in this case, higher metal and Pd contents (close to 30 wt % and Pd:Ag of 80:20) were necessary for achieving this performance. Therefore, it is possible to postulate rhodium as the suitable metal to be alloyed with Pd, in carbon-supported catalysts for anodes in DMFCs.

2.2. Non Carbon-Supported Pd-Alloys

Although carbon materials have been traditionally used as supports for catalysts in low temperature fuel cells, some problems associated with its corrosion suggest the necessity for design novel supports able to supply the properties of carbon materials. In the carbon corrosion process, carbon surface (C_s) is oxidized, forming intermediates and electrons (Equation (5)) [46]:

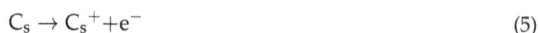

$$C_s \rightarrow C_s^+ + e^- \tag{5}$$

These electroactive species are susceptible to be hydrolyzed (Equation (6)):

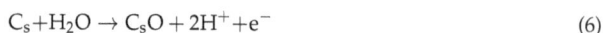

$$C_s + H_2O \rightarrow C_sO + 2H^+ + e^- \tag{6}$$

The final step of the corrosion process is the formation of carbon dioxide (Equation (7)):

$$2C_sO + H_2O \rightarrow C_sO + CO_{2(g)} + 2H^+ + e^- \tag{7}$$

Accordingly, some possible alternatives for substituting the carbon materials have been explored, looking for improving the activity of electrodes in fuel cells. For instance, Pd nanoparticles supported on Ni foam, with different Pd contents (Pd-2-Ni and Pd-4-Ni) were synthesized by galvanic replacement of Ni with $PdCl_4^{2-}$ and $PdCl_6^{2-}$ [47]. The activities of these materials were better than that exhibited by a commercial Pd/C, possibly due to the re-arrangement of Pd atoms and their improved utilization, which depended of the employed precursor during the synthesis. Moreover, a major exposition of Ni atoms probably promoted the oxidation of methanol, considering the above-mentioned role of this metal in this reaction.

On the other hand, titanium oxides have demonstrated attractive properties as low cost, commercial availability and high stability [48–50]. Since TiO_2 is recognized by its semiconductor properties, it is important to improve its electronic conductivity by means of the introduction of electron-donor dopants [51]. Nonetheless, there are few reports about the use of TiO_2 as support for Pd-based catalysts. In this sense, some researchers attempted to deposit Pd nanoparticles on TiO_2 nanotubes, finding remarkable catalytic activities in comparison with those observed for a pure Pd electrode and those Pd-nanostructures supported on TiO_2 nanoparticles [52]. The authors justified this improved activity from the well-dispersed nanoparticles on titanium tubes, which possessed a small diameter, thus promoting the obtaining of high specific surface areas. In these catalysts, the CO poisoning was also diminished, possibly due to some modifications in the Pd electronic structure induced by TiO_2, improving the CO oxidation process. In other work, Hosseini et al. [53] prepared Pd nanoparticles supported on an auto-assembled TiO_2 nanotubes/titanium support with a highly porous structure. This composite favored the diffusion of electroactive species toward the Pd nanoparticles, enhancing the methanol oxidation currents and conferring high stability. Another important property displayed by this support was its low electrochemical charge transfer resistance during the methanol oxidation activation stage.

Regarding the Pd-alloys supported on these TiO_2 supports, the most recent progresses are related with the presence of Ni and Ag on them, in agreement with some reports from Ju and co-workers. In the case of nickel [54], they supported Pd-Ag nanoparticles on TiO_2 nanotubes by surface reductive deposition of $PdCl_2$ and $NiSO_4$ on TiO_2 nanotubes. Large surface area and porous structure was observed for the Ti supports, improving the dispersion of Pd-Ni nanoparticles, whereas the electrochemical characterization exhibited high hysteresis peaks in the forward and backward scans, being this fact a proof of the high efficiency for the methanol intermediates oxidation. Again, the authors suggested the synergistic effect between Pd-Ni alloy and the TiO_2 nanotube support as responsible of the improved activity of these materials. On the other hand, the Pd-Ag alloys [54] showed remarkable electrocatalytic activity with high tolerance to the poisoning when tested towards the methanol electrochemical oxidation, although in this case, no hysteresis peaks were observed. Moreover, the dispersion of Pd-Ag nanoparticles was worse than that of Pd-Ni nanoparticles. In a recent work, Cao et al. [55] prepared nanoporous catalysts conformed by Pd, CuO and TiO_2, following a chemical dealloying process of amorphous Cu-Pd alloy ribbons in 5 M hydrochloric acid solution.

Then, they deposited TiO_2 nanoparticles, forming composites as np-Pd/CuO/160TiO$_2$ (with 160 as the TiO_2 moles present in the composite), which exhibited high activity toward the methanol oxidation and enhanced current densities (close to 2.86 mA cm^{-2}). The improved activity of this material was attributed to the ability of TiO_2 nanoparticles to form oxygen species able to promote the oxidation of methanol poisoning intermediates. Furthermore, these materials presented an outstanding long-term stability. The effect of some TiO_2 supports in the catalysis of methanol oxidation can be verified through the kinetic parameters shown in Table 5.

Table 5. Kinetic parameters for the methanol oxidation on Pd-based catalysts supported on TiO_2 materials.

Catalyst	Electrochemical Parameter		Reference
	Onset Potential (V vs. RHE)	Anodic Peak Current (A)	
Pd (pure)	0.22	1.0	
Pd/TiO$_2$ (nanoparticles)	0.22	4.5	[52]
Pd/TiO$_2$ (nanotubes)	0.23	9.0	
Catalyst	Onset Potential (V vs. RHE)	Anodic Peak Current Density (mA cm^{-2})	Reference
np-Pd/CuO/80TiO$_2$	0.55	1.5	
np-Pd/CuO/160TiO$_2$	0.51	2.8	
np-Pd/CuO/240TiO$_2$	0.55	1.4	[55]
np-Pd/CuO	0.60	0.8	
np-Pd	0.55	0.7	

From the works described above, it is possible to conclude that novel supports as nickel foams and TiO_2 nanotubes offer good properties to increase the performance of palladium nanoparticles toward the MOR. Basically, Ni foam facilitates the formation of OH_{ads}, and thus, increases to the CO poisoning tolerance, as reported by Niu et al. [47], who employed a surface metal loading of Pd close to 1 mg cm^{-2}, obtaining current densities close to 180 mA mg^{-1} Pd. In the case of the TiO_2-supported catalysts, the nanotube structure displayed the better performance [52] as a consequence of the improved diffusion of electroactive species toward the nanoparticles. The main drawback of this material is related with the low conductivity of this support, a fact that encourages the researches toward the doping of this support with other metals in order to increase its conductivity.

3. Oxygen Reduction on Pd-Based Catalysts

3.1. Carbon-Supported Pd-Alloys

Essentially, it is possible to consider similar problems in the cathode in comparison with those observed for the anode, in terms of cost issues and activity/technical limitations. This reaction starts with the adsorption of oxygen on catalyst surface, but different pathways can be followed (see Figure 1), as the direct forming process of water, which implies the maximum production of electrons (four electrons). However, production of hydrogen peroxide is also possible, a fact that results in the increase of the corrosive operation conditions of the DMFCs and the decrease in their performance, considering the production of two electrons during the reduction from molecular oxygen to H_2O_2 [56].

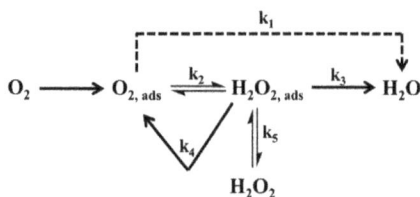

Figure 1. Mechanism and pathways for the oxygen reduction reaction.

Other typical drawbacks in the cathodes of direct methanol fuel cells are related with the adsorption of oxygen on the catalytic nanoparticles, the kinetics associated with the O-O cleavage, the crossover of methanol and the reduction of surface metal oxides once the oxygen has been transformed in water [56]. In this sense, significant progress has been achieved by researchers studying the ORR reaction on ruthenium chalcogenides [6,7,57], heat-treated macrocyclic compounds of transition metals [58–60], and palladium-based catalysts [8,61,62]. Regarding the catalyst supported on carbon materials different than carbon blacks, Zheng et al. supported palladium nanoparticles on carbon nanofibers and activated carbons in order to determine the influence of the carbon support on the ORR [63]. They found that this reaction is controlled by surface phenomena if the catalyst is supported on activated carbons, whereas diffusion of electroactive species is the factor controlling this reaction if the catalyst is supported on carbon nanotubes. Furthermore, carbon supports played a crucial role in the onset potentials for the ORR, being the Pd nanoparticles supported on activated carbons which displayed the most negative onset potential (0.50 V vs. RHE in acid media), whereas the catalysts supported on carbon nanofibers displayed values close to 0.70 V for the fishbone carbon nanofibers and 0.72 V for the platelet carbon nanofibers. In other study, Chakraborty and co-workers synthesized spherical and rod-like shapes nanosized Pd particles supported on multiwall carbon nanotubes (MWCNTs) [64] employing an electroless procedure. The results of this study indicated that these MWCNT-supported nanoparticles have remarkable catalytic activity towards the oxygen reduction, controlled by the surface morphology and coverage of particles on the carbon nanotubes, with a reaction mechanism that promoted the formation of hydrogen peroxide. The most important facts related with the electrocatalytic activity of these materials were related with the positive onset potentials for the ORR, the high stability of the catalysts as well as the definition of several peaks, which indicated the production of H_2O_2 in the first step of the ORR mechanism. Subsequently, this hydrogen peroxide is reduced to H_2O (second stage of the mechanism). Following the line of the shaped-controlled synthesis of Pd nanoparticles, Lusi et al. [65] recently reported the preparation of Pd nanocubes supported on Vulcan carbon black, finding several sizes such as 30, 10 and 7 nm. The authors assessed the electrocatalytic activity of these materials towards the ORR in alkaline media, and observed a four-electron pathway with the transfer of the first electron to adsorbed O_2 as the rate-limiting step. Furthermore, the specific activity of these nanocubes overcame that of the spherical Pd nanoparticles, finding a correlation between the enhanced activities and the increase of the particle size. Although this study demonstrated the achievement of high efficiencies in the ORR with big particle sizes, another recent research also suggested that it is possible to obtain high performances with ultra small thiolate-protected Pd nanoclusters [66], with ORR onset potential of −0.09 V (vs. Ag/AgCl) in alkaline media and improved durability. The authors verified the effects of thiolated ligands removing, finding a more positive onset potential (close to −0.02 V) and higher mass activity than that of a Pt/C catalyst, a fact explained from an activation of the nanoclusters after the thiolated ligands removing process.

Another study about the catalysis of the ORR in acid and alkaline media on Pd nanoparticles supported on MWCNTs was reported by Jukk and co-workers [67]. In this work, they employed Nafion® and polyvinylpyrrolidone as surfactants during the synthesis procedure. These surfactants conditioned the final performance of the materials, with the Nafion®-synthesized material exhibiting higher current densities and more positive onset potentials during the ORR tests, whereas the catalyst prepared in presence of polyvinylpyrrolidone displayed an equivalent behavior to that of a bulk Pd electrode. The authors suggested that these surfactants modified the amount and morphology of the as-formed Pd oxides, altering the adsorption oxygen and the reaction pathway.

Regarding the Pd alloys, Pd-Ni supported on carbon blacks exhibited high activity, as shown by Li and co-workers, who found high ORR activities, even better than those produced for a Pd/C catalyst [68]. The number of transferred electrons was also calculated, and it was demonstrated that a high content of surface Ni atoms promote the yielding of H_2O_2. This behavior was modified with a heat treatment at 800 °C, which induced the segregation of Pd atoms to the nanoparticles surface, generating a major number of produced electrons and thus, promoting the formation of water and increasing the current densities. Ramos-Sánchez et al. [69] also synthesized these nanoparticles supported on Vulcan carbon black by a borohydride reduction in THF. These materials developed power densities close to 120 mW cm^{-2} when used as cathodes in a direct methanol fuel single cell. Other works from the same authors reported bigger activities for the ORR on Pd-Ni catalysts compared with those obtained for Pd/C catalysts, with onset potential shifts of 100 mV towards positive values [70]. About other carbon supports different than carbon blacks, recently Calderon et al. supported Pd-Ni nanoparticles on previously chemically-treated carbon nanofibers [71]. The metal contents were close to 25 wt %, with 1:2 as Pd:Ni atomic ratio. The carbon nanofibers were enriched with surface oxygen and/or nitrogen groups after the chemical treatment. Although the onset potentials for ORR were more negative than that of the commercial catalyst Pd/C E-TEK, as well as this material displayed the highest diffusional current densities, higher mass activities i_k were determined for the Pd-Ni materials at 0.85 V, indicating the suitability of this alloy as possible replacement of Pt in DMFC cathodes. On the other hand, the catalyst supported on the carbon nanofibers with surface nitrogen groups displayed low production of hydrogen peroxide, suggesting an inhibiting effect of these functional groups toward the formation of this intermediate. The opposite behavior was observed for the catalysts supported on the O-modified carbon nanofibers (catalyst Pd-Ni/CNFO).

Similar to the case of platinum cathodes, cobalt has also been alloyed with palladium for cathode electrode applications in DMFCs. A recent work from Arroyo-Ramírez et al. [72] reported the use of organometallic complexes as precursors for synthesizing palladium-cobalt nanoparticles on Vulcan XC-72R carbon black, obtaining different ratios between Pd and Co (Pd_2Co/C and $PdCo_2/C$). A thermal reduction method was employed, which produced catalysts with high activity towards the ORR, even overcoming the performances of the commercial Pd/C catalyst. On the other hand, the tolerance to the methanol crossover of the synthesized materials was also higher than that of a Pt/C catalyst. The small particle diameters were obtained with the chemical reduction, being a crucial factor that conditioned the activity of the as-synthesized catalysts. Golmohammadi and co-workers proposed a novel alternative carbon support, consisting of the mixture between Vulcan XC-72R carbon black (VC) and multiwalled carbon nanotubes (MWCNTs), for supporting Pd_3Co nanoparticles [73]. The lowest potential drops and highest power densities were achieved with a MWCNTs/VC mass ratio close to 25:75. Presence of MWCNTs generated high surface area, major conductivity, and thus, fast ORR kinetics. Durability of membrane electrode assembly (MEA) with the mentioned mass ratio in the carbon support was also evaluated, finding both, high stability and improved durability, concluding that synergism between MWCNTs and VC contributed to the high performance of this catalyst.

Pd-Cu alloys are another promising alternative for carry out this reaction. Wu et al. [74] reported the wet chemical synthesis of Pd_nCu_{100-n} nanoalloys with controlled composition, finding that size, Pd/Cu ratio and phase structure of the nanoparticles are the key factors in the electrocatalytic activity and stability of these materials. The observed crystalline domains of the catalysts depended on the preparation method and influenced the performance of the catalysts, being the Pd_nCu_{100-n} catalysts with a Pd/Cu ratio close to 50:50, the most efficient toward the ORR, a fact explained from the body-centered cubic type ordered domains. The results from the structural and electrochemical characterizations suggested different correlations between the catalytic activity, atomic-scale structure and composition of the materials, bearing in mind the changes presented for $Pd_{50}Cu_{50}/C$ after the durability potential cycling. Other work reported the preparation of ternary catalyst with Pd, Cu and Co, increasing their activity and durability with the incorporation of gold traces [75] this noble metal galvanically replaced Co and Cu and entered into the Pd-Co-Cu lattice. The catalyst with the major durability (close to 100 h) was the $Au\text{-}Pd_6CoCu/C$, according with the results displayed by the single fuel-cell test, a result attributed to the homogeneous distribution of gold into the nanoparticles and the enhanced charge transfer between Pd and Au.

Fe is another metal with interesting properties in the catalysis of oxygen reduction. Neergat et al. investigated Pd-Fe alloys supported on carbon black, obtaining materials with high activity toward the ORR in comparison with the behavior observed for a Pd/C. Particularly, the Pd_3Fe/C catalyst displayed a half-wave potential more positive (>100 mV compared with that from Pd/C) and even similar than that of Pt/C with the same noble metal content [76]. Moreover, this catalyst showed both, a similar hydrogen peroxide production than that of Pt/C, but lower than that from Pd/C. The high tolerance to the methanol poisoning was also observed, making it useful in an eventual methanol crossover condition. Another example of this alloy could be the study made by Abo-Zeid et al. [77], who synthesized Pd-Fe alloys by a combined ethylene glycol-sodium borohydride procedure in presence of PDDA ionic polycation. Some heat treatments at 300, 500 and 700 °C were applied to these catalysts, which exhibited a better performance in terms of activity and stability. In fact, the most important improvements were obtained with an aging temperature of 300 °C. The effects of heat treatment were associated with the increase of crystallite sizes and alloying degrees. Within the context of Pd-Fe alloys, Rivera and co-workers studied the activity of Pd, PdFe, PdIr and PdFeIr alloys toward the ORR [78]. Main results indicated that insertion of Fe and Ir in the Pd lattice enhanced the activity toward the ORR, as well as increased the methanol tolerance, depending of the redox state of the transition metals. For example, the RDE tests demonstrated that absence of Ir and presence of Fe^{2+} and Fe^{3+} and Ir^{3+} and Ir^{4+} coincide with a promoted activity in presence of methanol. This behavior was verified with the membrane electrode assembly tests performed for a cathode containing the PdFeIr/C catalyst, which yielded the highest performance. Thus, material surface structure has a high impact on the activity of catalysts, as well as the iridium introduction into the Pd alloy promoted the methanol tolerance and inhibited the water dissociation, diminishing the oxidation of catalyst surface.

Finally, we can cite a very recent work regarding an alloyed Pd-W catalysts and its performance toward the ORR [79]. This material showed a good dispersion of their nanoparticles on Vulcan XC-72R carbon black support, a fact that increased the electrochemical activity, as demonstrated from the electrochemical tests performed for $Pd_{19}W/C$, which displayed a two-fold superior mass activity in comparison with that of Pd/C This result postulated this catalyst as hopeful for cathode application in DMFCs. Table 6 summarizes the ORR onset potential values determined for some carbon-supported Pd-based catalysts.

Table 6. Onset potential values for the ORR performed on carbon-supported Pd-based catalysts.

Catalyst	Onset Potential (V vs. RHE)	Reference
Pd/AC	0.584	
Pd/CNF$_{fisbone}$	0.624	[63]
Pd/CNF$_{platelet}$	0.764	
Pd/MWCNTs	1.014	[64]
Pd/MWCNTs-Nafion composite	0.900	[67]
Pd/MWCNTs-PVP composite	0.870	
Pd$_{nanocubes}$/Vulcan XC-72R	1.000	[65]
Pd$_{nanoclusters}$/XC-72R (Ligand on)	0.897	
Pd$_{nanoclusters}$/XC-72R (Ligand off)	1.017	[66]
Pt/XC-72R	0.987	
Pd/C	1.085	
Pd-Ni(3:1)/C	1.085	[68]
Pd-Ni(1:1)/C	1.105	
Pd-Ni(1:3)/C	1.005	
Pd/C E-TEK	1.050	
Pd-Ni/CB 1:2	0.960	
Pd-Ni/CNF 1:2	0.955	[71]
Pd-Ni/CNFO 1:2	0.960	
Pd-Ni/CNFN 1:2	0.940	
Pd/C (commercial)	0.728	
Pd$_2$Co/C	0.735	
PdCo$_2$/C	0.731	[72]
PtCo/C (commercial)	0.836	
Pt/C (commercial)	0.844	
Pd-Cu(36:64)/C	0.799	
Pd-Cu(54:46)/C	0.919	[74]
Pd-Cu(75:25)/C	0.799	
Pd/C (commercial)	0.879	
Pd-Fe/C (non-heat treated)	0.655	
Pd-Fe/C (300 °C)	0.865	[77]
Pd-Fe/C (500 °C)	0.815	
Pd-Fe/C (700 °C)	0.805	
Pd/C	0.940	
Pd$_{19}$W/C	0.950	
Pd$_9$W/C	0.950	[79]
Pd$_3$W/C	0.950	
JM Pt/C	0.950	

From the data presented in the Table 6, it is possible to conclude that the more active palladium catalysts toward the ORR are those alloyed with nickel, bearing in mind they showed the more positive onset potentials for this reaction [68]. The metal content of these materials was close to 12 wt %, suggesting that a small content of Pd is enough for achieving a good performance. The most positive onset potential was presented by the catalyst Pd-Ni (1:1)/C, demonstrating that composition is a determining factor to produce high ORR current densities with the most positive onset potential. Moreover, alloys between palladium and tungsten also presented positive onset potentials (values close to 0.95 V vs. RHE) [79], especially in the case of the catalyst Pd$_{19}$W/C. Only a few amount of tungsten alloyed with palladium was enough to obtain a remarkable increase in the catalytic activity in comparison with the results displayed for the Pd/C catalyst. Therefore, Ni and W are suitable metals to be alloyed with palladium to be used as cathodes for DMFCs.

3.2. Non Carbon-Supported Pd-Alloys

Pd catalysts supported on non-carbonaceous materials for the ORR have also been investigated, reaching an increased interest because of the possibilities for obtaining materials with novel properties and synergistic effects between metal nanoparticles and support, enabling the activity and the performance of cathodes in DMFCs. One alternative could be the doping of perovskites with Pd atoms at different oxidation states, as exposed by Zhu et al. [80], who reported that incorporation of the rare $Pd^{3/4+}$ oxidation state favorably affects the catalytic activity of perovskites. This fact resulted in the obtaining of a mass activity 2.5—higher than that of a Pt/C commercial catalyst. Moreover, this material displayed better durability and tolerance to methanol crossover. The authors suggested that this increase can be attributed to the reduction in the atomic distances between the atoms conforming the perovskite and the modifying of the d-band center, generating a decrease in the electron density, a weak linking between Pd and OH_{ads} and the availability of more active sites.

In the previous section, we mentioned some works related with the use of TiO_2 as support in Pd-alloyed anode catalysts. Some reports indicate that TiO_2 could be also employed as support for cathode catalysts. Lo Vecchio and co-workers [81] reported the preparation of Pd catalysts supported on titanium sub-oxides with formula Pd/Ti_nO_{2n-1}. This support conferred to the catalyst high stability and resistance after 1000 potential cycles, as well as better tolerance to the methanol poisoning, as corroborated by means of the onset potential shifts, which were displaced toward negative values for the Pt/C catalyst in presence of methanol, whereas in the case of the Pd/Ti_nO_{2n-1} this potential was constant.

Tungsten and molybdenum have also been employed for design supports. Ko et al. [82] synthesized mesoporous tungsten carbide in order to test them as catalysts and also as support for Pd nanoparticles. The materials were built in a CH_4/H_2 atmosphere using tungsten nitride nanoparticles as starting material, varying the temperatures (700, 800 and 900 °C). The authors detected a correlation between the pore structures and the reaction temperature, considering that 900 °C was the temperature that produced the tungsten carbide with the best electrocatalytic properties (most positive onset potentials, highest oxygen reduction current densities and lowest production of hydrogen peroxide). When the Pd nanoparticles were supported on these materials, the new catalysts displayed high durability and stability after 2000 cycles. In fact, the corrosion resistance was higher than that of a Pt/C catalyst with the same metal contents. In the case of molybdenum, Yan and co-workers [83] synthesized bimetallic carbide Fe_2MoC for anchoring Pd nanoparticles with high activity and stability compared to a Pd/C and a commercial Pt/C in the ORR. The authors attributed the remarkable performance of $Pd/GC-Fe_2MoC$ to an electron-donating effect from the Fe_2MoC to the supported Pd nanoparticles, which promoted the linkage strength between them. The Koutecky-Levich analysis of this material suggested water as major product during the ORR, after finding a four-electron transfer. In other kind of composites, Zuo et al. [84] prepared MoS_2/Pd catalysts by a sonochemical method. This method consisted of the ultrasonic exfoliation of bulk MoS_2 into single and few layers nanosheets, which then were employed as support for Pd nanoparticles. Electrochemical measurements made by cyclic voltammetry and rotating disk electrode showed a direct four-electron pathway for the ORR, with high electrocatalytic activity and long operation stability than a commercial Pt/C catalyst. In fact, mass and specific activities of this composite overcame that from the commercial catalyst. Catalysts containing Ti also have been synthesized from dealloying processes, displaying good results in the ORR. Chen and co-workers [85] applied this method for developing a nanoporous PdCuTiAl (np-PdCuTiAl) electrocatalyst with a three-dimensional network conformed by ultrafine channels. This composite displayed a half-wave potential more positive than those exhibited by Pd/C and Pt/C catalysts, with a four-electron reaction pathway. Moreover, the tolerance toward the methanol poisoning was major than that observed for the carbon supported-Pd and Pt catalysts.

Unsupported Pd-Ni alloyed catalysts have also been prepared and tested towards ORR. Xu et al. [86] dealloyed a PdNiAl precursor alloy to make a rich-nanoporous PdNi alloy. The as-synthesized electrode displayed a uniform structure with an interconnected network of hollow channels with a 5 nm-diameter approximately. The electrochemical tests for this material demonstrated a high activity towards the ORR, in terms of the bigger specific and mass activities in comparison with a commercial Pt/C, besides a longer durability and tolerance to methanol. The characteristics of the three-dimensional network and the synergistic effect between Ni and Pd were stated by the authors as key factors for the improved performance of this material. Dealloying process was also employed by Chen et al. [87], who fabricated a nanoporous PdNi catalyst by electrochemical dealloying. The starting material was a $Pd_{20}Ni_{80}$ alloy, which was putted in an acid solution to be dealloyed, obtaining a final Ni content of 9 at %. This material displayed higher electrocatalytic performances during the oxygen reduction than both, a commercial Pd/C and nanoporous Pd catalysts, with more positive onset potentials and enhanced mass activities. As mentioned above, the remarkable behavior of this alloy was attributed to the bimetallic synergistic effect and the well-connected porosity of the material. Within the line of the non-supported materials, Xiong and co-workers [88] synthesized a self-supported Pd-Cu catalyst with 3D porous structure, using an electrochemical deposition of copper from a $CuSO_4$ solution in acid media, in order to promote a galvanic replacement with the Pd^{2+} present in a Na_2PdCl_4 solution. The most important properties of the electrochemically synthesized catalyst were related with its long-term stability and the high oxygen reduction current densities, overcoming those of a Pt/C catalyst. The onset potentials on these materials were close to 1.0 V vs. RHE, while Pt/C exhibited a value closer to 0.9 V. Table 7 shows some of the parameters observed during the ORR on non-carbon supported catalysts.

The onset potential differences between the commercial catalysts and those tested during the ORR indicated that the alloying process of PdNi catalysts is a promising technique to obtain cathodes for fuel cells. Xu et al. [86] demonstrated that it is possible to dealloy PdNiAl composites to obtain PdNi structures with similar activity to a Pt/C commercial catalyst, this fact representing a decrease in the overall production costs of DMFCs. The three dimensional nanostructure and the changes induced by Ni in the electronic structure of palladium seems to be the factors influencing the outstanding performance of these materials.

Table 7. Onset potential values for the ORR performed on non-carbon-supported Pd-based catalysts.

Catalyst	Onset Potential (V vs. RHE)	Reference
Pd/LF	0.662	
Pd/LFP0.05	0.792	[80]
Pd/LFP0.05-R	0.722	
Pd/LFP0.05-RO	0.772	
Pd/WC-700-m	0.692	
Pd/WC-800-m	0.812	[82]
Pd/WC-900-m	0.872	
Pt/C	1.08	
Pd/C	0.92	[83]
Pd/C-MoC	0.95	
Pd/C-Fe$_2$MoC	1.08	
Pt/C (commercial)	1.04	
PdNi dealloyed	1.04	[86]
Pd dealloyed	0.90	
Pt/C (commercial)	0.90	[88]
Pd-Cu(nanodendrites)	1.00	

4. Conclusions

In this review, a revision about the state of the art of Pd-based catalysts has been described, in terms of the efficiency and performance towards both the methanol oxidation and oxygen reduction reactions, as fundamental chemical processes in direct methanol fuel cells (DMFCs). The most important conclusion in the case of the catalysts suggested as anodes for DMFCs is related with the role of second metal in the forming of OH_{ads}, bearing in mind its importance in the oxidizing of CO and others intermediates generated during the oxidation of methanol. Different authors have reported that these metals (Ag, Ni, and Rh) are able to form OH_{ads} at lower potentials than those determined for Pd or Pt, whereas novel carbon materials increase the diffusion of electroactive species and facilitate the electronic transference. On the other hand, novel non-carbon materials as supports create synergistic effects that improve the performance of anodes, basically by means of some changes in the electronic structure of catalytic metals, favoring the oxidation of methanol and the other species related with this reaction. In the case of the cathode Pd-based catalysts, the decrease of the onset potentials and the enhance current densities detected during the ORR were the most important parameters considered to assess the performance of the catalysts. In this sense, the low content of surface Pd oxides and the structure of support were started as crucial factors to obtain good values for the mentioned parameters. Crystalline structure also seems to promote the formation of more active sites and upgrade the tolerance of methanol towards the methanol crossover. Pd-Ni and Pd-W seems to be the most promising cathode catalysts, bearing in mind their positive onset potentials (some of them, overcoming those observed for commercial Pd and Pt catalysts), indicating the beneficial effects of the second metal incorporation on the structure of palladium. From the works here cited and revised, it is possible to conclude that replacement of platinum in the electrodes for direct methanol fuel cells could be carried out, using Pd and Pd-alloys catalysts supported on both, carbon materials and non-carbon materials, in order to decrease the consumption of this metal as catalytic phase in direct methanol fuel cells.

Acknowledgments: The authors want to thank the Spanish Ministry of Economy and Competitiveness and FEDER for financial support under the projects ENE2014-52518-C2-1-R.

Author Contributions: Juan Carlos Calderón Gómez wrote the paper; Rafael Moliner and Maria Jesus Lázaro revised the paper.

Conflicts of Interest: The authors declare no conflict of interest.

Abbreviations

The following abbreviations are used in this manuscript:

DMFC	Direct Methanol Fuel Cells
CNT	Carbon Nanotubes
MWCNT	Multi-Walled Carbon Nanotubes
RGO	Reduced Graphene Oxide (by sodium borohydride)
MOR	Methanol oxidation reaction
MEA	Membrane electrode assembly
DEMS	Differential Electrochemical Mass Spectrometry
ORR	Oxygen Reduction Reaction
VC	Vulcan Carbon
TKK	Tanaka Kikinzoku Kogyo®

References

1. Lee, J.B.; Park, Y.K.; Yang, O.B.; Kang, Y.K.; Jun, K.W.; Lee, Y.J.; Kim, H.Y.; Lee, K.H.; Choi, W.C. Synthesis of porous carbons having surface functional groups and their application to direct-methanol fuel cells. *J. Power Sources* **2006**, *158*, 1251–1255. [CrossRef]
2. Brouzgou, A.; Song, S.Q.; Tsiakaras, P. Low and non-platinum electrocatalysts for PEMFCs: Current status, challenges and prospects. *Appl. Catal. B* **2012**, *127*, 371–388. [CrossRef]
3. Fournier, J.; Faubert, G.; Tilquin, J.Y.; Cote, R.; Guay, D.; Dodelet, J.P. High-performance, low Pt content, catalysts for the electroreduction of oxygen in polymer electrolyte fuel cells. *J. Electrochem. Soc.* **1997**, *144*, 145–154. [CrossRef]
4. Lefèvre, M.; Dodelet, J.P.; Bertrand, P. Molecular oxygen reduction in PEM fuel cells: Evidence for the simultaneous presence of two active sites in Fe-based catalysts. *J. Phys. Chem. B* **2002**, *106*, 8705–8713. [CrossRef]

5. Matter, P.H.; Zhang, L.; Ozkan, U.S. The role of nanostructure in nitrogen-containing carbon catalysts for the oxygen reduction reaction. *J. Catal.* **2006**, *239*, 83–96. [CrossRef]
6. Zaikovskii, V.I.; Nagabhushana, K.S.; Kriventsov, V.V.; Loponov, K.N.; Cherepanova, S.V.; Kvon, R.I.; Bönnemann, H.; Kochubey, D.I.; Savinova, E.R. Synthesis and structural characterization of Se-modified carbon-supported Ru nanoparticles for the oxygen reduction reaction. *J. Phys. Chem. B* **2006**, *110*, 6881–6890. [CrossRef] [PubMed]
7. Serov, A.A.; Min, M.; Chai, G.; Han, S.; Kang, S.; Kwak, C. Preparation, characterization, and high performance of RuSe/C for direct methanol fuel cells. *J. Power Sources* **2008**, *175*, 175–182. [CrossRef]
8. Mustain, W.E.; Prakash, J. Kinetics and mechanism for the oxygen reduction reaction on polycrystalline cobalt-palladium electrocatalysts in acid media. *J. Power Sources* **2007**, *170*, 28–37. [CrossRef]
9. Assiongbon, K.A.; Roy, D. Electro-oxidation of methanol on gold in alkaline media: Adsorption characteristics of reaction intermediates studied using time resolved electro-chemical impedance and surface plasmon resonance techniques. *Surf. Sci.* **2005**, *594*, 99–119. [CrossRef]
10. Hu, F.P.; Shen, P.K. Ethanol oxidation on hexagonal tungsten carbide single nanocrystal-supported Pd electrocatalyst. *J. Power Sources* **2007**, *173*, 877–881. [CrossRef]
11. Li, R.S.; Wei, Z.; Huang, T.; Yu, A.S. Ultrasonic-assisted synthesis of Pd–Ni alloy catalysts supported on multi-walled carbon nanotubes for formic acid electrooxidation. *Electrochimica Acta* **2011**, *56*, 6860–6865. [CrossRef]
12. Shao, M.H. Palladium-based electrocatalysts for hydrogen oxidation and oxygen reduction reactions. *J. Power Sources* **2011**, *196*, 2433–2444. [CrossRef]
13. Nguyen, S.T.; Tan, D.S.L.; Lee, J.M.; Chan, S.H.; Wang, J.Y.; Wang, X. Tb promoted Pd/C catalysts for the electrooxidation of ethanol in alkaline media. *Int. J. Hydrogen Energy* **2011**, *36*, 9645–9652. [CrossRef]
14. Grden, M.; Łukaszewski, M.; Jerkiewicz, G.; Czerwinski, A. Electrochemical behavior of palladium electrode: Oxidation, electrodissolution and ionic adsorption. *Electrochimica Acta* **2008**, *53*, 7583–7598. [CrossRef]
15. Bianchini, C.; Shen, P.K. Palladium-based electrocatalysts for alcohol oxidation in half cells and in direct alcohol fuel cells. *Chem. Rev.* **2009**, *109*, 4183–4206. [CrossRef] [PubMed]
16. Wei, Y.C.; Liu, C.W.; Wang, K.W. Improvement of oxygen reduction reaction and methanol tolerance characteristics for PdCo electrocatalysts by Au alloying and CO treatment. *Chem. Commun.* **2011**, *47*, 11927–11929. [CrossRef] [PubMed]
17. Calderón, J.C.; Mahata, N.; Pereira, M.F.R.; Figueiredo, J.L.; Fernandes, V.R.; Rangel, C.M.; Calvillo, L.; Lazaro, M.J.; Pastor, E. Pt–Ru catalysts supported on carbon xerogels for PEM fuel cells. *Int. J. Hydrogen Energy* **2012**, *37*, 7200–7211.
18. Yu, X.W.; Ye, S.Y. Recent advances in activity and durability enhancement of Pt/C catalytic cathode in PEMFC: Part I. Physico-chemical and electronic interaction between Pt and carbon support, and activity enhancement of Pt/C catalyst. *J. Power Sources* **2007**, *172*, 133–144. [CrossRef]
19. Zhang, W.Y.; Yao, Q.S.; Wu, X.D.; Fu, Y.S.; Deng, K.M.; Wang, X. Intimately coupled hybrid of graphitic carbon nitride nanoflakelets with reduced graphene oxide for supporting Pd nanoparticles: A stable nanocatalyst with high catalytic activity towards formic acid and methanol electrooxidation. *Electrochim Acta* **2016**, *200*, 131–141. [CrossRef]
20. Zhang, X.; Zhu, J.X.; Tiwary, C.S.; Ma, Z.Y.; Huang, H.J.; Zhang, J.F.; Lu, Z.Y.; Huang, W.; Wu, Y.P. Palladium nanoparticles supported on nitrogen and sulfur dual-doped graphene as highly active electrocatalysts for formic acid and methanol oxidation. *ACS Appl. Mater. Interfaces* **2016**, *8*, 10858–10865. [CrossRef] [PubMed]
21. Zhang, W.Y.; Huang, H.J.; Li, F.; Deng, K.M.; Wang, X. Palladium nanoparticles supported on graphitic carbon nitride-modified reduced graphene oxide as highly efficient catalysts for formic acid and methanol electrooxidation. *J. Mat. Chem. A* **2014**, *2*, 19084–19094. [CrossRef]
22. Yin, Z.; Zhang, Y.N.; Chen, K.; Li, J.; Li, W.J.; Tang, P.; Zhao, H.B.; Zhu, Q.J.; Bao, X.H.; Ma, D. Monodispersed bimetallic PdAg nanoparticles with twinned structures: Formation and enhancement for the methanol oxidation. *Sci. Rep.* **2014**, *4*, 4288–4296. [CrossRef] [PubMed]
23. Wang, Y.; Sheng, Z.M.; Yang, H.B.; Jiang, S.P.; Li, C.M. Electrocatalysis of carbon black- or activated carbon nanotubes-supported Pd–Ag towards methanol oxidation in alkaline media. *Int. J. Hydrogen Energy* **2010**, *35*, 10087–10093. [CrossRef]

24. Luo, B.; Liu, S.M.; Zhi, L.J. Chemical approaches toward graphene-based nanomaterials and their applications in energy-related areas. *Small* **2012**, *8*, 630–646. [CrossRef] [PubMed]

25. Li, L.Z.; Chen, M.X.; Huang, G.B.; Yang, N.; Zhang, L.; Wang, H.; Liu, Y.; Wang, W.; Gao, J.P. A green method to prepare Pd–Ag nanoparticles supported on reduced graphene oxide and their electrochemical catalysis of methanol and ethanol oxidation. *J. Power Sources* **2014**, *263*, 13–21. [CrossRef]

26. Jurzinsky, T.; Cremers, C.; Pinkwart, K.; Tübke, J. On the influence of Ag on Pd-based electrocatalyst for methanol oxidation in alkaline media: A comparative differential electrochemical mass spectrometry study. *Electrochimica Acta* **2016**, *199*, 270–279. [CrossRef]

27. Shen, P.K.; Xu, C.W.; Zeng, R.; Liu, Y.L. Electro-oxidation of methanol on NiO-promoted Pt/C and Pd/C catalysts. *Electrochem. Solid-State Lett.* **2006**, *9*, A39–A42. [CrossRef]

28. Li, G.L.; Jiang, L.H.; Jiang, Q.; Wang, S.L.; Sun, G.Q. Preparation and characterization of Pd_xAg_y/C electrocatalysts for ethanol electrooxidation reaction in alkaline media. *Electrochimica Acta* **2011**, *56*, 7703–7711. [CrossRef]

29. Qi, Z.; Geng, H.; Wang, X.; Zhao, C.; Ji, H.; Zhang, C.; Xu, J.; Zhang, Z. Novel nanocrystalline PdNi alloy catalysts for methanol and ethanol electro-oxidation in alkaline media. *J. Power Sources* **2011**, *196*, 5823–5828. [CrossRef]

30. Amin, R.S.; Abdel Hameed, R.M.; El-Khatiba, K.M. Microwave heated synthesis of carbon supported Pd, Ni and Pd–Ni nanoparticles for methanol oxidation in KOH solution. *Appl. Catal. B* **2014**, *148-149*, 557–567. [CrossRef]

31. Amin, R.S.; Abdel Hameed, R.M.; El-Khatib, K.M.; Elsayed Youssef, M. Electrocatalytic activity of nanostructured Ni and Pd–Ni on Vulcan XC-72R carbon black for methanol oxidation in alkaline medium. *Int. J. Hydrogen Energy* **2014**, *39*, 2026–2041. [CrossRef]

32. Liu, Z.L.; Zhang, X.H.; Hong, L. Physical and electrochemical characterization of nanostructured Pd/C and PdNi/C catalysts for methanol oxidation. *Electrochem. Commun.* **2009**, *11*, 925–928. [CrossRef]

33. Calderón, J.C.; Nieto-Monge, M.J.; Pérez-Rodríguez, S.; Pardo, J.I.; Moliner, R.; Lázaro, M.J. Palladium–nickel catalysts supported on different chemically-treated carbon blacks for methanol oxidation in alkaline media. *Int. J. Hydrogen Energy* **2016**. [CrossRef]

34. Singh, R.N.; Singh, A.; Anindita. Electrocatalytic activity of binary and ternary composite films of Pd, MWCNT and Ni, Part II: Methanol electrooxidation in 1 M KOH. *Int. J. Hydrogen Energy* **2009**, *34*, 2052–2057. [CrossRef]

35. Zhao, Y.C.; Zhan, L.; Tian, J.N.; Nie, S.L.; Ning, Z. MnO_2 modified multi-walled carbon nanotubes supported Pd nanoparticles for methanol electro-oxidation in alkaline media. *Int. J. Hydrogen Energy* **2010**, *35*, 10522–10526. [CrossRef]

36. Sasahara, A.; Tamura, H.; Tanaka, K.I. Catalytic activity of Pt-deposited Rh(110) bimetallic surface for $NO + H_2$ reaction. *J. Phys. Chem. B* **1997**, *101*, 1186–1189. [CrossRef]

37. Ross, P.N.; Kinoshita, K.; Scarpellino, A.J.; Stonehart, P. Electrocatalysis on binary alloys: I. Oxidation of molecular hydrogen on supported Pt–Rh alloys. *J. Electroanal. Chem. Interfacial Electrochem.* **1975**, *59*, 177–189. [CrossRef]

38. Shen, S.Y.; Zhao, T.S.; Xu, J.B. Carbon supported PtRh catalysts for ethanol oxidation in alkaline direct ethanol fuel cell. *Int. J. Hydrogen Energy* **2010**, *35*, 12911–12917. [CrossRef]

39. de Souza, J.P.I.; Queiroz, S.L.; Bergamaski, K.; Gonzalez, E.R.; Nart, F.C. Electro-oxidation of ethanol on Pt, Rh, and PtRh electrodes. A study using DEMS and in-situ FTIR techniques. *J. Phys. Chem. B* **2002**, *106*, 9825–9830. [CrossRef]

40. Bergamaski, K.; Gonzalez, E.R.; Nart, F.C. Ethanol oxidation on carbon supported platinum-rhodium bimetallic catalysts. *Electrochimica Acta* **2008**, *53*, 4396–4406. [CrossRef]

41. Delpeuch, A.B.; Asset, T.; Chatenet, M.; Cremers, C. Electrooxidation of ethanol at room temperature on carbon-supported Pt and Rh-containing catalysts: A DEMS study. *J. Electrochem. Soc.* **2014**, *161*, F918–F924. [CrossRef]

42. Jurzinsky, T.; Bär, R.; Cremers, C.; Tübke, J.; Elsner, P. Highly active carbon supported palladium-rhodium Pd_xRh/C catalysts for methanol electrooxidation in alkaline media and their performance in anion exchange direct methanol fuel cells (AEM-DMFCs). *Electrochimica Acta* **2015**, *176*, 1191–1201. [CrossRef]

43. Hsieh, C.T.; Yu, P.Y.; Tzou, D.Y.; Hsu, J.P.; Chiu, Y.R. Bimetallic Pd–Rh nanoparticles onto reduced graphene oxide nanosheets as electrocatalysts for methanol oxidation. *J. Electroanal. Chem.* **2016**, *761*, 28–36. [CrossRef]

44. Alvi, M.A.; Akhtar, M.S. An effective and low cost Pd–Ce bimetallic decorated carbon nanofibers as electrocatalyst for direct methanol fuel cells applications. *J. Alloys Compd.* **2016**, *684*, 524–529. [CrossRef]

45. He, L.L.; Song, P.; Feng, J.J.; Fang, R.; Yu, D.X.; Chen, J.R.; Wang, A.J. Porous dandelion-like gold@ palladium core-shell nanocrystals in-situ growth on reduced graphene oxide with improved electrocatalytic properties. *Electrochimica Acta* **2016**, *200*, 204–213. [CrossRef]

46. Zhang, S.S.; Yuan, X.Z.; Wang, H.J.; Mérida, W.; Zhu, H.; Shen, J.; Wu, S.H.; Zhang, J.J. A review of accelerated stress tests of MEA durability in PEM fuel cells. *Int. J. Hydrogen Energy* **2009**, *34*, 388–404. [CrossRef]

47. Niu, X.H.; Zhao, H.L.; Lan, M.B. Palladium deposits spontaneously grown on nickel foam for electrocatalyzing methanol oxidation: Effect of precursors. *J. Power Sources* **2016**, *306*, 361–368. [CrossRef]

48. Park, K.W.; Han, S.B.; Lee, J.M. Photo (UV)-enhanced performance of Pt–TiO_2 nanostructure electrode for methanol oxidation. *Electrochem. Commun.* **2007**, *9*, 1578–1581. [CrossRef]

49. Xue, X.D.; Gu, L.; Cao, X.B.; Song, Y.Y.; Zhu, L.W.; Chen, P. One-pot, high-yield synthesis of titanate nanotube bundles decorated by Pd (Au) clusters for stable electrooxidation of methanol. *J. Solid State Chem.* **2009**, *182*, 2912–2917. [CrossRef]

50. Guo, X.; Guo, D.J.; Qiu, X.P.; Chen, L.Q.; Zhu, W.T. Excellent dispersion and electrocatalytic properties of Pt nanoparticles supported on novel porous anatase TiO_2 nanorods. *J. Power Sources* **2009**, *194*, 281–285. [CrossRef]

51. Dvoranova, D.; Brezova, V.; Mazur, M.; Malati, M.A. Investigations of metal-doped titanium dioxide photocatalysts. *Appl. Catal. B* **2002**, *37*, 91–105. [CrossRef]

52. Wang, M.; Guo, D.J.; Li, H.L. High activity of novel Pd/TiO_2 nanotube catalysts for methanol electro-oxidation. *J. Solid State Chem.* **2005**, *178*, 1996–2000. [CrossRef]

53. Hosseini, M.G.; Momeni, M.M.; Khalilpur, H. Synthesis and characterization of palladium nanoparticles immobilized on TiO_2 nanotubes as a new high active electrode for methanol electro-oxidation. *Int. J. Nanosci.* **2012**, *11*, 1250016. [CrossRef]

54. Ju, J.; Shi, Y.; Wu, D. TiO_2 nanotube supported PdNi catalyst for methanol electro-oxidation. *Powder Technol.* **2012**, *230*, 252–256. [CrossRef]

55. Cao, S.; Xu, W.; Zhu, S.; Liang, Y.; Li, Z.; Cui, Z.; Yang, X.; Inoue, A. Synthesis of TiO_2 nanoparticles loaded Pd/CuO nanoporous catalysts and their catalytic performance for methanol, ethanol and formic acid electro-oxidations. *J. Electrochem. Soc.* **2016**, *163*, E263–E271. [CrossRef]

56. Fernandez, J.L.; Walsh, D.A.; Bard, A.J. Thermodynamic guidelines for the design of bimetallic catalysts for oxygen electroreduction and rapid screening by scanning electrochemical microscopy. M–Co (M: Pd, Ag, Au). *J. Am. Chem. Soc.* **2005**, *127*, 357–365. [CrossRef] [PubMed]

57. Shukla, A.K.; Raman, R.K. Methanol-resistant oxygen-reduction catalysts for direct methanol fuel cells. *Annu. Rev. Mater. Res.* **2003**, *33*, 155–168. [CrossRef]

58. Baranton, S.; Coutanceau, C.; Roux, C.; Hahn, F.; Leger, J.M. Oxygen reduction reaction in acid medium at iron phthalocyanine dispersed on high surface area carbon substrate: tolerance to methanol, stability and kinetics. *J. Electroanal. Chem.* **2005**, *577*, 223–234. [CrossRef]

59. Wagner, A.J.; Wolfe, G.M.; Fairbrother, D.H. Reactivity of vapor-deposited metal atoms with nitrogen-containing polymers and organic surfaces studied by in situ XPS. *Appl. Surf. Sci.* **2003**, *219*, 317–328. [CrossRef]

60. Matter, P.H.; Zhang, L.; Ozkan, U.S. The role of nanostructure in nitrogen-containing carbon catalysts for the oxygen reduction reaction. *J. Catal.* **2006**, *239*, 83–96. [CrossRef]

61. Mustain, W.E.; Kepler, K.; Prakash, J. Investigations of carbon-supported $CoPd_3$ catalysts as oxygen cathodes in PEM fuel cells. *Electrochem. Commun.* **2006**, *8*, 406–410. [CrossRef]

62. Serov, A.A.; Cho, S.Y.; Han, S.; Min, M.; Chai, G.; Nam, K.H.; Kwak, C. Modification of palladium-based catalysts by chalcogenes for direct methanol fuel cells. *Electrochem. Commun.* **2007**, *9*, 2041–2044. [CrossRef]

63. Zheng, J.S.; Zhang, X.; Li, P.; Zhu, J.; Zhou, X.G.; Yuan, W.K. Effect of carbon nanofiber microstructure on oxygen reduction activity of supported palladium electrocatalyst. *Electrochem. Commun.* **2007**, *9*, 895–900. [CrossRef]

64. Chakraborty, S.; Retna Raj, C. Electrocatalytic performance of carbon nanotube-supported palladium particles in the oxidation of formic acid and the reduction of oxygen. *Carbon* **2010**, *48*, 3242–3249. [CrossRef]

65. Lüsi, M.; Erikson, H.; Sarapuu, A.; Tammeveski, K.; Solla-Gullón, J.; Feliu, J.M. Oxygen reduction reaction on carbon-supported palladium nanocubes in alkaline media. *Electrochem. Commun.* **2016**, *64*, 9–13. [CrossRef]

66. Zhao, S.; Zhang, H.; House, S.D.; Jin, R.; Yang, J.C.; Jin, R. Ultrasmall palladium nanoclusters as effective catalyst for oxygen reduction reaction. *ChemElectroChem.* **2016**, *3*, 1225. [CrossRef]

67. Jukk, K.; Alexeyeva, N.; Johans, C.; Kontturi, K.; Tammeveski, K. Oxygen reduction on Pd nanoparticle/multi-walled carbon nanotube composites. *J. Electroanal. Chem.* **2012**, *666*, 67–75. [CrossRef]

68. Li, B.; Prakash, J. Oxygen reduction reaction on carbon supported palladium-nickel alloys in alkaline media. *Electrochem. Commun.* **2009**, *11*, 1162–1165. [CrossRef]

69. Ramos-Sanchez, G.; Santana-Salinas, A.; Vazquez-Huerta, G.; Solorza-Feria, O. Electrochemical impedance study and performance of PdNi nanoparticles as cathode catalyst in a polymer electrolyte membrane fuel cell. *J. New Mat. Electrochem. Syst.* **2010**, *13*, 213–217.

70. Ramos-Sánchez, G.; Yee-Madeira, H.; Solorza-Feria, O. PdNi electrocatalyst for oxygen reduction in acid media. *Int. J. Hydrogen Energy* **2008**, *33*, 3596–3600. [CrossRef]

71. Calderón, J.C.; Celorrio, V.; Nieto-Monge, M.J.; Fermín, D.J.; Pardo, J.I.; Moliner, R.; Lázaro, M.J. Palladium-nickel materials as cathode electrocatalysts for alkaline fuel cells. *Int. J. Hydrogen Energy* **2016**. accepted for publication.

72. Arroyo-Ramírez, L.; Montano-Serrano, R.; Luna-Pineda, T.; Román, F.R.; Raptis, R.G.; Cabrera, C.R. Synthesis and characterization of palladium and palladium-cobalt nanoparticles on Vulcan XC-72R for the oxygen reduction reaction. *ACS Appl. Mater. Interfaces* **2013**, *5*, 11603–11612. [CrossRef] [PubMed]

73. Golmohammadi, F.; Gharibi, H.; Sadeghi, S. Synthesis and electrochemical characterization of binary carbon supported Pd₃Co nanocatalyst for oxygen reduction reaction in direct methanol fuel cells. *Int. J. Hydrogen Energy* **2016**, *41*, 7373–7387. [CrossRef]

74. Wu, J.; Shan, S.; Luo, J.; Joseph, P.; Petkov, V.; Zhong, C.J. PdCu nanoalloy electrocatalysts in oxygen reduction reaction: role of composition and phase state in catalytic synergy. *ACS Appl. Mater. Interfaces* **2015**, *7*, 25906–25913. [CrossRef] [PubMed]

75. Wang, D.; Liu, S.; Wang, J.; Lin, R.; Kawasaki, M.; Rus, E.; Silberstein, K.E.; Lowe, M.A.; Lin, F.; Nordlund, D.; et al. Spontaneous incorporation of gold in palladium-based ternary nanoparticles makes durable electrocatalysts for oxygen reduction reaction. *Nat. Commun.* **2016**. [CrossRef] [PubMed]

76. Neergat, M.; Gunasekar, V.; Rahul, R. Carbon-supported Pd–Fe electrocatalysts for oxygen reduction reaction (ORR) and their methanol tolerance. *J. Electroanal. Chem.* **2011**, *658*, 25–32. [CrossRef]

77. Abo Zeid, E.F.; Kim, Y.T. Effect of heat treatment on nanoparticle size and oxygen reduction reaction activity for carbon-supported Pd–Fe alloy electrocatalysts. *Am. J. Nano. Res. Appl.* **2015**, *3*, 71–77.

78. Rivera Gavidia, L.M.; García, G.; Anaya, D.; Querejeta, A.; Alcaide, F.; Pastor, E. Carbon-supported Pt-free catalysts with high specificity and activity toward the oxygen reduction reaction in acidic medium. *Appl. Catal. B* **2016**, *184*, 12–19. [CrossRef]

79. Dai, Y.; Yu, P.; Huang, Q.; Sun, K. Pd-W alloy electrocatalysts and their catalytic property for oxygen reduction. *Fuel Cells* **2016**, *16*, 165–169. [CrossRef]

80. Zhu, Y.; Zhou, W.; Chen, Y.; Yu, J.; Xu, X.; Su, C.; Tadé, M.O.; Shao, Z. Boosting oxygen reduction reaction activity of palladium by stabilizing Its unusual oxidation states in perovskite. *Chem. Mater.* **2015**, *27*, 3048–3054. [CrossRef]

81. Lo Vecchio, C.; Alegre, C.; Sebastián, D.; Stassi, A.; Aricò, A.S.; Baglio, V. Investigation of supported Pd-based electrocatalysts for the oxygen reduction reaction: performance, durability and methanol tolerance. *Materials* **2015**, *8*, 7997–8008. [CrossRef]

82. Ko, A.R.; Lee, Y.W.; Moon, J.S.; Han, S.B.; Cao, G.Z.; Park, K.W. Ordered mesoporous tungsten carbide nanoplates as non-Pt catalysts for oxygen reduction reaction. *Appl. Catal. A* **2014**, *477*, 102–108. [CrossRef]

83. Yan, Z.; Zhang, M.; Xie, J.; Zhu, J.; Shen, P.K. A bimetallic carbide Fe₂MoC promoted Pd electrocatalyst with performance superior to Pt/C towards the oxygen reduction reaction in acidic media. *Appl. Catal. B* **2015**, *165*, 636–641. [CrossRef]

84. Zuo, L.X.; Jiang, L.P.; Zhu, J.J. A facile sonochemical route for the synthesis of MoS₂/Pd composites for highly efficient oxygen reduction reaction. *Ultrason. Sonochem.* **2016**. [CrossRef] [PubMed]

85. Chen, X.; Si, C.; Wang, Y.; Ding, Y.; Zhang, Z. Multicomponent platinum-free nanoporous Pd-based alloy as an active and methanol-tolerant electrocatalyst for the oxygen reduction reaction. *Nano Res.* **2016**, *9*, 1831–1843. [CrossRef]

86. Xu, C.; Liu, Y.; Hao, Q.; Duan, H. Nanoporous PdNi alloys as highly active and methanol tolerant electrocatalysts towards oxygen reduction reaction. *J. Mater. Chem. A* **2013**, *1*, 13542–13548. [CrossRef]

87. Chen, L.; Guo, H.; Fujita, T.; Hirata, A.; Zhang, W.; Inoue, A.; Chen, M. Nanoporous PdNi bimetallic catalyst with enhanced electrocatalytic performances for electro-oxidation and oxygen reduction reactions. *Adv. Funct. Mater.* **2011**, *21*, 4364–4370. [CrossRef]
88. Xiong, L.; Huang, Y.X.; Liu, X.W.; Sheng, G.P.; Li, W.W.; Yu, H.Q. Three-dimensional bimetallic Pd–Cu nanodendrites with superior electrochemical performance for oxygen reduction reaction. *Electrochimica Acta* **2013**, *89*, 24–28. [CrossRef]

catalysts

MDPI

Review

Non-Precious Electrocatalysts for Oxygen Reduction Reaction in Alkaline Media: Latest Achievements on Novel Carbon Materials

Angeliki Brouzgou [1], Shuqin Song [2,*], Zhen-Xing Liang [3,*] and Panagiotis Tsiakaras [1,4,*]

[1] Laboratory of Alternative Energy Conversion Systems, Department of Mechanical Engineering, University of Thessaly, Pedion Areos, Volos 38334, Greece; brouzgou@gmail.com
[2] Key Laboratory of Low-Carbon Chemistry and Energy Conservation of Guangdong Province, School of Materials Science and Engineering, Sun Yat-sen University, Guangzhou 510275, China
[3] Key Laboratory on Fuel Cell Technology of Guangdong Province, School of Chemistry and Chemical Engineering, South China, University of Technology, Guangzhou 510641, China
[4] Laboratory of Electrochemical Devices Based on Solid Oxide Proton Electrolytes, Institute of High Temperature Electrochemistry, Russian Academy of Sciences, Yekaterinburg 620137, Russia
*   Correspondence: stsssq@mail.sysu.edu.cn (S.S.); zliang@scut.edu.cn (Z.-X.L.); tsiak@uth.gr (P.T.); Tel.: +86-20-8411-3253 (S.S.); +86-20-8711-3584 (Z.-X.L.); +30-24210-74065 (P.T.)

Academic Editors: Vincenzo Baglio and David Sebastián
Received: 1 August 2016; Accepted: 26 September 2016; Published: 19 October 2016

Abstract: Low temperature fuel cells (LTFCs) are considered as clean energy conversion systems and expected to help address our society energy and environmental problems. Up-to-date, oxygen reduction reaction (ORR) is one of the main hindering factors for the commercialization of LTFCs, because of its slow kinetics and high overpotential, causing major voltage loss and short-term stability. To provide enhanced activity and minimize loss, precious metal catalysts (containing expensive and scarcely available platinum) are used in abundance as cathode materials. Moreover, research is devoted to reduce the cost associated with Pt based cathode catalysts, by identifying and developing Pt-free alternatives. However, so far none of them has provided acceptable performance and durability with respect to Pt electrocatalysts. By adopting new preparation strategies and by enhancing and exploiting synergetic and multifunctional effects, some elements such as transition metals supported on highly porous carbons have exhibited reasonable electrocatalytic activity. This review mainly focuses on the very recent progress of novel carbon based materials for ORR, including: (i) development of three-dimensional structures; (ii) synthesis of novel hybrid (metal oxide-nitrogen-carbon) electrocatalysts; (iii) use of alternative raw precursors characterized from three-dimensional structure; and (iv) the co-doping methods adoption for novel metal-nitrogen-doped-carbon electrocatalysts. Among the examined materials, reduced graphene oxide-based hybrid electrocatalysts exhibit both excellent activity and long term stability.

Keywords: oxygen reduction reaction; low temperature fuel cells; non-precious carbon-based electrocatalysts; hybrid electrocatalysts; macro-mesoporous structure; co-doping

1. Introduction

The design and the development of highly efficient, stable and low cost oxygen reduction reaction (ORR) electrocatalysts is one of the main determining steps for the fabrication of more commercially viable electrochemical devices (fuel cells, electrolyzers, metal-air batteries, etc.).

During the last two decades, a joint effort has been made by the international research community towards: (i) the development of low-loading precious metal electrocatalysts; (ii) the development of non-Pt electrocatalysts; (iii) the development of new supporting materials; and (iv) the investigation of novel preparation methods [1].

Among the various developed non-Pt electrocatalysts, Ir- and Pd-based ones have attracted much more attention, as they exhibited comparable activity with Pt-based electrocatalysts, though in the presence of methanol or ethanol due to alcohol crossover, they exhibited much lower stability and less activity than Pt-based ones [1]. Moreover, different supports such as tungsten carbides, carbon nanotubes, and nitrogen-modified carbon were investigated and adopted, but without the expected results [2].

During the last decade, along with the development of alkaline membranes, their promising performance, in major cases better than that of the acidic membranes, significantly boosted further investigation on the non-Pt electrocatalysts [3]. Among the non-Pt electrocatalysts, heteroatom-doped carbon materials are currently considered as the most potential Pt substitutes for the oxygen reduction reaction (ORR) [4]. Various heteroatoms, such as N [5–8], B [7,8], S [9–11], Fe [12,13], and P [7,14–16], with different electronegativity, have been doped for nanotubes [17], graphene [18], graphite [19], etc.

The doped-carbons show excellent ability to adsorb and bond OOH$^\bullet$, significantly enhancing the formation of H_2O_2. This facile ORR activity has been attributed to the fact that carbon atoms next to heteroatoms are the catalytically active sites and to the modification of carbon electronegativity. For instance, in N-doped carbon, a charge separation occurs due to the different electronegativity values, in which the positive pole is favorable for oxygen adsorption, enhancing the ORR. In terms of alcohol crossover, the carbon-based catalyst shows an extremely high selectivity to the ORR and inactivity to alcohol oxidation reaction. The reason mostly lies in the high activation energy for the chemical dissociative adsorption of the hydrocarbon compounds on such catalysts.

Furthermore, it has been observed that co-doping with two heteroatoms creates a higher number of catalytically active sites than single-doped counterparts. Thus, many of the ORR investigations in the last two years were devoted to N-doped carbon catalysts with another element [8]. For example, Fe-N doped carbon systems have shown excellent ORR performance in both alkaline and acid electrolytes, due to the metal-nitrogen synergetic effect [13]. The significant contribution of nitrogen when being doped in carbon has been reported since 2009. Niwa et al. [20] investigated different electronic structures of nitrogen (pyridinic, graphitic, pyrrolic, oxidic, proposing that graphitic-like nitrogen is the main reason for the enhanced ORR activity of N-doped carbon electrocatalysts.

The mesoscopic morphology, the high surface area, the pore size and the order degree reflect the available active sites of a carbon material, and consequently affect its electrocatalytic activity. Most of the synthesis strategies of doped carbon material with high surface area include multi-step chemical processes, with some of them to be of high cost. Consequently, sometimes due to the complexity of the synthesis procedure of hetero-atom doped carbon materials it is difficult to designate the nature of catalytically active sites for ORR, which is of high importance for understanding the whole ORR process [21].

The key issue to obtain optimal ORR activities by using doped-carbons lies in: (i) uniform doping of heteroatoms (ii) unique nanostructure with high surface area; and (iii) suitable mesopore distribution. In most of the preparation methods, carbon support is either post-treated with heteroatom substances [7,8,15], or obtained through direct co-pyrolysis of the heteroatom-containing organic compounds with carbon under inert gas atmosphere [22,23].

However, a sufficiently doped and finely controlled structure is rather difficult to be simultaneously achieved. Therefore, a uniform distribution of dopants into a well-designed carbon matrix with significantly enhanced ORR performance still remains a challenging goal. A second target to be achieved concerns the choice of the appropriate precursor in combination with the pyrolysis conditions, to obtain large surface area, high nitrogen content and, thus, high performance to be attained.

The deliberated defects on carbon structures lead to the destruction of its conductivity, which is the most essential characteristic in electrocatalytic processes [4]. For this reason, in the last years, researchers have turned their attention to carbons with graphitic structure, such as graphene; the conductivity of which is not influenced by introducing heteroatoms.

The current review focuses on different strategies very recently developed by different research groups in order to enhance the activity of non-precious metal, namely carbon based electrocatalysts towards ORR. Mainly, four strategies have been distinguished: (i) the fabrication of 3D structure including interconnected networks by modifying carbon structure; (ii) the development of novel hybrid materials; (iii) the adoption of alternative precursor for a more homogeneous distribution of the heteroatoms into carbon matrix; and (iv) the co-doping technique of two heteroatoms, mainly into highly ordered mesoporous materials.

2. 2D and 3D Doped-Carbon Electrocatalysts

Two-dimensional (2D) and three-dimensional (3D) porous structures with controllable compositions can provide high activity and stability for ORR due to their interconnected open-pore structure, which can facilitate electron and mass transfer. In addition, the offered large active area of mesopore structure that can provide more active sites, in combination with microporous structure that reduces the electrolyte diffusion distances to the interior surfaces and serves as an electrolyte buffering reservoir, significantly enhances the ORR activity [24,25]. The addition of nitrogen at this advantageous carbon structure further facilitates ORR.

Recently, it has been reported that hierarchically porous materials offer greater active surface area via their interconnected macro-porosity [26]. Liu et al. [27], for the N-doped hierarchically porous carbon pyrolysed at 1000 °C, reported a kinetic-limiting current density (J_k) of 20.0 mA·cm^{-2} at 0.853 V (vs. RHE), a half wave-potential ($E_{1/2}$) of 0.773 V (vs. RHE), an excellent methanol tolerance (0.5 M) and 8.1% of current loss after 3.5 operational hours at −0.5 V vs Ag/AgCl. The observed high ORR activity was attributed to the existence of 3D-hierarchically interconnected porous frameworks with high surface area, large pore volume and high conductivity.

N-doped hierarchically macro/mesoporous carbon pyrolysed at the optimal temperature (850 °C) was also examined by Tao et al. [28]. By the aid of rotating disk electrode (RDE) measurements, they recorded a limiting current density close to 4.4 mA·cm^{-2} (1600 rpm), and a half-wave potential of 0.753 V (vs. RHE); in the presence of 3.0 M methanol in 0.1 M KOH the activity was almost unchanged. Chronoamperometric test at −0.55 V vs. Ag/AgCl reveals that after 12.5 h durability test, the electrocatalyst lost only 5% of its activity, while a dramatic degradation of 15% was observed for Pt/C after 2.5 h.

The 3D dual-doped highly ordered macro-mesoporous structure is observed to provide more active ORR electrocatalysts than the single-doped one. Namely, 3D Co and N-doped highly ordered macro-mesoporous carbon electrocatalyst with the optimum content of cobalt displayed high activity towards ORR due to the presence of a large number of active sites. According to linear sweep voltammetry curves (LSV), a limiting current density of 5.8 mA·cm^{-2} in oxygen saturated alkaline environment and a half-wave potential ($E_{1/2}$) of 0.83 V (vs. RHE), was observed [29]. The kinetic current density at 0.70 V vs. RHE was calculated close to 23.2 mA·cm^{-2}, while in the presence of 1.0 M of methanol, its ORR activity remained satisfactorily stable, indicating its high tolerance to the methanol. Additionally, after 25 h of continuous operation at 0.5 V vs. RHE, there was negligible current loss, in comparison with Pt (20 wt %)/C that lost ~50% of its initial activity when examined under the same experimental conditions. [29].

Transition metal oxides, especially the doped-ones, have shown very good ORR activity [30]. The role of Ni-doping in MnO$_2$ nanoneedles for the ORR in alkaline media was recently investigated by Hao et al. [31]. The increment of Mn(III), which serves as electrochemical active site, according to the authors, is the main reason for its higher activity, in comparison to the other samples. The ORR activity is enhanced when transition metals are adopted in 3D hierarchical porous spinel hollow nanospheres, especially those with spinel structure [32,33]. Hollow structure also contributes to the increment of the number of active sites. More precisely, $CoFeO_2$ 3D hollow nanospheres benefited the formation of a large three-phase boundary (solid–liquid–gas), which is necessary for the transfer of reactants and products. The yielded limiting current density (5.3 mA·cm^{-2}) decreased by 34% after 12-h durability test; whereas the Pt/C electrocatalyst decreased by 54% under the same experimental conditions [34]. The enhanced stability was attributed to the combination of its hollow morphology with the hierarchically porous structure. However, according to LSV curves, its half-wave potential was 0.58 V vs. RHE, negatively shifted by ~0.30 V compared with Pt. In general, even without a special 3D structure, transition metal spinel oxide nanoparticles (NPs) have shown really good ORR performance [35–37]. 3D hollow-structured $NiCo_2O_4$/C NPs with interconnected pores as bifunctional electrocatalysts, which are transformed from solid $NiCo_2$ alloy NPs, have also been investigated for ORR [35]. From the RDE measurements, a limiting current density of 5.7 mA·cm^{-2} and a half-wave potential of 0.68 V vs. RHE were recorded (Table 1). In addition, $NiCo_2O_4$/C showed a superior long-term stability over Pt/C electrocatalyst. At a constant potential of 0.6 V vs. RHE, a current decay of only 9% after 10 h occurred, whereas a gradual decrease of 52% was recorded for Pt/C under the same conditions.

Table 1. Non-precious 2D and 3D-structured doped-carbon ORR (oxygen reduction reaction) electrocatalysts.

Catalyst	Preparation Method and Experimental Conditions	Limiting Current (J_d; d = disc), Kinetic Current Density (J_k), Half Wave Potential ($E_{1/2}$))	Reference
N-HCS (hierarchically mesoporous spheres)-900	Nanocasting method, 0.1 M KOH, oxygen saturated 20 mV·s^{-1}, 1600 rpm	J_d = 4.7 mA·cm^{-2} J_k = 20.0 mA·cm^{-2} @ 0.3 V vs. Ag/AgCl sat. KCl (0.85 V vs. RHE) $E_{1/2}$ = 0.748 V vs. RHE	[27]
3D-HPC-N (N-doped-3D-hierarchically porous carbon materials)-850	Hierarchically macro/mesoporous silica as a hard template followed by a simple N-doping procedure and 0.1 M KOH, oxygen saturated 5 mV·s^{-1}, 1600 rpm (0.1 M KOH + 3.0 M MeOH)	J_d = 4.4 mA·cm^{-2} (Inactive in MeOH presence) $E_{1/2}$ = 0.753 V vs. RHE	[28]
3D Co-N-OMMC-0.6 (Co(NO$_3$)$_2$·6H$_2$O) (ordered macro-mesoporous carbon)	Dual-templating synthesis approach in a one-pot controllable procedure by the use of silica colloidal crystal (opal) as a macroporous mold and triblock copolymer Pluronic F127 as a mesoporous template and 0.1 M KOH, oxygen saturated; 5 mV·s^{-1}, 1600 rpm (0.1 M KOH + 1.0 M MeOH)	J_d = 5.8 mA·cm^{-2} (Inactive in MeOH presence), J_k = 23.2 mA·cm^{-2} @ 0.7 V vs. RHE, $E_{1/2}$ = 0.83 V vs. RHE	[29]
3D hierarchical porous-CoFe$_2$O$_4$ hollow nanospheres	Hydrothermal method and 0.1 M KOH, oxygen saturated, 10 mV·s^{-1}, 1600 rpm	J_d = 5.3 mA·cm^{-2} $E_{1/2}$ = 0.58 V vs. RHE	[34]
3D hollow NiCo$_2$O$_4$/C	Transformation from solid NiCo$_2$ alloy nanoparticles through the Kirkendall effect and 0.1 M KOH, oxygen saturated, 5 mV·s^{-1}, 1600 rpm	J_d = 5.7 mA·cm^{-2} $E_{1/2}$ = 0.68 V vs. RHE	[35]
3D Nanosheet Co$_3$O$_4$-doped-graphene	Microwave argon-plasma synthesis approach and 0.1 M KOH, oxygen saturated, 5 mV·s^{-1}, 1600 rpm	J_d = 5.7 mA·cm^{-2} J_k = 34mAcm^{-2} @ 0.75 V vs. RHE $E_{1/2}$ = 0.832 V vs. RHE	[38]
2D-CoAl-LDH@ZIF-67-800 (LDH: layered double hydroxides, ZIF: zeolitic imidazolate framework)	In situ nucleation and directed growth of MOFs arrays on the surface of LDHs nanoplatelets followed by a subsequent pyrolysis process and 0.1 M KOH, oxygen saturated 10 mV·s^{-1}, 1500 rpm (0.1 M KOH + 2.0 M MeOH)	J_d = 5.2 mA·cm^{-2} (Inactive in MeOH presence) $E_{1/2}$ = 0.675 V vs. RHE	[39]
3D-NCNT-900 (N-doped carbon nanotubes)	PPy nanotubes were synthesized by the chemical oxidative polymerization of pyrrole, in the presence of FeCl$_3$ as an oxidant, and p-toluene sulfonic acid (TsOH) as a dopant and 0.1 M KOH, oxygen saturated, 10 mV·s^{-1}, 1600 rpm (0.1 M KOH + 5.0 M MeOH)	J_d = 5.2 mA·cm^{-2} (Inactive in MeOH presence) $E_{1/2}$ = 0.707 V vs. RHE	[40]
3D-N-doped-TTF-700 (thermalized triazine-based framework)	A nitrogen-containing molecule, terephthalonitrile, as the basic building block and through first trimerization into a 2D covalent triazine-based framework and 0.1 M KOH, oxygen saturated; 10 mV·s^{-1}, 1600 rpm	J_d = 4.0 mA·cm^{-2} $E_{1/2}$ = 0.767 V vs. RHE	[41]

Among all carbon materials, graphene is the most promising for accommodating various nanoparticles to achieve high electron transport rate, electrolyte contact area and structural stability, all of which lead to markedly improved ORR performance. Replacing carbon with graphene in

3D structures combined with M-N-doped-carbon further enhances their electrocatalytic activity and stability (M-N-doped-3D graphene). The effective dispersion of Co_3O_4 nanosheets among 3D graphene sheets was made via argon-plasma synthesis. The Co_3O_4 nanosheets formed strong bonds with graphene sheets enhancing its electrical conductivity and electrochemical activity. The kinetic analysis resulted a kinetic current density of 34 mA·cm^{-2} at -0.75 V vs. RHE and a half-wave potential of 0.832 V vs. RHE. Additionally, a high stability was obtained as after 5.5 operational hours only 2% of its initial limiting current density was lost, while in the case of Pt (20 wt %)/C, ca 21% loss was detected [38].

The kinetic parameters (the available ones) of the electrocatalysts reported in the current review, along with the respective preparation methods are summarized in Table 1. It is observed that most of the electrocatalysts have also been investigated for the ORR in presence of methanol, as its crossover to cathode has been considered as one of the major reasons for the decay of ORR activity [3]. It is remarkable that each of them shows excellent tolerance to methanol, considering that just few years ago the main disadvantage of the non-precious electrocatalysts was related with their instability due to alcohol crossover [1,3].

It should be pointed out that the data given in literature are often problematic or misleading. For example, the limiting current density from RDE measurement varies considerably under the same conditions. According to Levich equation, the steady-state limiting current density for the 4e$^-$ ORR is ca 6.0 mA·cm^{-2} in a diluted electrolyte aqueous solution. The deviation observed in Table 1 should possibly be originated from several reasons. A negative deviation is caused by at least, but not limited, to the following reasons: (i) the poor quality of the fabricated electrode with an incomplete coverage of the electrocatalyst on the substrate; (ii) a mixed reaction pathway with both 2e$^-$ and 4e$^-$ charge transfer processes; (iii) poor electrocatalytic activity, and not mass transfer controlled steady state. A positive deviation, the other side of the coin, can also be found in literature (see Table 2), which is understandable as follows. The loading of the non-precious metal catalyst is generally pretty high, and thereby, the thickness of the "thin-film" electrode cannot be neglected any more. As such, the radial mass transfer should be considered for such "column-like" electrodes. Therefore, the comparison is sometimes tricky and should be cautious in practice. In our opinion, the normalized kinetic current density, other than the mass transfer-controlled current, should be used to benchmark the activity in future. Nevertheless, the data are summarized in this review (Tables 1–3), which are phenomenally used for reference.

On 2016, porous N-doped carbon nanotubes (NCNTs) with high surface area (>1000 m^2·g^{-1}) are fabricated by KOH activation and pyrolysis of polypyrrole nanotubes. They exhibited excellent ORR electrocatalytic (those annealed at 900 °C) performance and yielded a more positive onset potential, higher current density, and long-term operation stability in alkaline media, when compared with a commercially available 20 wt % Pt/C catalyst; this behaviour is mostly due to the synergetic effect between the dominant pyridinic/graphitic-N species and the porous tube structures. These metal-free porous nitrogen-doped carbon nanomaterials could be considered as potential alternatives to Pt/C catalysts, for electrochemical energy conversion and storage [40].

2D materials, as the 3D ones, have also attracted increased interest, with graphitic carbon nitride (g-C_3N_4) as a derivative graphene to be identified also as a good ORR electrocatalyst. More precisely, the development of porous organic networks, which stems from the polymerization of rigid organic molecules, offers the opportunity of 3D controllable structured electrocatalysts preparation [42]. Recently, Hao et al. [41] exploited the 2D structure of triazine-based framework (metal organic) and convert it to a 3D porous electrocatalyst. Its stability at 0.6 V vs. RHE was excellent as kept stable by 100% for 11 h of operation, as well as its tolerance in presence of 3.0 M methanol. Moreover, its catalytic performance is considered high compared with the results reviewed in the present work, as its half-wave potential reached 0.767 V vs. RHE.

Zeolitic imidazolate framework (ZIF) constitutes another precursor used often for the preparation of 2D/3D controllable structures More precisely, this is a zeolite-type nanoscale metal-organic framework (MOF), which is used as a self-sacrificing template as well as both carbon and nitrogen source [43]. A 2D carbon-based network electrocatalyst (Co-Al-LDH@ZIF67-800 °C), consisting of well-defined metal-organic framework (MOF) arrays on layered double hydroxides (LHDs) exhibited a limiting current density of 5.2 mA·cm^{-2}, with half-wave potential of 0.675 V vs. RHE; no reaction was observed in the presence of 2.0 M methanol [39]. Moreover, the as-prepared electrocatalyst presented superior durability as it lost only 1% of its activity after 5.5 h of operation at −0.5 V vs. RHE; under the same operation conditions, Pt/C lost 40% of its activity [39].

Miner et al. [44] using as MOF hexa-iminotriphenylene (HITP) synthesized a tunable highly ordered 2D-structured ORR electrocatalyst material, Ni$_3$(HITP)$_2$. However, according to RDE results it did not reach a plateau when examined in oxygen saturated 0.1 M KOH solution, at 2000 rpm. Therefore, for the enhancement of its ORR activity further investigation should be carried out.

In Figure 1, the long term stability comparison between the above-mentioned electrocatalysts is depicted. All the examined electrocatalysts are characterized by relatively good stability; after 10 h, they lose less than 8% of their initial current density. Longer-term stability for 25 h is observed for 3D-structured Co-N-doped highly ordered macro-mesoporous carbon electrocatalyst [29]. To the best of our knowledge, this is the highest value of long-term stability, which is mostly attributed to the 3D as well as to the macro and mesoporous structures.

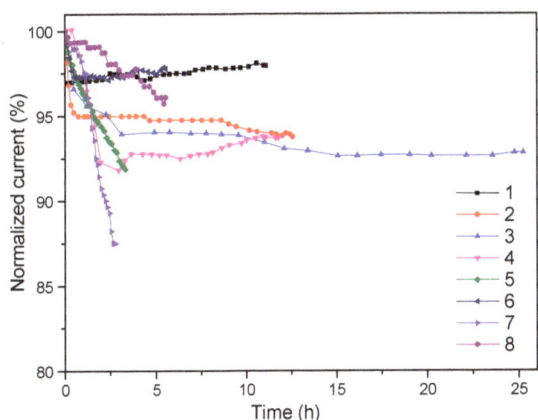

Figure 1. Comparison of stability of 3D-structured non-precious carbon based electrocatalysts: (1) NHCs (N-doped hierarchically carbon) (at 0.5 V vs. RHE) [27]; (2) 3D-HPC-N (N-doped-3D-hierarchically porous carbon materials)-850 (at 0.55 V vs. RHE) [28]; (3) 3D Co-N-OMMC-0.6 (Co(NO$_3$)$_2$·6H$_2$O) (ordered macro-mesoporous carbon) (at 0.6 V vs. RHE) [29]; (4) 3D hierarchical porous-CoFe$_2$O$_4$ hollow nanospheres (at 0.56 V vs. RHE) [34]; (5) Mn$_x$Co$_{3x}$O$_4$ (x = 0.4) [36]; (6) Nanosheet Co$_3$O$_4$-doped-graphene-3D [38]; (7) 2D-CoAl-LDH@ZIF-67-800 (LDH: layered double hydroxides, ZIF: zeolitic imidazolate framework [39]; and (8) 3D-N-doped-TTF (thermalized triazine-based framework)-700 (at 0.6 V vs. RHE) [41].

The half-wave potential ($E_{1/2}$) is conventionally used for the comparison of the ORR activity between different electrocatalysts [36]. To evaluate the catalytic activities of the electrocatalysts listed in Table 1, the half-wave potentials are depicted in Figure 2. As can be seen, 3D-structured Co-N-doped highly ordered macro-mesoporous carbon electrocatalyst [29] and 3D nanosheet Co$_3$O$_4$-doped-graphene [38] exhibit the same half-wave potential values (~0.83 V vs. RHE). The specific half-wave value is almost the same as that of Pt (ca 0.8 V vs. RHE). However, they differ in their kinetic current values: the 3D nanosheets Co$_3$O$_4$-doped-graphene [38] exhibited a kinetic current density of 34 mA·cm^{-2} at 0.75 V vs. RHE, while the 3D-structured Co-N-doped highly ordered macro-mesoporous carbon of 23.2 mA·cm^{-2}.

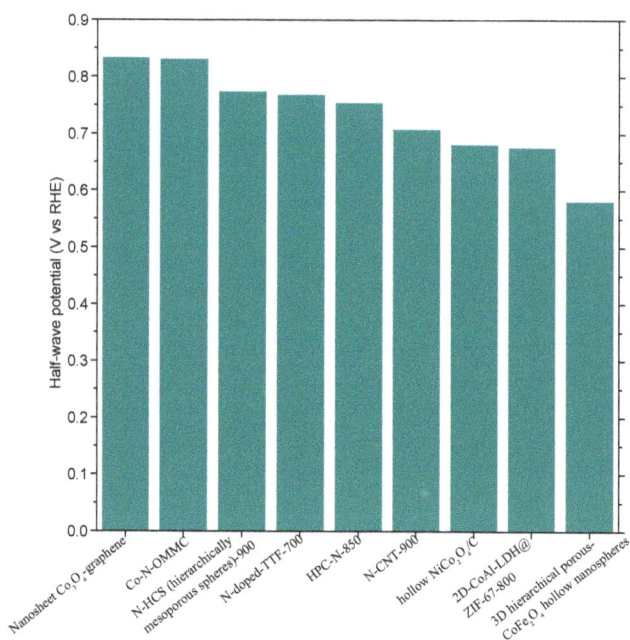

Figure 2. Comparison of half-wave potential (V vs. RHE). (References from left to right column: [38], [29], [27], [41], [28], [40], [35], [39], [34]).

3. Hybrid (Metal Oxide-Nitrogen-Carbon) Electrocatalysts

Despite the fact that, in acidic media, the transition metal oxides exhibit very low ORR activity and insufficient stability, in alkaline media, they exhibit high activity, mostly due to the oxygen-containing groups [45]. The key for their enhanced activity is the amplification of their conductivity by integrating their conductive matrix, forming the as-called hybrid materials; metal oxide-N-C [46].

In Table 2, along with their respective kinetic parameters, the most active ORR hybrid electrocatalysts are listed. Most of them seem to present higher ORR activity than the 2D or 3D electrocatalysts (Table 1). Characteristic is that all authors report higher limiting current and more positive onset and half-wave potential values than Pt/C electrocatalyst (Table 2). Additionally, most of those that were examined for methanol tolerance were found totally unaffected by methanol's presence.

Table 2. Non-precious hybrid ORR electrocatalysts.

Catalyst	Preparation Method and Experimental Conditions	Limiting Current (J_d; d = disc), Kinetic Current Density (J_k), Half Wave Potential ($E_{1/2}$))	Reference
$Co_{0.03}$@CoO-N-doped graphene carbon shells-800	Introduction of metal precursor (cobalt nitrate) to sucrose and urea followed by pyrolyzing and 0.1 M KOH, oxygen saturated, 10 mV·s^{-1}, 1600 rpm, (0.1 M KOH + 0.5 M MeOH)	J_d = 4.1 mA·cm^{-2} (Inactive in MeOH presence) $E_{1/2}$ = 0.81 V vs. RHE	[47]
BCN-2.5 at. %-1000	CVD synthesis of BCN sheets by thermally decomposing solid B C- and N-containing precursors at normal pressure and 0.1 M KOH, oxygen saturated, 10 mV·s^{-1}, 1600 rpm, (0.1 M KOH + 2.0 M MeOH)	J_d = 6.0 mA·cm^{-2} (Inactive in MeOH presence) $E_{1/2}$ = 0.707 V vs. RHE J_k = 26.62 mA·cm^{-2}	[48]
CoAl-LDHs (layered double hydroxide)/rGO (reduced graphene oxide)	Grow CoAl-LDHs on the surface of GO in-situ via coprecipitation and subsequently hydrothermal treatment and 0.1 M KOH, oxygen saturated, 10 mV·s^{-1}, 1600 rpm	J_d = 4.8 mA·cm^{-2} $E_{1/2}$ = 0.853 V vs. RHE	[49]

Table 2. *Cont.*

Catalyst	Preparation Method and Experimental Conditions	Limiting Current (J_d; d = disc), Kinetic Current Density (J_k), Half Wave Potential ($E_{1/2}$)	Reference
Co^{II}-A-rG-O (hybrid-ammonium) hydroxide-reduced graphene)	Synthesis at room temperature of archetypical hybrid materials consisting of cobalt-based organometallic complexes ([Co(acac)$_2$], acac = acetylacetonate) coordinated to N-doped graphenes and 0.1 M KOH, oxygen saturated, 10 mV·s^{-1}, 1600 rpm	J_d = 5.4 mA·cm^{-2} $E_{1/2}$ = 0.81 V vs. RHE J_k = 8.9 mA·cm^{-2} at 0.8 V vs. RHE	[50]
Co/N-HCOs (Co/N-co-doped hollowed-out carbon octahedrons)	Octahedral Co(II) complex with 2,6-bis(benzimidazol-2-yl)pyridine (BBP) as the precursor and 0.1 M KOH, oxygen saturated, 10 mV·s^{-1}, 1600 rpm	J_d = 4.9 mA·cm^{-2} $E_{1/2}$ = 0.81 V vs. RHE	[51]
Co(OH)$_2$-nanoplate/ N-RGO (N-doped reduced graphene oxide)	Hydrothermal method and 0.1 M KOH, oxygen saturated, 5 mV·s^{-1}, 1600 rpm	J_d = 4.7 mA·cm^{-2} $E_{1/2}$ = 0.66 V vs. RHE	[52]
N/Co-doped PCP(porous carbon polyhedron)/NRGO	Pyrolysis of graphene oxide-supported cobalt-based zeolitic imidazolate-framework and 0.1 M KOH, oxygen saturated, 5 mV·s^{-1}, 1600 rpm (0.1 M KOH + 3.0 M MeOH)	J_d = 7.8 mA·cm^{-2} (Inactive even after of 48h) J_k = 11.6 mA·cm^{-2} at 0.7 V vs. RHE $E_{1/2}$ = 0.93 V vs. RHE	[53]
CoCN@CoO$_x$(18)/NG (cobaltcarbonitride/ nitrogen doped graphene)	High temperature ammonia nitridation method and 0.1 M KOH, oxygen saturated, 10 mV·s^{-1}, 1600 rpm, (0.1 M KOH + 1.0 M MeOH)	J_d = 5.9 mA·cm^{-2} (Inactive in MeOH presence) $E_{1/2}$ 0.763 V vs. RHE	[54]
C-CZ-4(N-CNTs)-1000	In situ growth of metal–organic frameworks (ZIF-8) on carbon nanotubes, followed by pyrolysis and 0.1 M KOH, oxygen saturated, 5 mV·s^{-1}, 1600 rpm (0.1 M KOH + 1.0 M MeOH)	J_d = 6.0 mA·cm^{-2} (Inactive in MeOH presence) $E_{1/2}$ = 0.887 V vs. RHE	[55]
Bamboo-like CNT/Fe$_3$C nanoparticle hybrids-800	Annealing a mixture of PEG-PPG-PEG Pluronic P123, melamine, and Fe(NO$_3$)$_3$ at 800 °C in N$_2$ and 0.1 M KOH, oxygen saturated, 10 mV·s^{-1}, 1600 rpm	J_d = 4.0 mA·cm^{-2} $E_{1/2}$ = 0.861 V vs. RHE	[56]
Co$_3$O$_4$/NG (nitrogen-doped graphene)	Hydrothermal reaction of GO, MR, and CoCl$_2$ followed by a two-step heat treatment and 0.1 M KOH, oxygen saturated, 10 mV·s^{-1}, 1600 rpm	J_d = 4.6 mA·cm^{-2} $E_{1/2}$ = 0.74 V vs. RHE	[57]
CoNPs@NG (nitrogen-doped graphene)	Thermal condensation of biomass and corresponding metal salts and 0.1 M KOH, oxygen saturated, 10 mV·s^{-1}, 1600 rpm (0.1 M KOH + 1.0 M MeOH)	J_d = 7.2 mA·cm^{-2} $E_{1/2}$ = 1.01 V vs. RHE	[58]
C(PANI)/Mn$_2$O$_3$	Surface protected calcination processes and 0.1 M KOH, oxygen saturated, 10 mV·s^{-1}, 1600 rpm	J_d = 5.61 mA·cm^{-2} $E_{1/2}$ = 0.784 V vs. RHE	[59]

In general, Metal-N-doped carbon (especially graphene)-based electrocatalysts are preferred as conductive matrix materials. They are more positive than the others' hybrid materials, and half-wave potential as well as onset potential (Figure 3) make them more plausible candidate for cathodes. Namely, N/Co-doped graphene-type electrocatalyst (CoNPs@NG) [58] performed half-wave potential ($E_{1/2}$) of 1.01 V vs. RHE; about 0.19 V more positive than Pt/C's, the value of which is usually around 0.8 V vs. RHE. Its high activity was attributed to the growth of N-Co nanocrystal through the graphene plane and not at its edges, permitting the formation of a higher amount of active sites available to the reactants. Its special structure also offered very good stability performance as it is indicated from Figure 4. After ca 10 h of operation, only about 8% of current decay was observed. Remarkably, the same catalyst also presented good long-term stability in acidic environment, preserving 91% of its initial activity after 8.5 h of operation.

A novel hybrid electrocatalyst with reduced nitrogen-doped graphene/cobalt embedded porous carbon polyhedron was developed by Hou et al. [53]. A four-electron pathway was activated, performing $E_{1/2}$ = 0.93 V vs. RHE. Except for its excellent activity, this catalyst exhibited also a good durability, as after 5.5 h operation, its initial activity decreased by 14.5% in comparison with Pt/C, whose initial activity was reduced more than 30% for the same operation period. Its excellent activity was mainly attributed to the following factors: (i) the doping effect of N and Co; (ii) the porous structure; and (iii) the good contact between N/Co-doped carbon polyhedron and nitrogen-doped reduced graphene oxide.

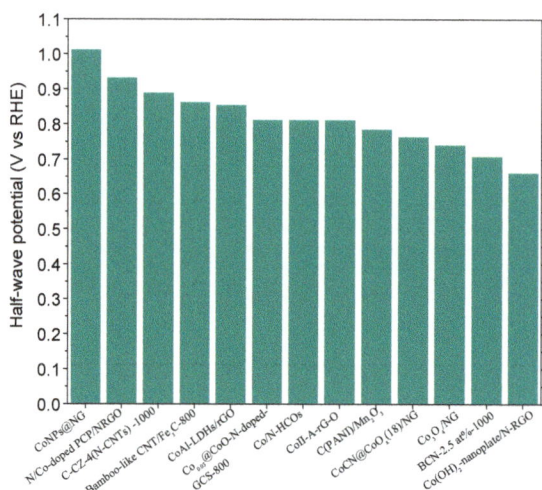

Figure 3. Comparison of half-wave potentials. (References from left to right bar: [58], [53], [55], [56], [49], [47], [50], [51], [59], [54], [57], [48], [52]).

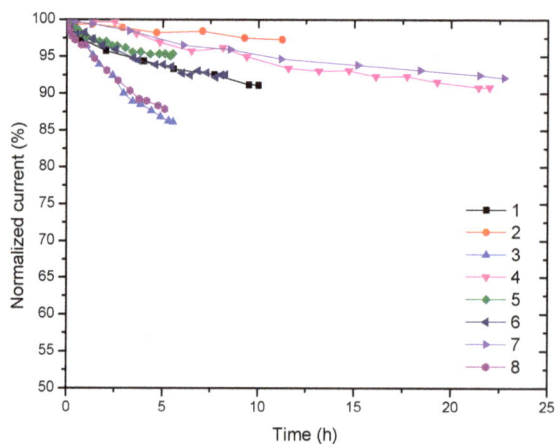

Figure 4. Long-term stability curves of hybrid materials: (1) $Co_{0.03}$@CoO-N-doped graphene carbon shells-800 (at −0.4 V vs. SCE) [47]; (2) Two-dimensional materials based on ternary system of B, C and N (B(2 at%)N(5 at%)G-1000 °C) [48]; (3) N/Co-doped PCP//NRGO (at −0.25 V vs. Ag/AgCl) [53]; (4) CoCN@CoO$_x$(18)/NG (cobalt carbonitride/nitrogen doped graphene) [54]; (5) Co_3O_4/NG (nitrogen-doped graphene) (at −0.36 V vs. SCE) [57]; (6) CoNPs@NG (nitrogen-doped graphene) [58]; (7) C(PANI)/Mn_2O_3 (hybrid) (at 0.8 V vs. RHE) [59]; and (8) CoO/MnO_2/RGO (reduced graphene oxide) (at 0.66 V vs. RHE) [60].

N-containing metal organic frameworks, like in the case of three-dimensional category, can deliver really good ORR electrocatalytic activity. An in in-situ grown metal-organic framework (MOF)/carbon nanotube (CNT) composite, instead of a surface grown one, was prepared and tested by Ge et al. [55], using ZIF-8 as precursor and by pyrolyzing the catalyst at 1000 °C. The as-prepared electrocatalyst reached a half-wave potential of 0.887 V vs. RHE and a limiting current of 6.0 mA·cm^{-2}, remaining stable even in the presence of 1 M CH_3OH.

Yang et al. [56] used, in their turn, melamine as natural source of nitrogen, synthesizing bamboo-like carbon nanotube Fe_3C NPs hybrid electrocatalyst, which also exhibited $E_{1/2} = 0.861$ V vs. RHE as well as excellent tolerance towards methanol. The Co/N-doped graphene hybrid catalyst, prepared by using urea as natural nitrogen source, in combination with sucrose and cobalt nitrate performed a half-wave potential of 0.81 V vs. RHE [47]. Moreover, the as-prepared electrocatalyst after 10 h operation at 0.4 V lost only 8% of its initial activity.

Urea with boric acid and polyethylene glycol were used for the preparation of B-N-co-doped graphite layers [48]. The obtained ORR kinetic current density was ca 27.0 mA·cm^{-2}, while its half-wave potential was very satisfactory; ca 0.707 V vs. RHE. Its activity remained unaffected in the presence of methanol (2.0 M) and its stability was very good, unchanged for 8.5 h, at 0.853 V vs. RHE, while Pt (20 wt %)/C, examined under the same experimental conditions, decreased by 20%. In hybrid-BCNs the extra electrons of nitrogen and the empty orbital of boron are connected into the carbon matrix, changing the electron density distribution of carbon [60–62].

Moreover, sandwich-like hybrid porous Co_3O_4 nanospheres and N-doped graphene using melamine as precursor were prepared by Zhang et al. [57]. By using melamine instead of urea or ammonia as nitrogen source, nitrogen NPs helped being attached not only at the edges of the graphene, but also in its inert areas; cobalt in each turn is intruded in the inert part of graphene providing much more active area. However, Co_3O_4/NG (nitrogen-doped graphene) exhibited good and close to Pt's half-wave potential, ca 0.74 V vs. RHE, indicating that it possesses high ORR activity. Meanwhile, its stability was good since after 5.5 h durability test at 0.913 V vs. RHE lost only 6% of its activity, while Pt/C lost 30% [57].

Additionally, Chao et al. [51] tried to make more environmentally friendly the preparation processes of hybrid materials by using inexpensive $Co(CH_3COO)_2 \cdot 4H_2O$ and 2,6-bis(benzimidazol-2-yl)pyridine (BBP) as raw materials. As described above, most of the fabrication methods of hybrid materials include multiple and relatively expensive processes. The result was a hollowed-out octahedral Co/N-co-doped carbon electrocatalyst that also performed a half-wave potential close to 0.81 V vs. RHE, and a limiting current density of 4.9 mA·cm^{-2}. Its activity remained almost constant even after 8000 cycles from 0 to 1.2 V, showing additionally remarkable methanol tolerance [51].

Among the investigated N-containing precursors, polyaniline (PANI) is commonly used for the preparation of N-doped carbon electrocatalysts. The combination of PANI's aromatic rings and the N-containing groups result in highly efficient ORR hybrid materials [63]. Recently, Cao et al. [59] reported half-wave potential of 0.784 V vs. RHE and 23 h long-term stability, at mesoporous hybrid shell of carbonized polyaniline Mn_2O_3 (C_{PANI}/Mn_2O_3) electrocatalyst. Their high activity and excellent stability was believed to be related with: (i) the high specific surface area; (ii) the surface oxidation state of Mn; and (iii) the composition codependent behavior (Mn_2O_3 and C_{PANI} are strongly dependent on each other). Hybrid materials have also been developed in order to improve the density of N-doped carbon electrocatalysts. The low density, observed in many N-doped carbons, leads to a thick electrode layer with reduced mass transfer ability [63,64]. By increasing the density of the carbon structure the volumetric ORR activity increases too.

Recently, N-doped reduced graphene oxide materials have attracted much interest due to their ability to increase electrocatalytic activity of the simple N-doped graphene electrocatalyst [65–67]. In N-doped graphene, it has been found that oxygen groups are responsible for the formation of the C-N bonds, increasing the conductivity of carbons. The reduction of N-doped graphene increases the oxygen groups and consequently the electronic-conductivity; reduced graphene oxide exhibit n-type electron doping behavior.

The introduction of metal oxides into N-doped graphene sheets, in order to form the hybrid material, can be processed through the following two-ways: (i) on the surface of the sheets; or (ii) through the plane of the graphene nanosheets. The second approach results in a better dispersion and higher stability of the NPs, giving the opportunity of exploiting both sides of graphene sheet [67,68].

According to Figure 3, Co^{II}-A-rG-O (hybrid-ammonium hydroxide-reduced graphene) performed the same ORR activity, giving also 0.81 V vs. RHE half-wave potential. Han et al. [50] proved that the high hybrid materials' activity toward ORR is attributed to the coordination of the modified carbon with the organometallic molecules. More precisely, the Co atom is coordinated in a square planar arrangement and an additional N-atom (or N containing aromatic ring) at an axial position, to two acetylacetonate ligands.

Co_3Mn-CO_3-LDH/RGO electrocatalyst exhibited that its relative activity maintained 87% of its initial value, while in the case of Pt/C at 82% [60].

A different geometrical hybrid electrocatalyst was developed by Zhan et al. [52], compounding hexagonal $Co(OH)_2$ nanoplates with nitrogen-doped reduced graphene oxide. According to the authors, its good performance (0.66 V vs. RHE) is attributed to the nanosizing of $Co(OH)_2$ and the synergistic interaction between $Co(OH)_2$ and N-rGO. After 4 h of operation at (-0.7 V), its activity remained stable and the same with that of (20 wt %) Pt/C.

Additionally, the combination of layered double hydroxides (LDHs) with metal oxides results in highly efficient ORR electrocatalysts. The LDHs' more positive charge and their good long-term stability in alkaline environment make them advantageous and promising. The CoAl-LDHs on reduced graphene oxide support at the optimum ratio ($w_{CoAl-LDHs}/w_{GO}$ = 1:5) performed 0.853 V vs. RHE half-wave potential, presenting no reaction in the presence of methanol (3.0 M) [49]. Moreover, it exhibited very stable behavior after 2.5 h operation at -0.4 V without losing its activity. It was found that the presence of LDHs improved the electrons transfer, while the reduced graphene oxide has been considered as the hydrogen peroxide cleaner. Thus, the synergetic effect of those two factors lead to a really effective ORR electrocatalyst [49].

Zhao et al. [69] prepared and tested, a new type of cost-effective artificial oxygen-evolving complex (OEC) rather active for the ORR catalysts based on cobalt-phosphate (Co-Pi). More precisely, Co-Pi OEC anchoring on reduced graphite oxide (rGO) nanosheet is shown to possess dramatically improved electrocatalytic activity. It was suggested that rGO serves as "peroxide cleaner" in enhancing the electrocatalytic behavior.

A new hybrid catalyst in the form of core-shell type, CoCN(carbonnitirde)@CoO_x NPs, supported on N-doped graphene was designed for the ORR in an alkaline solution [54]. The oxide 2 nm thin layer (CoO_x) played a protective role preventing the oxidation corrosion of CoCN. This electrocatalyst presented a low half-wave potential ($E_{1/2}$ = 0.763 V vs. RHE) compared to the other hybrid materials, but significantly high stability, losing only 10% of its initial reduction activity after 20 h of operation; under the same operational conditions, Pt/C lost ~30% of its activity. Its methanol tolerance was also high, as no any reaction was observed in 1.0 M methanol concentration.

By comparing Figures 1 and 4, it can be deduced that hybrid materials exhibit a higher long-term stability performance during the ORR process. Among the examined hybrid electrocatalysts, the mesoporous hybrid shell of carbonized polyaniline Mn_2O_3 (C_{PANI}/Mn_2O_3) [59] and the CoCN(carbonnitirde)@CoO_x supported on the graphene maintained more than 90% of their initial activity for about 20 h of operation [54]; both electrocatalysts have mesoporous core-shell structure offering more active sites and mass transfer. Thus, the strategy of intervening in carbon nanostructure by deliberately introducing in its main structure metal-organic frameworks develops hybrid materials/electrocatalysts. This strategy, according to the current review, has proven the most efficient one. Another, third, strategy of improving electrocatalytic activity of the non-precious metals concerns with the use of different nitrogen sources, as will be discussed below.

4. Alternative Raw Materials

Increasing homogeneity by the aid of doping constitutes another strategy adopted for the improvement of the characteristics of non-precious metal electrocatalysts. However, the control of the homogeneity is not easy since most of the synthesis procedures include multiple chemical steps. Especially when a precursor that includes an inbuilt homogeneous source of heteroatoms is adopted, the synthesis of a homogeneous doped carbon matrix is easier and more successful [70–72].

In Table 3, the electrochemical ORR activities in terms of half-wave potential (vs. RHE) for this class of electrocatalysts is reported. Among them, Prussian-blue, polyaniline (PANI), soya plant and others, have been investigated as alternative-raw precursors in order to facilitate the preparation procedure of electrocatalysts as well as to include a ready 3D-structure framework into the carbon matrix. These precursors contain M-N ready bonds responsible for the electrocatalysts activity enhancement [73].

Table 3. Non-precious alternative precursors' ORR electrocatalysts.

Catalyst	Preparation Method and Experimental Conditions	Limiting Current (J_d; d = disc), Kinetic Current Density (J_k), Half Wave Potential ($E_{1/2}$)	Reference
C-2PANI (polyaniline)/PBA (prussian blue analogue), 2-aniline/(aniline + PBA)	Mixing and pyrolysis & 0.1 M KOH, oxygen saturated, 10 mV·s^{-1}, 1600 rpm	J_d = 6.1 mA·cm^{-2} $E_{1/2}$ = 0.85 V vs. RHE	[73]
AFC-600 (ammonium ferric citrate, 600: treatment temperature)	A single-source molecular precursor containing carbon, nitrogen and transition metal & 1.0 M NaOH, oxygen saturated, 10 mV·s^{-1}, 2500 rpm	J_d = 2.6 mA·cm^{-2} $E_{1/2}$ = 0.881 V vs. RHE	[74]
N(okara source)-C-800	Nitrogen-doped carbon by okara treatment using also FeCl$_3$ & 0.1 M KOH, oxygen saturated, 5 mV·s^{-1}, 1600 rpm	J_d = 4.0 mA·cm^{-2} $E_{1/2}$ = 0.86 V vs. RHE	[75]
Fe/N-gCB (co-doped graphitic carbon bulb)	Prussian blue (PB) as the only precursor & 0.1 M KOH, oxygen saturated, 10 mV·s^{-1}, 1600 rpm	J_d = 5.0 mA·cm^{-2} $E_{1/2}$ = 0.81 V vs. RHE	[76]
FePhen@MOF-ArNH$_3$	Encapsulation synthesis and heat treatment in ammonia & 0.1 M KOH, oxygen saturated, 20 mV·s^{-1}, 1600 rpm	J_d = 5.6 mA·cm^{-2} $E_{1/2}$ = 0.86 V vs. RHE	[77]
S-P-N-doped graphitized carbon-1000	Soya (plant) as carbon source-graphitized product as support & 0.1 M KOH, oxygen saturated, 5 mV·s^{-1}, 1600 rpm (0.1 M KOH + 1.25 M MeOH)	J_d = 3.7 mA·cm^{-2} (MeOH tolerant) $E_{1/2}$ = 0.79 V vs. RHE	[78]
YS-Co/N-PCMs (yolk-shell/porous carbon microspheres)	Template-free hydrothermal method and a subsequent pyrolysis & 0.1 M KOH, oxygen saturated, 10 mV·s^{-1}, 1600 rpm	J_d = 5.0 mA·cm^{-2} $E_{1/2}$ = 0.706 V vs. RHE J_k = 16.0 mA·cm^{-2} at 0.3 V vs. SCE	[79]
N-S-co-doped-graphite	Pyrolysis of homogeneous mixture of exfoliated graphitic flakes and ionic liquid 1-butyl-3-methylimidazoliumbis(trifluoromethanesulfonyl) imide & 0.1 M KOH, oxygen saturated, 10 mV·s^{-1}, 2500 rpm	J_d = 6.5 mA·cm^{-2} $E_{1/2}$ = 0.768 V vs. RHE	[80]

According to the literature review, the composition of single-source molecular precursor affects the properties of any adopted metal and carbon and consequently the ORR electrocatalyst's activity. For example, Fe/Fe$_3$C NPs encapsulated in nitrogen-doped carbon using ammonium ferritic citrate as single molecular precursor exhibited the highest according to half-wave potential, $E_{1/2}$ = 0.881 V vs. RHE (Figure 5), ORR activity as well as an excellent methanol tolerance [74]. The conversion of amorphous carbon to graphitic carbon and the encapsulation of the iron nanoparticles into carbon's matrix were the main factors of its high ORR performance.

Another biomass product, okara, has been used as an alternative precursor, namely as nitrogen source. Okara is a byproduct from tofu and soy milk production, commonly used as feed for livestock or as a natural nitrogen fertilizer. The okara-derived nitrogen-doped carbon electrocatalyst exhibited the second highest activity, among the investigated samples, showing 0.86 V vs. RHE half-wave potential. This electrocatalyst was characterized by the high content of sp^{2}C, pyridinic-N and graphitic-N [75]. However, it is noted that such okara-derived catalysts are not easy to be reproduced as the composition of biomass varies considerably in reality, depending also on the factories and on the kind of soy beans.

Prussian blue is another precursor with ordered three-dimensional framework, which is consisted of coordinative bonded transition metal cations and cyanide groups. Such precursors contain M-N bonds and they are more active for ORR [76,77]. The observed high ORR activity 0.85 V vs. RHE is mainly attributed to the coordination of metal structure MN$_x$ [73]. Fe/N co-doped graphitic carbon bulb electrocatalyst was prepared using only Prussian blue as precursor; exhibiting 0.81 V vs. RHE half-wave potential [76].

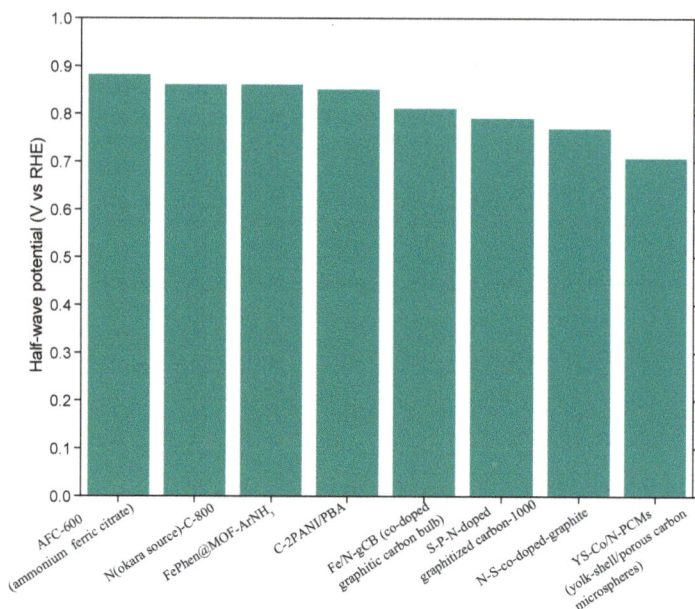

Figure 5. Comparison of half-wave potential of modified-alternative raw electrocatalytic materials. (Ref. from left to right bar: [74], [75], [77], [73], [76], [78], [80], [79]).

On the other hand, it has been also reported [77] a high ORR activity (Figure 5) for FePhen@MOF (metallic organic framework)-ArNH$_3$ electrocatalyst, with 0.86 V vs. RHE half-wave potential, without observing any direct coordination of iron with nitrogen. Its high activity was attributed also to the encapsulation of Fe into N-doped carbon's matrix and its synergetic effect with nitrogen via stabilizing the peroxide intermediate. Another kind of alternative-precursor, soya plant, which is rich in proteins, can lead to multiheteroatom-doped (S-N-P) carbon electrocatalysts after the appropriate treatment [78]. More precisely, the as prepared electrocatalyst exhibited 0.79 V vs. RHE half-wave potential for ORR and was inactive in presence of 1.25 M methanol. Additionally, at 0.4 V vs. RHE, its current density remained stable for almost one hour, while the corresponding commercial Pt/C's current density reduced more than 10%.

By pyrolyzing a homogeneous mixture of exfoliated graphitic flakes and ionic liquid, a very active ORR electrocatalyst was fabricated performing 0.768 V vs. RHE [80]. The great ORR activity was ascribed to the exfoliated graphite flakes which aid to the increased nitrogen doping and consequently the increased active surface area as well as to the co-doping of nitrogen and sulfur.

Hierarchically porous carbon materials are appropriate for accommodating more active sites and are synthesized via chemically complicated routes. A template-free and easier preparation method via using melamine, formaldehyde and cobalt acetate in combination with pyrolysis were adopted to prepare hierarchically structured yolk-shell Co and N co-doped porous carbon microspheres by Chao et al. [79]. This structure combined the high surface of the micropores, the active sites of Co-N$_x$ and graphitic N, yielding 0.706 V vs. RHE.

5. Dual Hetero-Atom Doped Electrocatalysts

Doping nitrogen can facilitate 4e⁻ ORR process leading to water formation, while boron, sulfur, and phosphorus are more prone to enhance 2e⁻ ORR process leading to hydrogen peroxide formation. Besides, nitrogen-doped carbon materials are cheaper and also offer long durability; for this reason, in the last 2–3 years, they have been studied more intensively. Among the investigated N-doped carbon materials ordered mesoporous carbon (OMC) attracts much more attention due to the unique regular arrays of uniform mesopores [81–83].

Sheng et al. [84] recently reported a highly active N-OMC electrocatalyst which exhibited 0.853 V vs. RHE half-wave potential (Figure 6). Furthermore, it presented very good stability, as after 5 h operation at −0.4 V vs. Ag/AgCl, its relative current density decreased only by 10% (Figure 7). The well-defined ordered mesoporous framework consisting of interconnected N-doped carbon rods with uniform size was the reason for the high ORR activity.

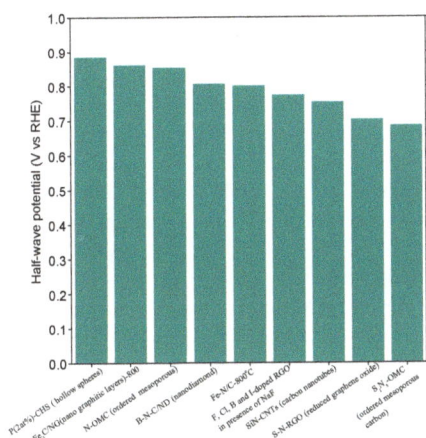

Figure 6. Comparison of ORR activity in terms of half-wave potential. (Ref. from left to right bars: [85], [86], [84], [87], [88], [89], [90], [91], [92]).

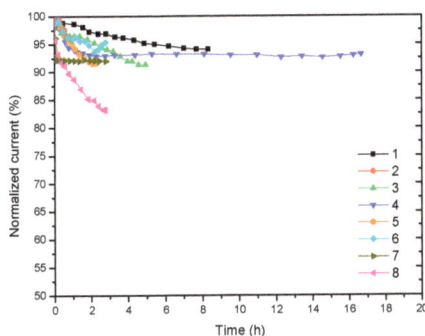

Figure 7. Long-term stability curves for modified with alternative precursor and mono or dual-doped electrocatalysts: (1) NCNT-900 (N-doped carbon nanotubes) (at −0.4 V vs. SCE) [40]; (2) S-P-N-doped graphitized carbon-1000 (at −0.3 V vs. SCE) [78]; (3) NOMC (N-doped ordered mesoporous) (at −0.3 V vs. Ag/AgCl) [84]; (4) P(2 at. %)-CHS (phosphorus-doped hollow spheres) (at 0.8 V vs. RHE) [85]; (5) shell core structural boron and nitrogen co-doped graphitic carbon/nanodiamond (BN-C/ND) [87]; (6) Fe-N/C-800 (at −0.3 V vs. Ag/AgCl) [88]; (7) SiN-CNTs (at −0.3 V vs. Ag/AgCl) [90]; and (8) S-doped/CNTs-900 (at −0.3 V vs. Ag/AgCl) [93].

The doping of N-doped-carbon materials with a second element such as silicon, boron, phosphorus, etc. favors the synergetic effect between the second element and the N-doped carbon. The resulting materials exhibit a better ORR activity than single N-doped carbon materials.

For instance, phosphorus has lower electronegativity than nitrogen and sulfur, leading to a higher modification extent of carbon and having as a result a more efficient electrocatalyst towards ORR. Wu et al. [85] prepared phosphorus-doped carbon hollow spheres (P(2 at.%)-CHS) electrocatalysts, which exhibited more positive than Pt's half-wave potential, 0.883 V vs. RHE (Figure 6). Its special structural design offered higher triple phase (solid–liquid–gas) area and consequently enhanced ORR activity. Moreover, the as-prepared electrocatalyst operating at -0.25 V for almost 17 h lost only 5% of its initial activity (Figure 7). On the contrary, under the same operation conditions, the activity loss on Pt/C recorded almost 20%.

Iron encapsulation into carbon's matrix, as preparation method of ORR electrocatalysts' fabrication, has been adopted from other research groups too [74,78,86]. According them, the observed enhanced ORR activity was attributed to the iron's encapsulation, which stabilized the produced peroxide intermediate, and to the coordination of iron with nitrogen.

Niu et al. [88] tried to enlarge the active surface area of N-doped mesoporous carbon by embedding rigid templates within precursor matrices. They fabricated Fe-N/C catalysts by the aid of controlled pyrolysis of a poly(2-fluoroaniline) (P_2FANI) matrix, within which FeO-(OH) nanorods were homogeneously embedded. They obtained a half-wave potential value of 0.8 V vs. RHE, the same with Pt's one. The Fe-N/C pyrolysed at 800 °C exhibited, additionally, remarkable long-term stability keeping 93.3% of its initial activity for 2.8 h; while under the same experimental conditions Pt/C lost half of its initial activity.

It is worth to be noticed that it is surprising that in some cases Pt/C is so much unstable compared to non-noble metal catalysts. Part of this difference (sometimes higher than 20%) could be due to the comparison of "normalized current" at different potentials. However, we could conclude here that most of the examined non-noble catalysts exhibit at least similar stability with Pt/C one.

Iron has been also investigated as a second co-doped with nitrogen element. A novel nitrogen-doped-graphene nanoplatelets of Fe@N-graphene nanoplatelet-embedded carbon nanofibers was proposed by Ju et al. [94], delivering half-wave potential value 0.8 V vs. RHE (Figure 6), which is the same as Pt. Its high activity was due to the high number of functional groups.

Sulfur constitutes another element that has been investigated a lot as a second element in N-doped electrocatalysts. According to Figure 6, sulfur-nitrogen-doped CNTs [92] and sulfur-nitrogen-doped reduced graphene oxide [91] electrocatalysts exhibited 0.685 and 0.703 V vs. RHE half-wave potential, respectively. According to the authors of both works, the synergetic effect between N and S favors the $4e^-$ ORR pathway. Additionally, S-N-CNTs electrocatalyst exhibited excellent methanol tolerance (10 wt %) even after 2.5 h operation at -0.3 V vs. Hg/HgO [93]. The advantages of the ordered mesoporous structure and of the co-doping method were exploited by Jiang et al. [92], fabricating a S-N-ordered mesoporous electrocatalyst. The authors also stated very good long-term stability, as the electrocatalyst maintained 91.6% of its relative current density after 2.5 h operation, while Pt/C maintained only 70% of its owns; excellent methanol (1.0 M) tolerance was recorded.

Additionally, silicon doped into nitrogen-doped carbon nanotubes (SiN-CNTs) matrix exhibited a very good activity, $E_{1/2} = 0.753$ V vs. RHE (Figure 6), remaining also inactive in presence of 1.0 M methanol. Moreover, for the first operational seconds its initial relative current decayed by 8% and it remained stable for 2.8 h, at -0.3 V vs. Ag/AgCl, while over Pt/C, the measured current decay was 30% [90].

Boron and nitrogen co-doping mostly on graphene and carbon nanotubes has been investigated thoroughly in international literature [95–98]. Liu et al. [87], recently proved that synergetic effect of co-doping and doping one by one element was the main factor for enhancing ORR activity. As depicted in Figure 6, the B-N co-doped over core shell graphitic carbon nanodiamonds electrocatalysts exhibited high half-wave potential, $E_{1/2} = 0.805$ V vs. RHE [87].

Finally, some research groups turned their attention to halogen doped-nitrogen carbon materials. By introducing a halogen element into graphene structure, its electronic structure changes, accommodating more active sites [89]. F-, Cl-, Br- and I-co-doped reduced graphene oxide electrocatalyst, according to LSV measurements, exhibited a half-wave potential of 0.773 V vs. RHE; just a little more negative from Pt's. After 2000 s, its current density remained the same, while for Pt/C it was decreased by 10%.

Taking into consideration the results listed in Table 4, a common characteristic of the dual-doped electrocatalysts is the hollow carbon nanospheres, as well as the mesoporous structure. According to Figure 7, the highest long-term stability was displayed by phosphorus-doped hollow sphere electrocatalyst, ca 18 h, maintaining 90% of its initial activity [85]. It has been stated that the hollow structure offers efficient mass transfer of the reactants. Finally, the development of ORR electrocatalysts with alternative precursors, by the aid of simpler preparation methods, is still of great challenge as deduced during the current review.

Table 4. Non-precious dual-doped carbon ORR electrocatalysts.

Catalyst	Preparation Method and Experimental Conditions	Limiting Current (J_d; d = disc), Kinetic Current Density (J_k), Half-Wave Potential ($E_{1/2}$))	Reference
N-OMC (N-doped ordered mesoporous)	Two-step nanocasting method (DHN as precursors) & 0.1 M KOH, oxygen saturated 10 mV·s⁻¹, 2500 rpm	J_d = 5.8 mA·cm⁻² J_k = 22 mA·cm⁻² (at 0.4 V vs. Ag/AgCl) $E_{1/2}$ = 0.853 V vs. RHE	[84]
P(2 at. %)-CHS (phosphorus-doped hollow spheres)	Hydrothermal method using glucose as a carbon source, tetraphenylphosphonium bromide as a P source and anionic surfactant sodium dodecyl sulfate as a soft template & 0.1 M KOH, oxygen saturated 10 mV·s⁻¹, 1600 rpm	J_d = 5.7 mA·cm⁻² $E_{1/2}$ = 0.883 V vs. RHE	[85]
Fe₃C/N-G(nano graphitic layers)-800	Pyrolysis of poly (1,8-diaminonapphthalene) (PDAN) using precursors of 1,8-diaminonaphthalene (DAN) and FeCl₃ & 0.1 M KOH, oxygen saturated, 5 mV·s⁻¹, 1600 rpm, (0.1 M KOH + 1.0 M MeOH)	J_d = 5.8 mA·cm⁻² $E_{1/2}$ = 0.86 V vs. RHE (Inactive in MeOH presence)	[86]
Shell core structural B and N co-doped graphitic carbon/nanodiamond (BN-C/ND)	One-step heat-treatment of the mixture with nanodiamond, melamine, boric acid and FeCl₃ & 0.1 M KOH, oxygen saturated, 10 mV·s⁻¹, 1600 rpm (0.1 M KOH + 1.0 M MeOH)	J_d = 6.0 mA·cm⁻² $E_{1/2}$ = 0.22 V vs. Hg/HgO (0.805 V vs. RHE) (Inactive in MeOH presence)	[87]
Fe-N/C-800	Thermally removable nanoparticle templates & 0.1 M KOH, oxygen saturated (0.1 M KOH + 1.0 M MeOH)	J_d = 4.9 mA·cm⁻² $E_{1/2}$ = 0.80 V vs. RHE (Inactive in MeOH presence)	[88]
F, Cl, B and I-doped RGO in presence of NaF	Halogenations of reduced graphene oxide (RGO) with simultaneously fluorine, chlorine, bromine and iodine by electrochemical exfoliation of GO and obtained XRGO in presence of IL and halogen salts (X = F, Cl, Br, I) & 0.5 M KOH, oxygen saturated, 10 mV·s⁻¹, 1600 rpm	J_d = 5.7 mA·cm⁻² $E_{1/2}$ = 0.773 V vs. RHE	[89]
SiN-CNTs	Thermolysis of 3-aminopropyl-triethoxysilane and dimethylsilicone oil, respectively, using FeMo/Al₂O₃ as catalysts & 0.1 M KOH, oxygen saturated 5 mV·s⁻¹, 1600 rpm (0.1 M KOH + 1.0 M MeOH)	J_d = 6.12 mA·cm⁻² $E_{1/2}$ = 0.753 V vs. RHE (Inactive in MeOH presence)	[90]
S-N-RGO (reduced graphene oxide)	Single-step non-hydrothermal chemical route Reflux in ethylene glycol at 180 °C for 3 h & 0.1 M KOH, oxygen saturated, 5 mV·s⁻¹, 1600 rpm	J_d = 5.1 mA·cm⁻² $E_{1/2}$ = 0.703 V vs. RHE J_k = 7.7 mA·cm⁻² (at 0.6 V vs. Hg/HgO)	[91]
S₁N₅-OMC (dual doped with S and N ordered mesoporous carbon)	Polythiophene (PTh) and polypyrrole (PPy) as the precursors, ordered mesoporous silica (SBA-15) as the hard template, and FeCl₃ as the catalyst & 0.1 M KOH, oxygen saturated, 10 mV·s⁻¹, 1600 rpm, (0.1 M KOH + 0.5 M MeOH)	J_d = 4.6 mA·cm⁻² $E_{1/2}$ = 0.685 V vs. RHE (Inactive in MeOH presence)	[92]

6. Conclusions

The present review mainly focuses on novel carbon based materials for ORR, including: (i) the development of three-dimensional structures; (ii) the synthesis of novel hybrid (metal oxide-nitrogen-carbon) electrocatalysts; (iii) the use of alternative raw precursors characterized from three-dimensional structure; (iv) the adoption of co-doping methods for novel metal-nitrogen-doped-carbon electrocatalysts.

Hybrid-materials exceed the other types of electrocatalysts in terms of ORR activity as well as long-term stability. Among them, nitrogen-doped-reduced graphene oxide with mesoporous structure is the most efficient support. However, the greatest challenge of those electrocatalysts is the simplification of their synthesis procedure. Towards this direction, non-precious ORR electrocatalysts via one step process utilizing alternative precursors, mainly Prussian-blue and/or polyaniline, were prepared. The main advantage of these kinds of alternative precursors is their ready (from their nature) three-dimensional structures offering enriched M-N bonds, which, after pyrolysis, form MN_x ORR active sites. The same precursors can serve also as carbon or nitrogen sources. This class of electrocatalysts exhibits very good ORR electrochemical activity; however, their main drawback is their short stability over time.

In their turn, *dual-doped ORR electrocatalysts* succeed to change the ORR mechanism in N-doped electrocatalysts from $2e^-$ to $4e^-$ process. The most efficient dual-doped electrocatalysts: (i) are characterized by either ordered mesoporous or hollow mesoporous graphene structure; (ii) have been prepared by co-doping metals; and (iii) the doped-metals have been embedded throughout carbon matrix and not at its edges. In terms of long-term stability, it can be deduced that they are less stable than hybrids, but more stable than the alternative precursors' electrocatalysts.

Finally, *3D interconnected structures* facilitate electron and mass transfer. The key for the development of the most active electrocatalyst (3D Co-N-OMMC) was the combination of: (i) co-doping of the two metals; (ii) highly ordered macro-mesoporous graphene's structure; and (iii) synergetic effect of the two metals with the support. Typically, their ORR activity is still lower than the other as-reported electrocatalysts. The degradation of their structure and, consequently, the much lower stability are the main challenges that need to be overcome in future.

Acknowledgments: The authors are grateful to the Sino-Greek Science and Technology Cooperation (Project 2013DFG62590), the National Natural Science Foundation of China (Grant No. 21575299, 21576300, and 21276290), and to the Guangdong Province Nature Science Foundation (2014A030313150) for the financial support. Panagiotis Tsiakaras is also grateful to the Ministry of Education and Science of the Russian Federation (Mega-grant contract No. 14.Z50.31.0001) for funding.

Author Contributions: Angeliki Brouzgou searched the literature papers, categorized the electrocatalysts in the present paper and wrote the main part of the current work; Shuqin Song and Zhen-Xing Liang corrected English, helped with data analysis and wrote part of this work; Panagiotis Tsiakaras wrote part of this work, corrected English grammar and syntax and supervised the whole work.

Conflicts of Interest: The authors declare no conflict of interest.

References

1. Brouzgou, A.; Song, S.Q.; Tsiakaras, P. Low and non-platinum electrocatalysts for pemfcs: Current status, challenges and prospects. *Appl. Catal. B Environ.* **2012**, *127*, 371–388. [CrossRef]
2. Wang, Y.; He, C.; Brouzgou, A.; Liang, Y.; Fu, R.; Wu, D.; Tsiakaras, P.; Song, S. A facile soft-template synthesis of ordered mesoporous carbon/tungsten carbide composites with high surface area for methanol electrooxidation. *J. Power Sources* **2012**, *200*, 8–13. [CrossRef]
3. Brouzgou, A.; Podias, A.; Tsiakaras, P. PEMFCS and AEMFCS directly fed with ethanol: A current status comparative review. *J. Appl. Electrochem.* **2013**, *43*, 119–136. [CrossRef]
4. Wan, K.; Long, G.-F.; Liu, M.-Y.; Du, L.; Liang, Z.-X.; Tsiakaras, P. Nitrogen-doped ordered mesoporous carbon: Synthesis and active sites for electrocatalysis of oxygen reduction reaction. *Appl. Catal. B Environ.* **2015**, *165*, 566–571. [CrossRef]
5. Gong, K.; Du, F.; Xia, Z.; Durstock, M.; Dai, L. Nitrogen-doped carbon nanotube arrays with high electrocatalytic activity for oxygen reduction. *Science* **2009**, *323*, 760–764. [CrossRef] [PubMed]
6. Lee, J.S.; Park, G.S.; Kim, S.T.; Liu, M.; Cho, J. A highly efficient electrocatalyst for the oxygen reduction reaction: N-doped ketjenblack incorporated into Fe/Fe$_3$C-functionalized melamine foam. *Angew. Chem.* **2013**, *125*, 1060–1064. [CrossRef]

7.   Choi, C.H.; Chung, M.W.; Kwon, H.C.; Park, S.H.; Woo, S.I. B, N-and P, N-doped graphene as highly active catalysts for oxygen reduction reactions in acidic media. *J. Mater. Chem. A* **2013**, *1*, 3694–3699. [CrossRef]

8.   Xue, Y.; Yu, D.; Dai, L.; Wang, R.; Li, D.; Roy, A.; Lu, F.; Chen, H.; Liu, Y.; Qu, J. Three-dimensional B, N-doped graphene foam as a metal-free catalyst for oxygen reduction reaction. *Phys. Chem. Chem. Phys.* **2013**, *15*, 12220–12226. [CrossRef] [PubMed]

9.   Zhang, L.; Niu, J.; Li, M.; Xia, Z. Catalytic mechanisms of sulfur-doped graphene as efficient oxygen reduction reaction catalysts for fuel cells. *J. Phys. Chem. C* **2014**, *118*, 3545–3553. [CrossRef]

10. Park, J.-E.; Jang, Y.J.; Kim, Y.J.; Song, M.-S.; Yoon, S.; Kim, D.H.; Kim, S.-J. Sulfur-doped graphene as a potential alternative metal-free electrocatalyst and Pt-catalyst supporting material for oxygen reduction reaction. *Phys. Chem. Chem. Phys.* **2014**, *16*, 103–109. [CrossRef] [PubMed]

11. Ma, Z.; Dou, S.; Shen, A.; Tao, L.; Dai, L.; Wang, S. Sulfur-doped graphene derived from cycled lithium–sulfur batteries as a metal-free electrocatalyst for the oxygen reduction reaction. *Angew. Chem. Int. Ed.* **2015**, *54*, 1888–1892. [CrossRef] [PubMed]

12. Peng, H.; Mo, Z.; Liao, S.; Liang, H.; Yang, L.; Luo, F.; Song, H.; Zhong, Y.; Zhang, B. High performance Fe-and N-doped carbon catalyst with graphene structure for oxygen reduction. *Sci. Rep.* **2013**. [CrossRef]

13. Liang, J.; Zhou, R.F.; Chen, X.M.; Tang, Y.H.; Qiao, S.Z. Fe–N decorated hybrids of CNTs grown on hierarchically porous carbon for high-performance oxygen reduction. *Adv. Mater.* **2014**, *26*, 6074–6079. [CrossRef] [PubMed]

14. Wu, J.; Yang, Z.; Li, X.; Sun, Q.; Jin, C.; Strasser, P.; Yang, R. Phosphorus-doped porous carbons as efficient electrocatalysts for oxygen reduction. *J. Mater. Chem. A* **2013**, *1*, 9889–9896. [CrossRef]

15. Zhang, C.; Mahmood, N.; Yin, H.; Liu, F.; Hou, Y. Synthesis of phosphorus-doped graphene and its multifunctional applications for oxygen reduction reaction and lithium ion batteries. *Adv. Mater.* **2013**, *25*, 4932–4937. [CrossRef] [PubMed]

16. Li, R.; Wei, Z.; Gou, X.; Xu, W. Phosphorus-doped graphene nanosheets as efficient metal-free oxygen reduction electrocatalysts. *RSC Adv.* **2013**, *3*, 9978–9984. [CrossRef]

17. Liang, Y.; Wang, H.; Diao, P.; Chang, W.; Hong, G.; Li, Y.; Gong, M.; Xie, L.; Zhou, J.; Wang, J.; et al. Oxygen reduction electrocatalyst based on strongly coupled cobalt oxide nanocrystals and carbon nanotubes. *J. Am. Chem. Soc.* **2012**, *134*, 15849–15857. [CrossRef] [PubMed]

18. Yang, Z.; Yao, Z.; Li, G.; Fang, G.; Nie, H.; Liu, Z.; Zhou, X.; Chen, X.; Huang, S. Sulfur-doped graphene as an efficient metal-free cathode catalyst for oxygen reduction. *ACS Nano* **2012**, *6*, 205–211. [CrossRef] [PubMed]

19. Liu, R.; Wu, D.; Feng, X.; Müllen, K. Nitrogen-doped ordered mesoporous graphitic arrays with high electrocatalytic activity for oxygen reduction. *Angew. Chem. Int. Ed.* **2010**, *49*, 2565–2569. [CrossRef] [PubMed]

20. Niwa, H.; Horiba, K.; Harada, Y.; Oshima, M.; Ikeda, T.; Terakura, K.; Ozaki, J.-I.; Miyata, S. X-ray absorption analysis of nitrogen contribution to oxygen reduction reaction in carbon alloy cathode catalysts for polymer electrolyte fuel cells. *J. Power Sources* **2009**, *187*, 93–97. [CrossRef]

21. Ferrandon, M.; Kropf, A.J.; Myers, D.J.; Artyushkova, K.; Kramm, U.; Bogdanoff, P.; Wu, G.; Johnston, C.M.; Zelenay, P. Multitechnique characterization of a polyaniline-iron-carbon oxygen reduction catalyst. *J. Phys. Chem. C* **2012**, *116*, 16001–16013. [CrossRef]

22. Wu, G.; More, K.L.; Johnston, C.M.; Zelenay, P. High-performance electrocatalysts for oxygen reduction derived from polyaniline, iron, and cobalt. *Science* **2011**, *332*, 443–447. [CrossRef] [PubMed]

23. Zhang, P.; Sun, F.; Xiang, Z.; Shen, Z.; Yun, J.; Cao, D. Zif-derived in situ nitrogen-doped porous carbons as efficient metal-free electrocatalysts for oxygen reduction reaction. *Energy Environ. Sci.* **2014**, *7*, 442–450. [CrossRef]

24. Zhang, Y.; Chu, M.; Yang, L.; Deng, W.; Tan, Y.; Ma, M.; Xie, Q. Synthesis and oxygen reduction properties of three-dimensional sulfur-doped graphene networks. *Chem. Commun.* **2014**, *50*, 6382–6385. [CrossRef] [PubMed]

25. Fang, B.; Kim, J.H.; Kim, M.-S.; Yu, J.-S. Hierarchical nanostructured carbons with meso–macroporosity: Design, characterization, and applications. *Acc. Chem. Res.* **2013**, *46*, 1397–1406. [CrossRef] [PubMed]

26. Wang, J.G.; Zhou, H.J.; Sun, P.C.; Ding, D.T.; Chen, T.H. Hollow carved single-crystal mesoporous silica templated by mesomorphous polyelectrolyte-surfactant complexes. *Chem. Mater.* **2010**, *22*, 3829–3831. [CrossRef]

27. Liu, Y.-L.; Shi, C.-X.; Xu, X.-Y.; Sun, P.-C.; Chen, T.-H. Nitrogen-doped hierarchically porous carbon spheres as efficient metal-free electrocatalysts for an oxygen reduction reaction. *J. Power Sources* **2015**, *283*, 389–396. [CrossRef]

28. Tao, G.; Zhang, L.; Chen, L.; Cui, X.; Hua, Z.; Wang, M.; Wang, J.; Chen, Y.; Shi, J. N-doped hierarchically macro/mesoporous carbon with excellent electrocatalytic activity and durability for oxygen reduction reaction. *Carbon* **2015**, *86*, 108–117. [CrossRef]

29. Sun, T.; Xu, L.; Li, S.; Chai, W.; Huang, Y.; Yan, Y.; Chen, J. Cobalt-nitrogen-doped ordered macro-/mesoporous carbon for highly efficient oxygen reduction reaction. *Appl. Catal. B Environ.* **2016**, *193*, 1–8. [CrossRef]

30. Duan, J.; Zheng, Y.; Chen, S.; Tang, Y.; Jaroniec, M.; Qiao, S. Mesoporous hybrid material composed of Mn_3O_4 nanoparticles on nitrogen-doped graphene for highly efficient oxygen reduction reaction. *Chem. Commun.* **2013**, *49*, 7705–7707. [CrossRef] [PubMed]

31. Hao, J.; Liu, Y.; Shen, H.; Li, W.; Li, J.; Li, Y.; Chen, Q. Effect of nickel-ion doping in MnO_2 nanoneedles as electrocatalyst for the oxygen reduction reaction. *J. Mater. Sci. Mater. Electron.* **2016**, *27*, 6598–6605. [CrossRef]

32. Bian, W.; Yang, Z.; Strasser, P.; Yang, R. A $CoFe_2O_4$/graphene nanohybrid as an efficient bi-functional electrocatalyst for oxygen reduction and oxygen evolution. *J. Power Sources* **2014**, *250*, 196–203. [CrossRef]

33. Huo, R.; Jiang, W.J.; Xu, S.; Zhang, F.; Hu, J.S. Co/CoO/$CoFe_2O_4$/G nanocomposites derived from layered double hydroxides towards mass production of efficient Pt-free electrocatalysts for oxygen reduction reaction. *Nanoscale* **2014**, *6*, 203–206. [CrossRef] [PubMed]

34. Xu, Y.; Bian, W.; Wu, J.; Tian, J.-H.; Yang, R. Preparation and electrocatalytic activity of 3D hierarchical porous spinel $CoFe_2O_4$ hollow nanospheres as efficient catalyst for oxygen reduction reaction and oxygen evolution reaction. *Electrochim. Acta* **2015**, *151*, 276–283. [CrossRef]

35. Wang, J.; Wu, Z.; Han, L.; Lin, R.; Xin, H.L.; Wang, D. Hollow-structured carbon-supported nickel cobaltite nanoparticles as an efficient bifunctional electrocatalyst for the oxygen reduction and evolution reactions. *ChemCatChem* **2016**, *8*, 736–742. [CrossRef]

36. Lee, E.; Jang, J.-H.; Kwon, Y.-U. Composition effects of spinel $Mn_xCo_{3-x}O_4$ nanoparticles on their electrocatalytic properties in oxygen reduction reaction in alkaline media. *J. Power Sources* **2015**, *273*, 735–741. [CrossRef]

37. Ma, Y.; Wang, R.; Wang, H.; Key, J.; Ji, S. Control of MnO_2 nanocrystal shape from tremella to nanobelt for ehancement of the oxygen reduction reaction activity. *J. Power Sources* **2015**, *280*, 526–532. [CrossRef]

38. Odedairo, T.; Yan, X.; Ma, J.; Jiao, Y.; Yao, X.; Du, A.; Zhu, Z. Nanosheets Co_3O_4 interleaved with graphene for highly efficient oxygen reduction. *ACS Appl. Mater. Interfaces* **2015**, *7*, 21373–21380. [CrossRef] [PubMed]

39. Li, Z.; Shao, M.; Zhou, L.; Zhang, R.; Zhang, C.; Wei, M.; Evans, D.G.; Duan, X. Directed growth of metal-organic frameworks and their derived carbon-based network for efficient electrocatalytic oxygen reduction. *Adv. Mater.* **2016**, *28*, 2337–2344. [CrossRef] [PubMed]

40. Pan, T.; Liu, H.; Ren, G.; Li, Y.; Lu, X.; Zhu, Y. Metal-free porous nitrogen-doped carbon nanotubes for enhanced oxygen reduction and evolution reactions. *Sci. Bull.* **2016**, *61*, 889–896. [CrossRef]

41. Hao, L.; Zhang, S.; Liu, R.; Ning, J.; Zhang, G.; Zhi, L. Bottom-up construction of triazine-based frameworks as metal-free electrocatalysts for oxygen reduction reaction. *Adv. Mater.* **2015**, *27*, 3190–3195. [CrossRef] [PubMed]

42. Xu, Y.; Jin, S.; Xu, H.; Nagai, A.; Jiang, D. Conjugated microporous polymers: Design, synthesis and application. *Chem. Soc. Rev.* **2013**, *42*, 8012–8031. [CrossRef] [PubMed]

43. Zhang, L.; Su, Z.; Jiang, F.; Yang, L.; Qian, J.; Zhou, Y.; Li, W.; Hong, M. Highly graphitized nitrogen-doped porous carbon nanopolyhedra derived from ZIF-8 nanocrystals as efficient electrocatalysts for oxygen reduction reactions. *Nanoscale* **2014**, *6*, 6590–6602. [CrossRef] [PubMed]

44. Miner, E.M.; Fukushima, T.; Sheberla, D.; Sun, L.; Surendranath, Y.; Dincă, M. Electrochemical oxygen reduction catalysed by Ni_3(hexaiminotriphenylene)$_2$. *Nat. Commun.* **2016**, *7*, 10942. [CrossRef] [PubMed]

45. Verma, A.; Jha, A.K.; Basu, S. Manganese dioxide as a cathode catalyst for a direct alcohol or sodium borohydride fuel cell with a flowing alkaline electrolyte. *J. Power Sources* **2005**, *141*, 30–34. [CrossRef]

46. Rojas-Carbonell, S.; Babanova, S.; Serov, A.; Ulyanova, Y.; Singhal, S.; Atanassov, P. Hybrid electrocatalysts for oxygen reduction reaction: Integrating enzymatic and non-platinum group metal catalysis. *Electrochim. Acta* **2016**, *190*, 504–510. [CrossRef]

47. Zhang, G.; Lu, W.; Cao, F.; Xiao, Z.; Zheng, X. N-doped graphene coupled with Co nanoparticles as an efficient electrocatalyst for oxygen reduction in alkaline media. *J. Power Sources* **2016**, *302*, 114–125. [CrossRef]

48. Jin, J.; Pan, F.; Jiang, L.; Fu, X.; Liang, A.; Wei, Z.; Zhang, J.; Sun, G. Catalyst-free synthesis of crumpled boron and nitrogen Co-doped graphite layers with tunable bond structure for oxygen reduction reaction. *ACS Nano* **2014**, *8*, 3313–3321. [CrossRef] [PubMed]

49. Wang, Y.; Wang, Z.; Wu, X.; Liu, X.; Li, M. Synergistic effect between strongly coupled coal layered double hydroxides and graphene for the electrocatalytic reduction of oxygen. *Electrochim. Acta* **2016**, *192*, 196–204. [CrossRef]

50. Han, J.; Sa, Y.J.; Shim, Y.; Choi, M.; Park, N.; Joo, S.H.; Park, S. Coordination chemistry of [Co(acac)$_2$] with N-doped graphene: Implications for oxygen reduction reaction reactivity of organometallic Co-O$_4$-N species. *Angew. Chem. Int. Ed.* **2015**, *54*, 12622–12626. [CrossRef] [PubMed]

51. Chao, S.; Bai, Z.; Cui, Q.; Yan, H.; Wang, K.; Yang, L. Hollowed-out octahedral Co/N-codoped carbon as a highly efficient non-precious metal catalyst for oxygen reduction reaction. *Carbon* **2015**, *82*, 77–86. [CrossRef]

52. Zhan, Y.; Du, G.; Yang, S.; Xu, C.; Lu, M.; Liu, Z.; Lee, J.Y. Development of cobalt hydroxide as a bifunctional catalyst for oxygen electrocatalysis in alkaline solution. *ACS Appl. Mater. Interfaces* **2015**, *7*, 12930–12936. [CrossRef] [PubMed]

53. Hou, Y.; Wen, Z.; Cui, S.; Ci, S.; Mao, S.; Chen, J. An advanced nitrogen-doped graphene/cobalt-embedded porous carbon oolyhedron hybrid for efficient catalysis of oxygen reduction and water splitting. *Adv. Funct. Mater.* **2015**, *25*, 872–882. [CrossRef]

54. Wu, Y.; Shi, Q.; Li, Y.; Lai, Z.; Yu, H.; Wang, H.; Peng, F. Nitrogen-doped graphene-supported cobalt carbonitride@oxide core–shell nanoparticles as a non-noble metal electrocatalyst for an oxygen reduction reaction. *J. Mater. Chem. A* **2015**, *3*, 1142–1151. [CrossRef]

55. Ge, L.; Yang, Y.; Wang, L.; Zhou, W.; De Marco, R.; Chen, Z.; Zou, J.; Zhu, Z. High activity electrocatalysts from metal–organic framework-carbon nanotube templates for the oxygen reduction reaction. *Carbon* **2015**, *82*, 417–424. [CrossRef]

56. Yang, W.; Liu, X.; Yue, X.; Jia, J.; Guo, S. Bamboo-like carbon nanotube/Fe$_3$C nanoparticle hybrids and their highly efficient catalysis for oxygen reduction. *J. Am. Chem. Soc.* **2015**, *137*, 1436–1439. [CrossRef] [PubMed]

57. Zhang, M.; Li, R.; Chang, X.; Xue, C.; Gou, X. Hybrid of porous cobalt oxide nanospheres and nitrogen-doped graphene for applications in lithium-ion batteries and oxygen reduction reaction. *J. Power Sources* **2015**, *290*, 25–34. [CrossRef]

58. Lv, L.-B.; Ye, T.-N.; Gong, L.-H.; Wang, K.-X.; Su, J.; Li, X.-H.; Chen, J.-S. Anchoring cobalt nanocrystals through the plane of graphene: Highly integrated electrocatalyst for oxygen reduction reaction. *Chem. Mater.* **2015**, *27*, 544–549. [CrossRef]

59. Cao, S.; Han, N.; Han, J.; Hu, Y.; Fan, L.; Zhou, C.; Guo, R. Mesoporous hybrid shells of carbonized polyaniline/Mn$_2$O$_3$ as non-precious efficient oxygen reduction reaction catalyst. *ACS Appl. Mater. Interfaces* **2016**, *8*, 6040–6050. [CrossRef] [PubMed]

60. Huang, W.; Zhong, H.; Li, D.; Tang, P.; Feng, Y. Reduced graphene oxide supported CoO/MnO$_2$ electrocatalysts from layered double hydroxides for oxygen reduction reaction. *Electrochim. Acta* **2015**, *173*, 575–580. [CrossRef]

61. Zheng, Y.; Jiao, Y.; Ge, L.; Jaroniec, M.; Qiao, S.Z. Two-step boron and nitrogen doping in graphene for enhanced synergistic catalysis. *Angew. Chem.* **2013**, *125*, 3192–3198. [CrossRef]

62. Zhao, Y.; Yang, L.; Chen, S.; Wang, X.; Ma, Y.; Wu, Q.; Jiang, Y.; Qian, W.; Hu, Z. Can boron and nitrogen co-doping improve oxygen reduction reaction activity of carbon nanotubes? *J. Am. Chem. Soc.* **2013**, *135*, 1201–1204. [CrossRef] [PubMed]

63. Ding, W.; Wei, Z.; Chen, S.; Qi, X.; Yang, T.; Hu, J.; Wang, D.; Wan, L.J.; Alvi, S.F.; Li, L. Space-confinement-induced synthesis of pyridinic-and pyrrolic-nitrogen-doped graphene for the catalysis of oxygen reduction. *Angew. Chem.* **2013**, *125*, 11971–11975. [CrossRef]

64. Cao, Y.; Wei, Z.; He, J.; Zang, J.; Zhang, Q.; Zheng, M.; Dong, Q. α-MnO$_2$ nanorods grown in situ on graphene as catalysts for Li–O$_2$ batteries with excellent electrochemical performance. *Energy Environ. Sci.* **2012**, *5*, 9765–9768. [CrossRef]

65. Qu, L.; Liu, Y.; Baek, J.-B.; Dai, L. Nitrogen-doped graphene as efficient metal-free electrocatalyst for oxygen reduction in fuel cells. *ACS Nano* **2010**, *4*, 1321–1326. [CrossRef] [PubMed]

66. Lin, Z.; Waller, G.; Liu, Y.; Liu, M.; Wong, C.P. Facile synthesis of nitrogen-doped graphene via pyrolysis of graphene oxide and urea, and its electrocatalytic activity toward the oxygen-reduction reaction. *Adv. Energy Mater.* **2012**, *2*, 884–888. [CrossRef]

67. He, D.; Jiang, Y.; Lv, H.; Pan, M.; Mu, S. Nitrogen-doped reduced graphene oxide supports for noble metal catalysts with greatly enhanced activity and stability. *Appl. Catal. B Environ.* **2013**, *132*, 379–388. [CrossRef]

68. Li, X.-H.; Antonietti, M. Metal nanoparticles at mesoporous N-doped carbons and carbon nitrides: Functional mott–schottky heterojunctions for catalysis. *Chem. Soc. Rev.* **2013**, *42*, 6593–6604. [CrossRef] [PubMed]

69. Zhao, Z.-G.; Zhang, J.; Yuan, Y.; Lv, H.; Tian, Y.; Wu, D.; Li, Q.-W. In-situ formation of cobalt-phosphate oxygen-evolving complex-anchored reduced graphene oxide nanosheets for oxygen reduction reaction. *Sci. Rep.* **2013**. [CrossRef] [PubMed]

70. Zhao, H.; Hui, K.S.; Hui, K.N. Synthesis of nitrogen-doped multilayer graphene from milk powder with melamine and their application to fuel cells. *Carbon* **2014**, *76*, 1–9. [CrossRef]

71. Li, Z.; Xu, Z.; Tan, X.; Wang, H.; Holt, C.M.B.; Stephenson, T.; Olsen, B.C.; Mitlin, D. Mesoporous nitrogen-rich carbons derived from protein for ultra-high capacity battery anodes and supercapacitors. *Energy Environ. Sci.* **2013**, *6*, 871–878. [CrossRef]

72. Senoz, E.; Stanzione, J.F.; Reno, K.H.; Wool, R.P.; Miller, M.E.N. Pyrolyzed chicken feather fibers for biobased composite reinforcement. *J. Appl. Polym. Sci.* **2013**, *128*, 983–989. [CrossRef]

73. Wang, X.; Zou, L.; Fu, H.; Xiong, Y.; Tao, Z.; Zheng, J.; Li, X. Noble metal-free oxygen reduction reaction catalysts derived from prussian blue nanocrystals dispersed in polyaniline. *ACS Appl. Mater. Interfaces* **2016**, *8*, 8436–8444. [CrossRef] [PubMed]

74. Wang, J.; Wang, G.; Miao, S.; Jiang, X.; Li, J.; Bao, X. Synthesis of Fe/Fe$_3$C nanoparticles encapsulated in nitrogen-doped carbon with single-source molecular precursor for the oxygen reduction reaction. *Carbon* **2014**, *75*, 381–389. [CrossRef]

75. Wang, R.; Wang, H.; Zhou, T.; Key, J.; Ma, Y.; Zhang, Z.; Wang, Q.; Ji, S. The enhanced electrocatalytic activity of okara-derived N-doped mesoporous carbon for oxygen reduction reaction. *J. Power Sources* **2015**, *274*, 741–747. [CrossRef]

76. Zhou, R.; Qiao, S.Z. An Fe/N co-doped graphitic carbon bulb for high-performance oxygen reduction reaction. *Chem. Commun.* **2015**, *51*, 7516–7519. [CrossRef] [PubMed]

77. Strickland, K.; Miner, E.; Jia, Q.; Tylus, U.; Ramaswamy, N.; Liang, W.; Sougrati, M.-T.; Jaouen, F.; Mukerjee, S. Highly active oxygen reduction non-platinum group metal electrocatalyst without direct metal-nitrogen coordination. *Nat. Commun* **2015**, *6*, 7343. [CrossRef] [PubMed]

78. Rana, M.; Arora, G.; Gautam, U.K. N-and S-doped high surface area carbon derived from soya chunks as scalable and efficient electrocatalysts for oxygen reduction. *Sci. Technol. Adv. Mater.* **2016**, *16*, 014803. [CrossRef]

79. Chao, S.; Cui, Q.; Wang, K.; Bai, Z.; Yang, L.; Qiao, J. Template-free synthesis of hierarchical yolk-shell Co and N codoped porous carbon microspheres with enhanced performance for oxygen reduction reaction. *J. Power Sources* **2015**, *288*, 128–135. [CrossRef]

80. She, Y.; Lu, Z.; Ni, M.; Li, L.; Leung, M.K.H. Facile synthesis of nitrogen and sulfur codoped carbon from ionic liquid as metal-free catalyst for oxygen reduction reaction. *ACS Appl. Mater. Interfaces* **2015**, *7*, 7214–7221. [CrossRef] [PubMed]

81. Xi, J.; Xia, Y.; Xu, Y.; Xiao, J.; Wang, S. (Fe,Co)@nitrogen-doped graphitic carbon nanocubes derived from polydopamine-encapsulated metal-organic frameworks as a highly stable and selective non-precious oxygen reduction electrocatalyst. *Chem. Commun.* **2015**, *51*, 10479–10482. [CrossRef] [PubMed]

82. Wang, D.-W.; Su, D. Heterogeneous nanocarbon materials for oxygen reduction reaction. *Energy Environ. Sci.* **2014**, *7*, 576–591. [CrossRef]

83. Daems, N.; Sheng, X.; Vankelecom, I.F.; Pescarmona, P.P. Metal-free doped carbon materials as electrocatalysts for the oxygen reduction reaction. *J. Mater. Chem. A* **2014**, *2*, 4085–4110. [CrossRef]

84. Sheng, X.; Daems, N.; Geboes, B.; Kurttepeli, M.; Bals, S.; Breugelmans, T.; Hubin, A.; Vankelecom, I.F.; Pescarmona, P.P. N-doped ordered mesoporous carbons prepared by a two-step nanocasting strategy as highly active and selective electrocatalysts for the reduction of O$_2$ to H$_2$O$_2$. *Appl. Catal. B Environ.* **2015**, *176*, 212–224. [CrossRef]

85. Wu, J.; Jin, C.; Yang, Z.; Tian, J.; Yang, R. Synthesis of phosphorus-doped carbon hollow spheres as efficient metal-free electrocatalysts for oxygen reduction. *Carbon* **2015**, *82*, 562–571. [CrossRef]

86. Xiao, M.; Zhu, J.; Feng, L.; Liu, C.; Xing, W. Meso/macroporous nitrogen-doped carbon architectures with iron carbide encapsulated in graphitic layers as an efficient and robust catalyst for the oxygen reduction reaction in both acidic and alkaline solutions. *Adv. Mater.* **2015**, *27*, 2521–2527. [CrossRef] [PubMed]

87. Liu, X.; Wang, Y.; Dong, L.; Chen, X.; Xin, G.; Zhang, Y.; Zang, J. One-step synthesis of shell/core structural boron and nitrogen co-doped graphitic carbon/nanodiamond as efficient electrocatalyst for the oxygen reduction reaction in alkaline media. *Electrochim. Acta* **2016**, *194*, 161–167. [CrossRef]

88. Niu, W.; Li, L.; Liu, X.; Wang, N.; Liu, J.; Zhou, W.; Tang, Z.; Chen, S. Mesoporous N-doped carbons prepared with thermally removable nanoparticle templates: An efficient electrocatalyst for oxygen reduction reaction. *J. Am. Chem. Soc.* **2015**, *137*, 5555–5562. [CrossRef] [PubMed]

89. Kakaei, K.; Balavandi, A. Synthesis of halogen-doped reduced graphene oxide nanosheets as highly efficient metal-free electrocatalyst for oxygen reduction reaction. *J. Colloid Interface Sci.* **2016**, *463*, 46–54. [CrossRef] [PubMed]

90. Liu, Z.; Fu, X.; Li, M.; Wang, F.; Wang, Q.; Kang, G.; Peng, F. Novel silicon-doped, silicon and nitrogen-codoped carbon nanomaterials with high activity for the oxygen reduction reaction in alkaline medium. *J. Mater. Chem. A* **2015**, *3*, 3289–3293. [CrossRef]

91. Bag, S.; Mondal, B.; Das, A.K.; Raj, C.R. Nitrogen and sulfur dual-doped reduced graphene oxide: Synergistic effect of dopants towards oxygen reduction reaction. *Electrochim. Acta* **2015**, *163*, 16–23. [CrossRef]

92. Jiang, T.; Wang, Y.; Wang, K.; Liang, Y.; Wu, D.; Tsiakaras, P.; Song, S. A novel sulfur-nitrogen dual doped ordered mesoporous carbon electrocatalyst for efficient oxygen reduction reaction. *Appl. Catal. B Environ.* **2016**, *189*, 1–11. [CrossRef]

93. Li, W.; Yang, D.; Chen, H.; Gao, Y.; Li, H. Sulfur-doped carbon nanotubes as catalysts for the oxygen reduction reaction in alkaline medium. *Electrochim. Acta* **2015**, *165*, 191–197. [CrossRef]

94. Ju, Y.-W.; Yoo, S.; Kim, C.; Kim, S.; Jeon, I.-Y.; Shin, J.; Baek, J.-B.; Kim, G. Fe@N-graphene nanoplatelet-embedded carbon nanofibers as efficient electrocatalysts for oxygen reduction reaction. *Adv. Sci.* **2016**, *3*. [CrossRef] [PubMed]

95. Zhong, S.; Zhou, L.; Wu, L.; Tang, L.; He, Q.; Ahmed, J. Nitrogen-and boron-co-doped core–shell carbon nanoparticles as efficient metal-free catalysts for oxygen reduction reactions in microbial fuel cells. *J. Power Sources* **2014**, *272*, 344–350. [CrossRef]

96. Bai, J.; Zhu, Q.; Lv, Z.; Dong, H.; Yu, J.; Dong, L. Nitrogen-doped graphene as catalysts and catalyst supports for oxygen reduction in both acidic and alkaline solutions. *Int. J. Hydrog. Energy* **2013**, *38*, 1413–1418. [CrossRef]

97. Cheng, Y.; Tian, Y.; Fan, X.; Liu, J.; Yan, C. Boron doped multi-walled carbon nanotubes as catalysts for oxygen reduction reaction and oxygen evolution reactionin in alkaline media. *Electrochim. Acta* **2014**, *143*, 291–296. [CrossRef]

98. Pumera, M. Heteroatom modified graphenes: Electronic and electrochemical applications. *J. Mater. Chem. C* **2014**, *2*, 6454–6461. [CrossRef]

catalysts

MDPI

Review

Proton Exchange Membrane Fuel Cell Reversal: A Review

Congwei Qin [1,2], Jue Wang [1,2], Daijun Yang [1,2], Bing Li [1,2,]* and Cunman Zhang [1,2,]*

[1] School of Automotive Studies, Tongji University, Shanghai 201804, China; 1434427@tongji.edu.cn (C.Q.); 1410816@tongji.edu.cn (J.W.); yangdaijun@tongji.edu.cn (D.Y.)

[2] Clean Energy Automotive Engineering Center, Tongji University, Shanghai 201804, China

* Correspondence: libing210@tongji.edu.cn (B.L.); zhangcunman@tongji.edu.cn (C.Z.); Tel.: +86-21-6958-3891 (B.L.); +86-21-6958-3850 (C.Z.)

Academic Editors: David Sebastián and Vincenzo Baglio

Received: 19 July 2016; Accepted: 10 September 2016; Published: 8 December 2016

Abstract: The H_2/air-fed proton exchange membrane fuel cell (PEMFC) has two major problems: cost and durability, which obstruct its pathway to commercialization. Cell reversal, which would create irreversible damage to the fuel cell and shorten its lifespan, is caused by reactant starvation, load change, low catalyst performance, and so on. This paper will summarize the causes, consequences, and mitigation strategies of cell reversal of PEMFC in detail. A description of potential change in the anode and cathode and the differences between local starvation and overall starvation are reviewed, which gives a framework for comprehending the origins of cell reversal. According to the root factor of cell starvation, i.e., fuel cells do not satisfy the requirements of electrons and protons of normal anode and cathode chemical reactions, we will introduce specific methods to mitigate or prevent fuel cell damage caused by cell reversal in the view of system management strategies and component material modifications. Based on a comprehensive understanding of cell reversal, it is beneficial to operate a fuel cell stack and extend its lifetime.

Keywords: PEMFC; cell reversal; starvation; mitigation strategies

1. Introduction

At present, the proton exchange membrane fuel cell is the most promising energy system used in commercialized electric vehicles; it has the following advantages: low-temperature operation (50–100 °C), high power density (40%–60%), nearly zero pollutants compared to conventional internal combustion gasoline vehicles, simple structure, and so on [1,2]. However, during the PEMFC technological development, there are two primary barriers to commercialization—durability and cost [3].

Cell polarity reversal occurs frequently in standard fuel cell operating conditions. In such a condition, there is some irreversible damage to fuel cell stack material, including flow field plate, membrane electrode assembly (MEA), and other constructional elements, which seriously affects the durability of PEMFC. There are two principal approaches to solving this problem. The first is a system control strategy whereby specially designed software would monitor anode and cathode outlet exhaust gas composition, cell voltage, and local density variation response. The software would then regulate fuel/air stoichiometry, cell temperature, current density, water management, and any other operating parameters to minimize fuel cell reversal damage [4–6]. Although the system control strategy can be an effective and efficient method to extend a fuel cell's lifespan, it would require a peripheral sensor to monitor and feedback information, and even to regulate the system parameters for steady operation. In fact, it does not take long for cell reversal damage to occur. The extra equipment involved would not only increase system complexity and decrease cost efficiency, but also cannot address cell reversal in time. Therefore, there is a second more direct strategy, system hardware material modification. Because all forms of cell reversal damage eventually lead to system material degradation, optimized materials would minimize cell reversal and improve fuel cell durability in nature. This paper presents the causes, types of damage, and improvement approaches related to cell reversal in a fuel cell.

2. PEMFC Cell Reversal Description

In a single fuel cell or fuel cell stack system, electrode potential drives chemical reactions, including hydrogen oxidation reaction (HOR), oxygen reduction reaction (ORR), carbon corrosion, and water electrolysis. In normal operations, there are excess hydrogen and air supplied to the anode and cathode, respectively. HOR and ORR take place in the anode and cathode, respectively, to produce sufficient amounts of electrons and protons. Nevertheless, there are many extreme operations or conditions, such as insufficient supply of reactant [5–7], low catalyst performance, uneven gas distribution, drastic current load change, and startup or shutdown operations [8], which would result in potential change. In PEMFC cell reversal, the experiment polarity curve shows that the anode potential increases and the cathode potential decreases, narrowing the electric potential difference and even reversing for several seconds. The fuel cell stack cannot work normally anymore and will consume energy instead of supplying energy. As a result, hydrogen and oxygen would be present in the wrong electrode components of fuel cell, which induces a corresponding potential change. In each cavity, all kinds of matter spontaneously recombine and react with each other to make the system right. However, because of the wrong initial conditions, the result cannot match what the fuel cell should be.

As shown in Figure 1, the fuel cell was subjected to cell reversal test experiment (hydrogen starvation). Figure 1 illustrates the time-dependent change in the anode and cathode potentials versus RHE in hydrogen starvation. The cell terminal voltage quickly changed to a negative voltage, from 1.0 to -2.0 V [6]. This will be discussed in more detail in Section 3.1.1.

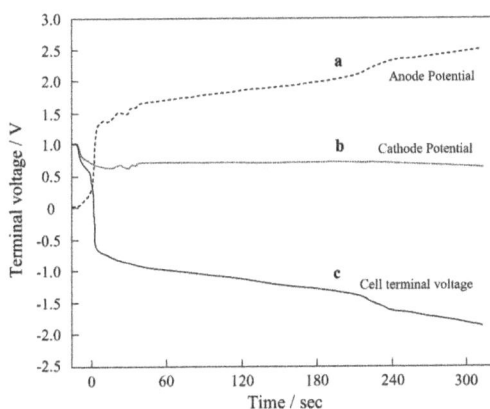

Figure 1. Anode and cathode potential change experienced cell reversal caused by hydrogen starvation. Reproduced with permission from [6]. Copyright Elsevier, 2004.

In the case of air starvation, there is another type of cell reversal detected in currency to voltage (I–V) image. In Figure 2, we can see that cell voltage decreased rapidly to become negative when the cell reversal experiment started (air starvation). After the initial drop, the anode potential, cathode potential, and cell voltage all reached constant values but remained near zero [5].

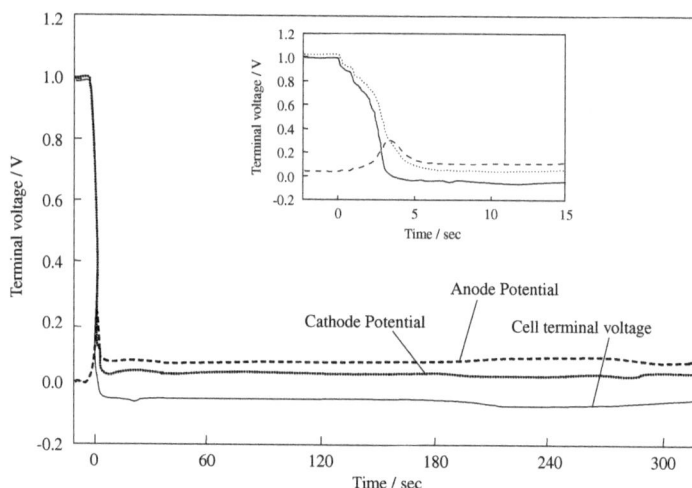

Figure 2. Anode and cathode potential change in cell reversal caused by oxygen starvation. Reproduced with permission from [5]. Copyright Elsevier, 2008.

Fuel starvation results in a markedly different level of cell reversal compared to air starvation. Cell terminal voltage is much larger with low fuel feed compared to low air feed. The damage to material under low potential is less serious than the high potential caused by hydrogen starvation. From this point, we can understand that the damage of fuel starvation could be caused by gas supply and flow field diffusion. In view of keeping reactant diffusion steady, and considering gas diffusion layer design, flow field design, and membrane gas permeation, the assembly of different layers should attract more attention. In the meantime, how to eliminate interface effect is also a point to be researched. Steadier system operation conditions will mean a longer lifetime.

Due to the carbon corrosion in cell reversal, polarity curve, gas chromatography (GC) and differential electrochemical mass spectrometry (DEMS) can be the main in situ measurements to analyze the extent of carbon corrosion, the exhaust gas compositions, and the anode or cathode outlet contents [9–11]. These techniques can realize online detection of carbon dioxide generated by carbon corrosion. After cell reversal, energy dispersive X-ray analysis (EDX), transmission electron microscopy (TEM) [6], scanning electron microscopy (SEM), X-ray diffraction (XRD), electrochemical impedance spectrum (EIS), inductively coupled plasma (ICP), and Raman Spectroscopy are the main means used to characterize the status of the catalyst layer (CL), catalyst nanoparticles sintering, and carbon corrosion extent [10,12–15]. These methods analyze CL morphology, catalyst distribution, elemental content, atomic structure, and impedance, respectively.

In a fuel cell reversal system experiment, there is a three-step protocol to test and categorize the electrode's reversal tolerance [16]. The point of this protocol is to build fuel starvation by replacing H_2 content with humidified N_2 to measure the performance of MEA after cell reversal. Specifically, Step 1, MEA should be in the condition of cell reversal for a long time, with a period of H_2 supply. Step 2, MEA is in the cycle between cell reversal and normal H_2 supply every 30 s. Step 3, MEA would be in cell reversal until the cell potential is -2.0 V. In the whole test procedure, the anode outlet air would be measured for the concentration of CO_2 and O_2.

To simulate and design cell reversal experiment, sub-stoichiometric reactant feeding is the first step. Then, regulating the fuel and air ratio should take a more important place for cell reversal data image. The data collection should include I–V image, different electrode potential of time-dependent like Figures 1 and 2, localized current and potential change across the electrode plane, and so on.

3. Causes and Consequences of Cell Reversal

3.1. Causes of Cell Reversal

The main cause of cell polarity reversal is that the fuel cell cannot satisfy the requisition of the electron and proton from the normal anode and cathode chemical reactions [17]. This key factor occurs in some improper or nonstandard operations, such as fuel and air sub-stoichiometric feeding, rapid load change, low catalyst performance, control module failure, and so on. During a sudden load change or incorrect operation, such as start-up or artificial reasons, fuel and oxidant starvation would occur. Li et al. [18] found that the cell performance loss was induced by the rapidly changing load frequencies resulting in catalyst particles' growth and agglomeration in the anode. This is a typical example where the mass transfer rate cannot match the load change rate.

In the fuel cell stack or galvanostatic operation of a single cell, other cells with sufficient reactant or the single cell system would electrolyze the sick cell for the specified electron and proton or location of saturated fuel supply electrolyze sub-saturated area. There are three key points to define the sick fuel cell: (1) low electrode performance; (2) low fuel supply; (3) low air or oxygen supply. By filling the current requirements, the sick cell would be self-abandoned to conduct some different reaction under high potential, like carbon oxidation, water oxidation in an anode, or hydrogen evolution in a cathode [17]. Hydrogen evolution occurs in the cathode catalyst layer, close to the proton exchange membrane; hydrogen re-oxidization is in the other part of the cathode catalyst layer; and ORR occurs between the cathode catalyst layer and the cathode gas diffusion layer.

Sanyo Electronics previously studied the performance degradation of fuel cell caused by fuel starvation [19]. Then, Kim et al. [20] researched the effect of stoichiometry on the dynamic behavior of a PEMFC during load change, and found a vacuum effect in fuel starvation. Hydrogen utilization was significant to overshoot or undershoot behaviors. Taniguchi et al. [6] designed an experiment using a single cell to simulate cell reversal in a stack.

An abrupt change in oxygen demand at the cathode, like start-up operation, load change, or water management failure, would induce air starvation. Rao et al. [21] developed a distributed dynamic model including three layers (the gas flow channel, the diffusion layer, and the reaction layer) in a cathode to discuss the problem of gas starvation. There are many reports about current distribution and degradation caused by air starvation [5,22,23]. In these papers, current distribution is relevant to matter diffusion. The different reactant combinations in different layers have various reactant reaction modules. Due to reaction spontaneity and external drive, electron transfer and high potential occur at the same time.

All of the experiments reported above show that reactant starvation would change anode or cathode reaction conditions and lead to abnormal chemical reactions in high potential. These reasons would complicate the situation of the electrode surface. The following section will discuss in detail the specific situation, causes, and effects in view of a PEMFC anode and cathode.

3.1.1. PEMFC Anode

Compared to air starvation, complete hydrogen starvation will induce high potential to lead to carbon corrosion. Generally speaking, its impact on fuel cell is more far-reaching and thorough. So we pay more attention to cell reversal caused by hydrogen starvation.

In normal operations, excess hydrogen and air are supplied to the anode and cathode, respectively, whereby hydrogen and air distribution in the two compartments meet the mass transportation requirement. However, local fuel starvation induced by heterogeneous fuel distribution and complete fuel starvation in the whole anode have different effects in the whole anode. These two hydrogen-deficient conditions are similar but the damage to the fuel cell is completely different, and in fact heterogeneous fuel distribution cannot cause cell reversal but current reversal [24]. Local fuel starvation, or partial hydrogen coverage, damages the air electrode and induces oxygen migration across the membrane from the cathode [25,26]. There are two mathematical guides to help us distinguish between and get a clear picture of these two situations by demonstrating electron diffusion and gas flow.

Complete hydrogen starvation:

$$N < I/(2F) \tag{1}$$

Equation (1) depicts the practical status of fuel cell operations in the whole fuel starvation. In this equation, N is the molar flow of hydrogen, I is the practical current, and F is Faraday's constant. Without enough hydrogen to support the required current, the cell has to oxidize other cell system materials such as water and carbon at the fuel electrode to produce electrons and protons. Obviously, the fuel cell must be a single cell during galvanostatic operation or in a fuel cell stack where other cells have excess hydrogen supply.

Partial hydrogen coverage:

$$N > I/(2F) \text{ but } \frac{I}{A} > I_{lim} \tag{2}$$

Equation (2) shows the local hydrogen starvation. Here, A is the active area and I_{lim} is the limiting current density on active area. According to the equation, enough hydrogen is fed to the anode to satisfy the necessary current, but the fuel distribution is uneven in the gas distribution layer (GDL). Finally, the partial electrode area is not fed with adequate fuel. Oxygen in the cathode is permitted to migrate to the anode across the membrane to form the potential in some fuel-deficit areas to produce electrons, resulting in the average current value being higher than the limiting current. Some active areas have the same cell potential difference as a normal fuel cell, about 0.8 V. Nevertheless, other negative regions' cell potential is lower than 0.8 V but more than 0. This potential difference between the anode and cathode can be due to partial pressure discrepancy. So this situation cannot induce cell reversal but it will cause current reversal. Patterson and Darling studied the damage caused by local fuel starvation to the cathode catalyst [27]. The experiment showed that the situation of local fuel starvation is not similar to cell reversal caused by fuel starvation.

Only when the low hydrogen stoichiometry is less than 1 can the fuel cell polarity be reversed in the condition of a single-cell galvanostatic operation or a fuel cell stack where other cells have excess hydrogen supply. Obviously, all mechanisms and consequences of local fuel starvation would also occur at some location of the CL during overall fuel complete starvation [15]. In fact, the relationship between local fuel starvation and complete hydrogen starvation is just a qualitative change induced by a different hydrogen degree.

Different fuel flow fed to the anode provokes different phenomena on its electrode surface. Fuel starvation induced uneven currency diffusion, which had an increasing trend in the anode inlet and a decreasing trend in the outlet [4]; hydrogen oxidation reaction and water oxidation occurred at different regions of the electrode simultaneously. Due to the lack of hydrogen supply, which cannot satisfy the fuel cell system's electron requirements, the fuel cell system would spontaneously increase the anode potential as a result of an abnormal reaction or cell performance degradation, such as water oxidation, carbon corrosion, and so on, to obtain extra electrons [28]. In a galvanostatic operation or fuel cell stack, it will decompose the electrode material or the sick cell, like an electrolytic cell.

Therefore, in this condition, oxygen is simultaneously reduced in the cathode while oxygen is produced in the anode. The final consequence is that oxygen is pumped from cathode to anode while water moves in the opposite direction. The cell is similar to an "oxygen pump". We can use GC to test the anode production online [23]. The electrode half reactions and the overall reaction are listed below:

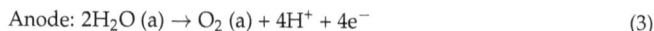

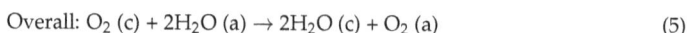

$$\text{Anode: } 2H_2O \text{ (a)} \rightarrow O_2 \text{ (a)} + 4H^+ + 4e^- \tag{3}$$

$$\text{Cathode: } O_2 \text{ (c)} + 4H^+ + 4e^- \rightarrow 2H_2O \text{ (c)} \tag{4}$$

$$\text{Overall: } O_2 \text{ (c)} + 2H_2O \text{ (a)} \rightarrow 2H_2O \text{ (c)} + O_2 \text{ (a)} \tag{5}$$

Water oxidation is the initial main reaction in a fuel cell, caused by fuel complete starvation in the anode. While the cell reversal experiment proceeds, the cell voltage becomes more and more negative. In Figure 1, we can see that the anode potential could be more than 1.5 V and the cathode is about 0.85 V at the first decreased stage. So the cell voltage is less than −0.65 V. At the next stage, the anode potential would become so high that carbon could be corroded gradually, especially with a platinum (Pt)-based catalyst, which would promote carbon corrosion. Finally, the cell voltage would get close to −2.0 V until the fuel cell cannot function.

According to Equations (3), (10), and (11), corrosion in the anode, water oxidation, and carbon would produce heat to dehydrate the anode ionomer and membrane electrolyte. When the cell voltage reaches −2.0 V, the membrane would be completely degraded by dielectric breakdown [16]. We can judge which cell reversal strategy is best based on the time it takes the cell to reach −2.0 V in order to avoid cell reversal causing irreversible damage. The better strategy should have a longer time to reach −2.0 V of cell potential.

Aside from the cell polarity change, local current density and temperature also have special distribution across the CL in cell reversal. Zhang and his team analyzed dynamically the local current densities and temperatures in proton exchange membrane fuel cells during reactant starvations [7]. The dynamic variations of local current densities with cell voltage and local temperatures during hydrogen starvation are studied. In this experiment, the hydrogen flow rate was held at 50 NmL·min^{-1} (equal to average current density of 450 mA·cm^{-2}). In the average current controlled operation, it was set to increase from 440 to 470 mA·cm^{-2} to mimic fuel starvation. We can see that at the initial stage in this experiment, about 10 to 20 s, carbon corrosion is less than water oxidation, as indicated by a relatively stable current density variation. Then the local current densities began to diverge. The closer the location is to the anode inlet, the larger the current density. The outlet area had the completely opposite case. The local temperature variation was similar to local current density. Without a stable −0.2 V cell voltage during fuel starvation [16,29], current density and temperature would be so divergent that the MEA would fail. The temperature and current density show a relatively clear picture to explain the electrode reaction.

In addition, cell reversal caused by hydrogen starvation would induce an abnormal oxidation reduction reaction at the electrode surface plane. There are different situations in different hydrogen feeding stoichiometry.

Akira Taniguchi et al. discussed and analyzed electrocatalyst degradation in PEMFC caused by cell reversal during fuel starvation [6]. The ruthenium dissolution occurred in the anode catalyst layer and the outlet had more severe degradation. There was also surface activity area loss in the cathode. The standard potential of transition metal couple is smaller than the electrode surface potential. Hence, some transition metal dissolution and redeposition are likely to occur. This is why the catalyst particle size becomes bigger than before cell reversal. In fuel starvation, the anode reaction is complex, involving hydrogen oxidation, water oxidation, carbon corrosion, and oxygen reduction. All of these reactions involve proton and electron transfer, which promote potential difference across the electrode plane. Reaction correspondence cannot be executed across the anode plane. It would simply get worse and worse until the system is dead.

Liang et al. focused their attention on a single cell under different degrees of fuel starvation [30]. They measured some electric parameters, such as cell voltage, current distribution, cathode and anode potentials, local interfacial potentials between anode and membrane, and so on, by means of a specially constructed segmented fuel cell. Experimental results showed that the current distribution was extremely uneven during fuel cell reversal, due to starvation or high local interfacial potential near the anode outlet. Hydrogen and water were oxidized at different areas of the anode. Anode carbon corrosion was proven to occur by monitoring its outlet's CO_2 concentration. Figure 3 shows that the anode potential became higher and higher with the decrease of hydrogen stoichiometry, and water oxidation gradually took on an important role in the current contribution. Simultaneously, carbon corrosion was more and more serious than before. The comparison between cell reversal voltage and anode potential under different hydrogen conditions is shown in Table 1. Different hydrogen stoichiometry creates different terminal cell potential.

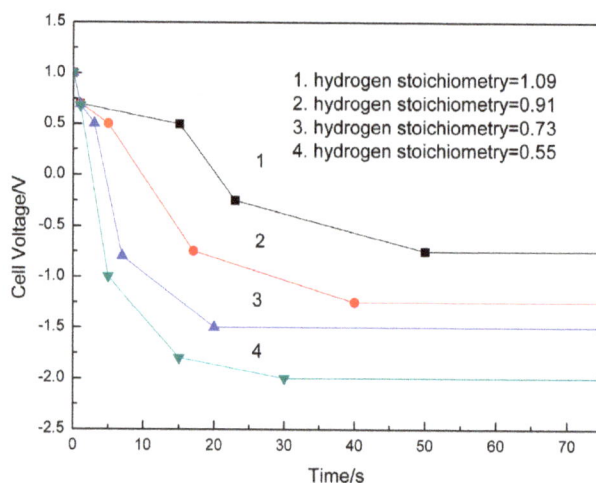

Figure 3. Changes of the anode potentials over time under different hydrogen stoichiometry. From 1 to 4, hydrogen stoichiometry is 1.09, 0.91, 0.73, and 0.55, respectively. Reproduced with permission from [30]. Copyright Elsevier, 2009.

Table 1. Different times for cell reversal in different hydrogen stoichiometry.

Hydrogen Stoichiometry	Time for Cell Reversal (s)	Cell Voltage (V)	Anode Potential (V)	Reference
0.8	Experiment started	About −0.7	1.3	[6]
0.8	About 300	−2.0	2.5	
1.09	26	−0.718	0.955	
0.91	10	−1.125	1.313	[30]
0.73	6	−1.689	1.821	
0.55	4	−1.951	2.058	

With the change of cell voltage over time, fuel cell system would begin to search reaction balance. In the anode, a compromise between hydrogen oxidation, water oxidation, and carbon corrosion move anode reaction from chaos to relative balance. Hydrogen stoichiometry has a different influence on the anode when it is lower than its satisfied supply. However, the experimental result is also affected by the fuel cell test device. Taniguchi's test result was not steady even 300 s after the experiment started, as compared to Liang's. It involves support material, test device design, sealing, and so on.

They also proposed that the hydrogen stoichiometry would shift the position of the lowest current [31]. With a comparatively high hydrogen feed, the position of the lowest current was closer to the anode outlet, where the fuel feed would encounter the gas drawn back from the outlet manifold and fuel starvation first occurred. The inadequate hydrogen supplied and "vacuum effect" would mix in the fuel cell. In this situation, researchers could analyze the area of fuel starvation in the anode according to the hydrogen stoichiometry to design a catalyst layer of anisotropy to avoid or mitigate the fuel starvation caused by inadequate fuel feed.

3.1.2. PEMFC Cathode

When the real-time current exceeds the limiting current for ORR, air starvation occurs. In this case, the cathode voltage approaches the potential of reversible hydrogen due to the existence of hydrogen evolution. The potential of the cell is slightly negative in this condition, which is smaller than the hydrogen starvation. At this comparatively low potential value, it is hard to corrode carbon material.

In the PEMFC cathode, oxidation supply and the obstruction of gas channels caused by water flooding or freezing are the main causes of cell reversal. The causes also include control system failure and incorrect operation. Water flooding or ice would block the access of the reactant gases to the electrocatalyst surface and result in a slower mass transport rate compared to the oxygen reduction reaction and a relative absence of cathode reactant. There are three ways for the PEMFC cathode to generate excess water, i.e., humidified hydrogen and air to optimize the interaction between the interface and reactant, the reaction product in the cathode, and water migrating across the membrane by electro-osmotic drag with protons [5].

Combined with Figure 2, it is evident that the final cell voltage in cell reversal caused by air starvation was close to -0.1 V, which is much lower than the fuel starvation including anode and cathode voltage. Because potential is the driving force of electrochemical reaction, with high potential it is easier to realize carbon corrosion, catalyst metal dissolution, and sinter. Fuel starvation at the PEMFC anode results in irreversible damage to the PEMFC after several minutes under higher potential and is more critical than improper gas supply or distribution at the cathode [15].

When oxidation starvation happens, protons and electrons produced in the anode enter into the cathode. Simultaneously, hydrogen is consumed in the anode and produced in the cathode. Protons act as a "hydrogen pump" [23]. The electrode half reactions and the overall reaction are listed below:

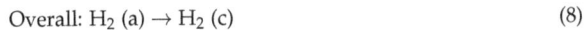

$$\text{Anode: } H_2 \text{ (a)} \rightarrow 2H^+ + 2e^- \tag{6}$$

$$\text{Cathode: } 2H^+ + 2e^- \rightarrow H_2 \text{ (c)} \tag{7}$$

$$\text{Overall: } H_2 \text{ (a)} \rightarrow H_2 \text{ (c)} \tag{8}$$

Regarding air starvation, researchers have focused on its causes, its dynamic characteristics across the anode plane, and the relationships between air feed and fuel cell degradation. Tüber et al. focused on the water flooding phenomenon at the fuel cell cathode caused by an accumulation of H_2O [32]. They analyzed the effect of air flow, humidity, temperature, and flow-field type on flooding.

Zhang et al. also detected the dynamic characteristics of local current densities and temperatures in proton exchange membrane fuel cells during air starvation under current controlled and voltage controlled operation [7]. Their experimental results were obtained at different times and relative positions. This paper demonstrated that the local temperature changes followed the local current density changes during experimental operation. It is similar with hydrogen starvation. Due to reactant starvation, anode and cathode chemical reactions cannot follow normal operations. It refers that a mixture of gas evolution, reactant oxidation, and reduction reaction occur in anode or cathode. According to different reactions' energy change, there must be a relationship between electrode temperature and current density.

Taniguchi et al. researched degradation in PEMFC caused by cell reversal during air starvation [5]. They detected catalyst samples by means of cell reversal experiment and analyzed electrochemical performance, such as electrochemical surface area (ECSA) and current-potential character, using TEM images to illustrate that the size of catalyst particles increased, agglomerated, and recrystallized. Obviously, this degradation effect caused by air starvation was smaller than fuel starvation. It is shown in Figure 4 that ECSA shrinks as a function of time. In high electrode potential, it is easier to make catalyst particles dissolve and agglomerate. Similarly, it also will impact fuel cell durability and performance. Improving catalyst electrode surface diffusion is a feasible method to mitigate its sinter.

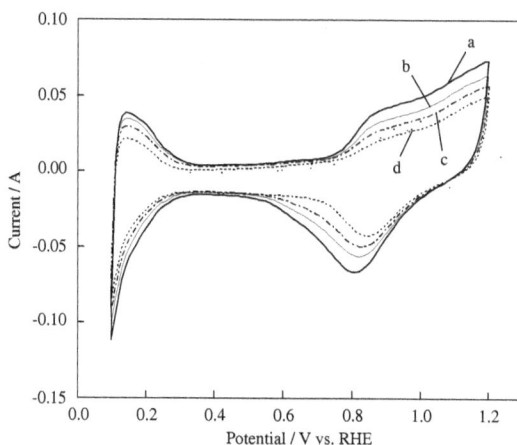

Figure 4. Change in the electrochemically active surface by a cell-reversal experiment: (a) before experiment; (b) after 5 min of experiment; (c) after 10 min of experiment; and (d) after 120 min of experiment. Reproduced with permission from [5]. Copyright Elsevier, 2008.

Natarajan et al. researched the relationship between cell temperature, anode and cathode gas streams humidity, and oxidant flow rate and current density distribution in a single-channel PEM fuel cell using a segmented collector setup [33]. The relationship between various oxidant flow rates and current densities in segment downstream was studied specifically. However, this setup also had its limitations in that this result did not include data from when the test temperature is over 55 °C. A local current density map at various test conditions is an accurate experimental measurement. With this method, we can analyze and understand the chemical reaction situation at the local electrode surface and the layer interface.

3.2. Consequences of Cell Reversal

There are some specific situations in anode and cathode above. Then, we put more attention on fuel starvation due to the permanent damage caused to the fuel cell. Hydrogen and air starvation are the main factors in cell reversal. Compared to air starvation, anode potential is increasing in fuel starvation. Once anode potential rises over 0.207 V, or further to over 1.23 V with fuel consumption, water electrolysis and carbon oxidation would occur at the anode to provide the protons and electrons required for the load and the ORR at the cathode [16].

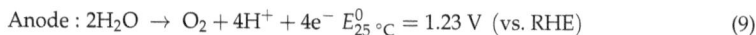

$$\text{Anode}: 2H_2O \rightarrow O_2 + 4H^+ + 4e^- \quad E^0_{25\,°C} = 1.23 \text{ V} \text{ (vs. RHE)} \tag{9}$$

$$\text{Anode}: C + 2H_2O \rightarrow CO_2 + 4H^+ + 4e^- \quad E^0_{25\,°C} = 0.21 \text{ V} \text{ (vs. RHE)} \tag{10}$$

There are some other reactions happening under higher potential, like carbon reacting with water [34].

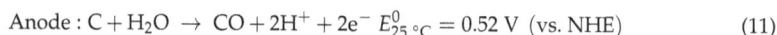

$$\text{Anode}: C + H_2O \rightarrow CO + 2H^+ + 2e^- \quad E^0_{25\,°C} = 0.52 \text{ V} \text{ (vs. NHE)} \tag{11}$$

As an indispensable component of the fuel cell system, carbon supports tend to corrode with water to produce CO_2 at high potential [28]. Thanks to a large number of micropores, catalyst particles can be physically segregated by carbon supports to prevent particle sintering and catalyst dissolution, which can improve catalyst particles' specific surface area [35,36]. On the other hand, under cell reversal conditions, the advantages presented above would be reversed due to the degradation of the catalyst layer structure. Without integrated carbon support structure, three degradation effects would occur: (1) Pt particle agglomeration and particle growth; (2) Pt loss and redistribution; and (3) poisonous effects caused by contaminants [13]. Due to the inherent tendency of catalyst nanoparticles to agglomerate to reduce the surface energy, without the limit of support carbon it would promote deterioration of catalyst performance. Low catalyst performance would inhibit electron and proton production for the anode. With the low chemical reaction rate, the anode would only seize carbon and water as reactant material for ORR. This creates a vicious cycle that leads to fuel cell degradation and obsolescence. In the end, the fuel cell would be in a dynamic balance for several chemical reactions on the cathode and anode plane or interface.

However, in air starvation conditions, because of the lower anode and cathode potential, 0.05 V at anode and 0.85 or 0.05 V at cathode (localized ORR and HOR, respectively), carbon corrosion is unlikely to occur [24]. Instead, the mixture of oxygen and hydrogen may damage the fuel cell, due to the large hydrogen explosion ranging from 4% to 75.6% that is likely to occur at room temperature in a narrow and airtight space. Elsewhere, with the reduction of water and high cell voltage, the membrane would be completely degraded and broken apart. Air starvation does not have a direct or serious effect on electrodes compared to fuel starvation.

4. Mitigation Strategies

In the case of cell reversal, there are many approaches to attenuate or solve this question, such as system management and material modification. This review will discuss two ways to summarize cell reversal failures. System strategies mainly monitor operation parameters and feed it back by adjusting gas flow and load change, and controlling gas humidity and system temperature. Because the system management method should consider material features, we mainly pay attention to the role of electrode material and catalyst layer design. The quality and design of fuel cell installation are key to solving fuel cell system failure. Thus, in terms of materials, we focus on three parts: high performance catalysts, carbon support materials, and water oxidation reaction. Because the damage caused by air starvation is less than that caused by fuel starvation, we focus on the design and structure of the anode CL. For the cathode, researchers [37] adjusted the porosity of GDL to improve gas diffusion and changed the supported material to optimize the transformation of oxygen and water.

4.1. System Management Strategies

System management mitigates catalyst degradation by regulating fuel cell system operation parameters such as pressures, cell temperature, humidity levels, and current density [38]. With auxiliary equipment, researchers can detect reverse cell potential by monitoring cell voltages of individual cells or groups of cells in the stack [39]. Combined with current density distribution and temperature distribution detected by electrode segmentation technology [40], researchers can also understand the situation of cell starvation specifically. Furthermore, by setting an alarm value on the monitored cell voltage, the system can adjust operation parameters such as reactant flow rate, operation load, and water management data, and so on to avoid cell reversal or system halt situations. For instance, a sensor in the cell test system activates a switch that plays the role of removing the stack load the instant the sensor detects a precipitous loss of voltage [41].

Vahidi et al. proposed a fast-responding auxiliary power source to prevent oxygen starvation in a fuel cell during rapid current transitions [42]. In the oxidation starvation experiment of increasing stack power, the reactant starvation was less than 1% in the hybrid installation. An image of the hybrid installation follows. In the schematic, a capacitor was used to provide deficit current when the power sensor gives a signal. With capabilities of model predictive control, it can mitigate cell reversal damage in air starvation. However, it is transitory for fuel starvation to make carbon corrosion. This auxiliary installation cannot support adequate current in fuel starvation.

4.2. System Material Modification

Compared to the system management strategy, the design and composition of a fuel cell structure are more important, especially in terms of preventing cell reversal. While there are additional accessory costs, improving corrosion resistance naturally increases the fuel cell lifespan. Due to the high potential, slow reaction rate, and reactant transfer rate, which cannot meet current density in fuel cell stacks and in galvanostatic operations, researchers focused on the three aspects previously discussed—catalyst, carbon material, and water oxidation reaction.

Specifically, there are four approaches used to attenuate performance degradation and extend fuel cell lifespan [11,15,37].

1. High catalyst activity and homogenous distribution;
2. Corrosion-resistant support material;
3. Water oxidation catalysts to promote water oxidation instead of carbon corrosion;
4. Increasing anode gas humidity.

4.2.1. High-Performance Catalyst

Low-performance catalysts would induce low anode and cathode reaction rates without sufficient protons and electrons supported in cell stacks and galvanostatic operations. However, in normal operation, Pt-based catalysts could meet the requirements of HOR and ORR due to their excellent catalyst performance [43].

Catalyst dissolution and agglomeration would occur via carbon support material corrosion [44], Pt loss [13,45], and surface energy reduction during standard fuel cell operations. Generally, the dissolution of Pt in the anode was more stable than in the cathode because of the electrode potential; the Pt particles' dissolution follows Equations (12) and (13) into Pt^{2+}, and the standard equilibrium potential is +0.88 V vs. RHE [46]. However, in cell reversal, the potential polarity is changed so that catalyst metals have high ORR potential with carbon support corrosion.

$$Pt + H_2O \rightarrow PtO + 2H^+ + 2e^- \tag{12}$$

$$PtO + 2H^+ + 2e^- \leftrightarrow Pt^{2+} + H_2O \tag{13}$$

Pt-based catalyst dissolution is more common during cell reversal in the anode. According to the metal couple potential, different catalyst metal dissolution and redeposition occur. Furthermore, the catalyst particle size becomes bigger as a result of the electrochemical surface area decreasing. Then, it also would induce activity degradation and the potential would grow higher and higher. Obviously, this is another vicious cycle.

High durability and catalyst activity are the keys to solving this problem. Pt–alloy catalysts have better activity and stability compared to traditional Pt/C catalysts, like Pt–Co and Pt–Cr–Ni [47]. The different alloy elements and the larger size of catalyst particles show excellent properties [48]. With high catalyst activity to decrease ORR or HOR activity energy, the basic electrode reaction rate can occur even in poor operating conditions such as bad fuel or air transfer operation. As to the durability, the non-noble metals in the Pt alloy catalysts are more sensitive to the ionomer phase by XRD analysis. These non-noble metals can partially lose to induce (111) Pt skin to increase catalyst activity [49]. It indirectly improves fuel cell catalyst durability.

Various kinds of alloy metals have already proved to be capable in a fuel cell operating environment, including Co, Cr, Fe, Ni, and V. Due to its advantage of stability, researchers paid more attention to Pt–Co/C recently [50,51]. Adzic et al. [52] showed that the durability of Pt-based catalyst could be improved with the addition of gold (Au) clusters. The above reports showed that the Pt alloy could not only use a second metal to replace unavailable Pt to reduce costs, but also could use other metals to stabilize the Pt skin and frame structure to increase its lifespan. However, in cell reversal, catalyst performance only adds brilliance to its present splendor. Its root is reactant supply and material corrosion resistance.

4.2.2. Carbon Support Material

As to material corrosion resistance, carbon is commonly used in PEMFC as a catalyst support, in GDLs, and as a bipolar-plate material [24]. Carbon support has many basic features, such as high specific surface area, good electrical conductivity, and chemical stability in a fuel cell system [53]. The stronger the interaction between carbon material and catalyst particles is, the better the performance of catalyst particles that is not prone to sinter is [36]. However, carbon oxidation reaction (COR) has a direct relationship with electrocatalytic material stability, and thermodynamically the reaction would happen at the cathode, because the reaction oxidation potential is 0.207 V (vs. RHE at $T = 298$ K). In the case of cell reversal, high potential in both the anode and cathode could electrolyze carbon material. Furthermore, with the existence of Pt-based catalyst, carbon corrosion would be easier [10,54].

Different kinds of carbon material have an impact on corrosion resistance. As a new carbon material, graphene has special optical, chemical, and mechanical properties. Due to its high surface area, high conductivity, and unique graphitized basal plane structure, it has potential to serve as a fuel cell catalyst support material to overcome the problem of corrosion [55]. Considering the strong interaction between graphene and nanoparticles with functional groups, it is hard for Pt-based catalyst nanoparticles to dissolve or sinter [56]. Graphene increases a Pt-based catalyst metal's dissolution potential. Guo and Sun developed a facile solution-phase self-assembly method to deposit FePt NPs on a graphene surface [43]. This experiment mainly shows that graphene as a carbon support could have an effect on carbon corrosion in harsh conditions. Although graphene has influence on improving catalyst activity and durability, easily stacking between graphene sheets, the effects of surface functional groups should be intensively investigated.

Generally, graphitic content is a point to be researched. Luis Castanheira and coworkers proposed that the structure of carbon material, including high-surface area carbon (HSAC), Vulcan XC72, or reinforced-graphite (RG) with identical Pt (40 wt %), was the key to carbon corrosion caused by the hydrogen/air interface [10]. Results showed that there were 60% and 80% ECSA loss for Pt/Vulcan and Pt/HSAC, respectively. However, Pt/RG only had 25%. Compared to other disorganized carbon material, graphitic carbon is more robust against carbon corrosion.

Similarly, Artyushkova et al. [57,58] found a relationship between carbon material structure and chemical parameters. Good electrochemical activity of ORR but poor activity of COR were seen with high amounts of graphite carbon, small specific surface area, less carbon surface oxides, and many large-sized pores. In fact, it is a contradiction that graphite carbon has less surface deficit, which makes it resistant to corrosion, and has weak interaction with catalyst particles. ORR and COR should be in balance only if we consider carbon material surface factor. Carbon material surface structure and function can be a method of improving this situation. This paper [58] introduced the relationship between different carbon material structural features, corrosion resistance, and hydrophobicity or hydrophilicity. According to this relationship, we have a path to manufacture and synthesize ideal carbon material.

Carbon surface porosity plays a role in corrosion resistance and wettability. Wlodarczyk [59] suggested that carbon could show good corrosion resistance with a porosity value in excess of 10%, because its wettability decreased with the value of porosity.

Apart from graphene, other carbon materials have also attracted researchers' attention. Oh et al. [60] focused on the performance of various carbon support materials including carbon black [44,61], carbon nanofiber [62,63], and carbon nanocage [64]. The electrochemical data about Pt-based catalyst supported by these three different types of carbon materials before and after corrosion are shown in Figure 5 and Table 2 [60]. Among these carbon materials, the amorphous carbon black was more susceptible to corrosion compared to carbon nanofiber (CNF) and carbon nanocage (CNC). In addition, CNC exhibited better performance than CNF in carbon corrosion resistance and preventing Pt particle aggregation, because of the balance between hydrophobicity and surface roughness [60]. Different three-dimensional carbon materials also lend corrosion resistance and physical segregation.

By combining these two points—carbon surface and spatial structure, we can design the desired carbon material structure to optimize a fuel cell.

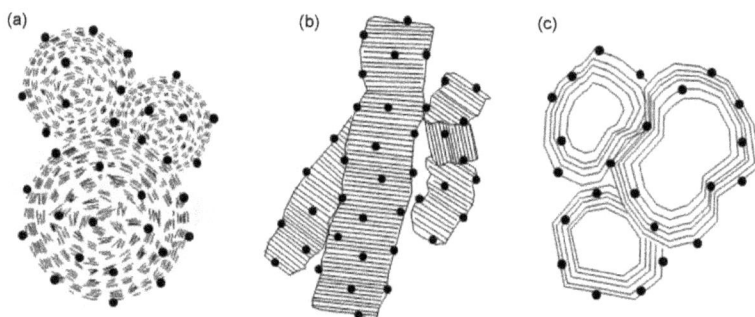

Figure 5. Three different types of carbon materials: (**a**) carbon black; (**b**) carbon nanofiber; (**c**) carbon nanocage. Reproduced with permission from [60]. Copyright Elsevier, 2009.

Table 2. Changes in different types of carbon materials' electrical data before and after corrosion experiments.

Condition	Active Surface Area (m²·g⁻¹)		Membrane Resistance		Charge Transfer Resistance		CO₂ Emission (μL)	Reference
	Before	After	Before	After	Before	After	30 min@1.4 V	
Pt/XC-72 (1200 cycle)	62.2	40.0	–	–	–	–	–	[44]
Pt/BP2000	59.9	26.7	–	–	–	–	–	–
Pt/Carbon Black (up to 4000 cycle)	41.7	15.2	0.016	0.025	0.039	0.328	651	[60]
Pt/CNF	30.1	27.7	0.016	0.016	0.054	0.058	49	–
Pt/CNC	33.6	32.9	0.016	0.016	0.04	0.041	4	–

There are some specific examples to show various kinds of carbon material durability performance. Vinod et al. made an effort to test graphite carbon support [65]. In the accelerated stress test (AST), graphite Pt/C catalyst had better carbon corrosion resistance relative to the untreated carbon sample. They tested ECSA, cell performance, charge transfer resistance by CV, cell polarization, and resistance, respectively. In the PEMFC cathode, graphite Pt/C catalyst electrical performance deteriorated by 10% after 70 h AST compared to 77% of untreated carbon support, as shown in Figure 6 [65].

Zhao et al. evaluated the change in nanostructure through the various heat treatments of carbon materials and their durability for the startup/shutdown operation of PEFC [66]. This paper had a similar result in that great catalyst support should have a high graphitization degree and good interaction with the catalyst. It gave an optimum heat treatment temperature of 1600 °C for best durability.

Generally, carbon material corrosion is caused by high electrode potential, which induces a reaction between carbon material and water. Carbon material surface defects are the most influential factor in corrosion. The purpose of carbon material graphite is to reduce surface defects for a more compact surface. Different types of carbon materials, like CNT, CNC, graphene, and so on, due to their unique surface structures, have different properties that have yet to be fully investigated. To achieve good ORR activity and carbon corrosion resistance, one must consider the balance between the graphitic level, the amount of surface oxide, surface roughness, pore size, and hydrophobicity. With the consideration of the important relationship among catalyst layer effective properties, ORR performance, and durability, it would prove useful to set design rules to produce and optimize the structure carbon material and MEA in order to satisfy various kinds of fuel cell operation conditions.

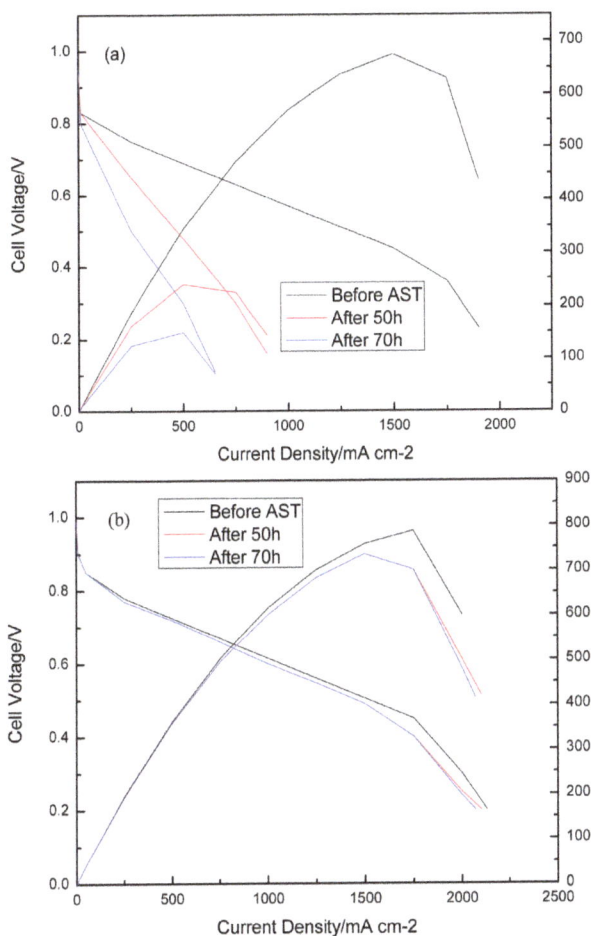

Figure 6. Steady-state performance curves for PEFCs with (**a**) Pt/Non-GrC and (**b**) Pt/GrC cathodes prior to AST and after 50 or 70 h of AST with H_2–O_2 at 60 °C. Reproduced with permission from [65]. Copyright WILEY-VCH, 2011.

4.2.3. Water Oxidation Reaction

In fuel cell reversal, water oxidation occurs in combination with carbon corrosion among abnormal reactions. Like the sacrificial anode preventing metal corrosion, water oxidation would occur more prevalently to minimize carbon oxidation. Some papers [11,37] also propose that the promotion of water oxidation over carbon through anode structure and component modification would mitigate the catalyst degradation. Obviously, there are two kinds of approaches to promoting chemical reaction: (1) increasing the amount of reactant and (2) reducing water oxidation reaction's activation energy. In Equations (9) and (10), we can see that reducing the water oxidation reaction's over-potential and increasing its kinetic rate instead of the protons and electrons produced by carbon corrosion can protect the fuel cell system. These two methods are reported in detail as follows.

Water Oxidation Catalyst

Catalysts can increase or decrease reaction activity energy to promote chemical reactions in directions we expect. Adding catalysts to boost water oxidation and reduce carbon corrosion would be an effective method of decreasing cell reversal.

Ralph et al. [16] added 20 wt % RuO_2 to 20% Pt and 10% Ru catalysts supported on Vulcan XC 72R with PTFE. The time to achieve −2.0 V increased from 22 to 32 min. As Table 3 shows, three water oxidation catalysts, RuO_2, RuO_2–TiO_2, and RuO_2–IrO_2, had different effects on cell reversal with the increasing reactivity RuO_2–IrO_2 > RuO_2–TiO_2 > RuO_2.

Table 3. The catalysts effect in anodes on the cell reversal tolerance in Step 3 of the protocol. Reproduced with permission from [16]. Copyright ECS, 2006.

Anode Design with Different Catalyst	Time (min)
20% Pt, 10% Ru/Shawinigan anode catalyst	0.25
Anode Catalyst* + RuO_2/Shawinigan	4.5
Anode Catalyst* + RuO_2–TiO_2 (90:10 atomic ratio Ru/Ti)/Shawinigan	24
Anode Catalyst* + RuO_2–IrO_2 (90:10 atomic ratio Ru/Ir)/Shawinigan	48
20% Pt, 10% Ru/Shawinigan+ RuO_2–IrO_2 (90:10 atomic ratio Ru/Ir)	74
40% Pt, 20% Ru/Shawinigan	167
40% Pt, 20% Ru/Shawinigan + RuO_2–IrO_2 (90:10 atomic ratio Ru/Ir)	1630

Ralph and coworkers made a series of investigations into electrochemical catalyst loading for water oxidation both unsupported and supported on various kinds of carbon materials. In the three-step protocol presented above, it is easy to recognize the performance of electrochemical catalysts according to the time it takes to get −2.0 V of cell potential, as listed in Table 3. The time to get to −2.0 V can partially demonstrate the effect of a water oxidation catalyst. Combined with CV, TEM, or SEM imaging before and after cell reversal experiment, we can analyze the morphology of carbon support and catalyst particles generally.

Jang et al. [67] studied the effect of water oxidation catalyst added into cathode to weaken carbon corrosion. In Figure 7, compared with commercialized Pt/C catalyst, the electrochemical carbon corrosion decreased by 76% with 2 wt % IrO_2 (0.016 mg·cm^{-2}) under 1.6 V (vs. NHE) and 70 °C. CO_2 monitoring showed that the water oxidation catalyst had the effect of mitigating carbon corrosion. Carbon dioxide is produced by carbon corrosion in a fuel cell. The decrease of carbon dioxide production means the amount of COR is reducing. Theoretically, a water oxidation catalyst could also be used in anodes to decrease carbon corrosion.

Jang et al. [68] also studied the performance of IrO_2 and IrO_2/C mixed with Pt/C on cathode durability during fuel starvation. In an accelerated fuel starvation test, with varying amounts of IrO_2 (2.5, 5, 10, or 20 wt %), the peak power densities of the cell cathode catalyst were decreased by 10.21%, 6.52%, 2.93%, and almost 0, respectively. This shows that IrO_2 plays an important role in maintaining cell performance under extreme operation conditions. However, due to the screening effect, IrO_2 particles in the cathode surface mixed with the commercial Pt/C would reduce the activity of the Pt-based catalyst. Therefore, IrO_2/C was chosen as the mixture of cathode catalyst. As a water electrolysis catalyst, with the IrO_2/C increasing, the durability of the fuel cell became better and better due to the rapid decomposition of water molecules. According to the screening effect and catalyst durability, 10 wt % IrO_2/C with Pt/C catalyst is a better choice.

In the case of water oxidation, the most important points are the design of catalyst layers mixed between Pt-based catalysts and water oxidation catalysts. Some papers [16,67,68] proposed that the water oxidation catalyst has a positive effect on water electrolysis over and above carbon corrosion and the degradation of carbon support material. It is obvious that water oxidation catalyst improves the cell reversal tolerance with increasing effect (RuO_2–IrO_2 > RuO_2–TiO_2 > RuO_2).

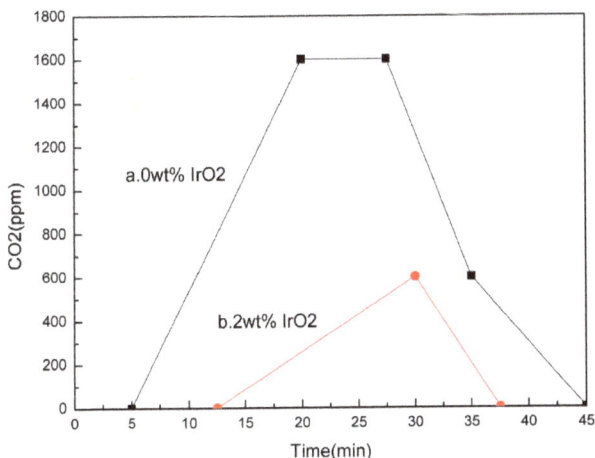

Figure 7. MEA–CO_2 mass spectrum (a) without IrO_2; (b) with 2 wt % IrO_2. Reproduced with permission from [67]. Copyright American Chemical Society, 2010.

Control Humidity and Holding Water Volume

Enhancing water retention is another way to promote water oxidation over carbon corrosion, due to the increase of reactant resources. However, humidity has a negative relationship with fuel cell performance in some ranges. More does not mean better. Some patents and papers [42,69,70] focused on cell reversal prevention over its performance and lifetime mitigation due to its irreversible damage. Therefore, a hydrophobic agent is a way to hold more water in the anode when cell reversal occurs induced by fuel starvation, which is always used in the gas diffusion layer to improve reactant gas diffusion and water management [71].

Ralph et al. [16] changed the water supply at the anode in their fuel cell reversal mimic experiment [72], in which the anode was fed with humidified N_2 and the cathode fed with O_2. When the cell current was set at 120 A, the current density was 0.5 A·cm^{-2} and the cell potential was more than -1.0 V (when the cell potential is less than -2.0 V, the polymer membrane can be completely degraded by dielectric breakdown). This experiment demonstrated that water supply played an important role in sustaining the cell potential and preventing MEA deterioration.

Besides increasing fuel humidity, the other method is to add hydrophobic materials into the catalyst layer, like PTFE. It could slow down the rate of water and increase the water contact angle by keeping water in the anode without shrinking the contact area between the catalyst layer and the membrane. Due to its key role in water retention, it is hoped that a hydrophobic agent would promote water oxidation reactions to solve the issues caused by cell reversal, resulting in less carbon component oxidation [16]. As Figure 8 depicted, Ralph et al. [16] put four Ballard cell stacks under cell reversal condition. When the cell potential got to -2.0 V, the one added PTFE was increased to 22 from 3 min, which is the time the standard anode took.

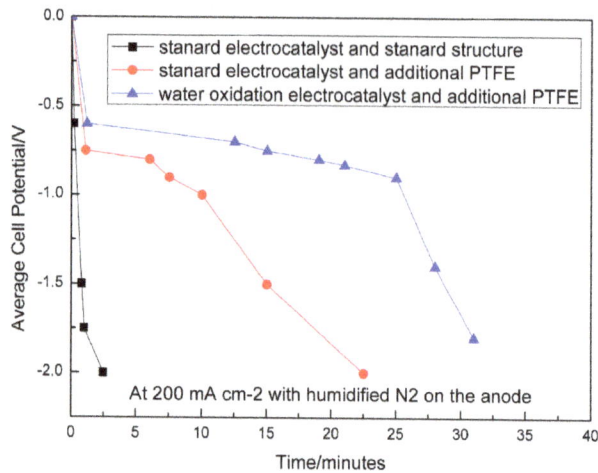

Figure 8. The effect of anode catalyst layer structure and composition in Ballard four-cell stack tests. Reproduced with permission from [16]. Copyright ECS, 2006. The anode contains 20 wt % Pt and 10 wt % Ru supported on Vulcan XC72R at an electrode loading rate of ca. $0.3 \, \text{mg·Pt·cm}^{-2}$ with the water oxidation electrocatalyst added at a loading rate of ca. $0.1 \, \text{mg·Ru·cm}^{-2}$.

Numerous research papers have studied the effect of hydrophobic agent content and species, including polytetrafluoroethylene (PTFE), polyvinylidene fluoride (PVDF), and fluorinated ethylene propylene (FEP), in GDL on the performance of fuel cells [73–75]. Given their excellent characteristics, hydrophobic agents can be used in the anode catalyst layer to hold more water.

5. Conclusions

Fuel starvation, inadequate air supply, water flooding, load change, low catalyst performance, etc. can induce cell reversal when the current rate exceeds the limiting mass transfer rate. Because these factors cannot produce sufficient electron and proton feeds to match current stack requirements and achieve current value balance, the fuel cell voltage could reverse and some other abnormal chemical reactions might occur to make up for the deficit. In these abnormal operating conditions, there is irreversible damage to the PEMFC system, including carbon support corrosion, catalyst and sinter agglomeration, and membrane degradation. In fact, gross fuel starvation is different from localized fuel starvation in terms of causes, carbon corrosion area, cell voltage, and current phenomena. Generally, cell reversal caused by overall fuel starvation is more critical than in a cathode and is induced by improper gas supply, flooding, and any other factors that obstruct the flow field and catalyst layer; therefore, the design of the anode catalyst layer and the catalytic performance have drawn more attention to cell reversal. Due to the existence of hydrogen in cathodes or oxygen in anodes, cell reversal can cause permanent damage to a fuel cell.

To solve this problem or mitigate the harm to fuel cells, two approaches have been used in recent years. First is the system management strategy, where auxiliary monitoring equipment is used to determine necessary adjustments to operation parameters to avoid cell reversal, such as adding reactant flow rate, change operation load, water management, and so on. The second is system material modification to resist cell reversal. Specifically, there are three approaches to attenuate performance loss and extend fuel cell lifespan. Researchers have placed attention on high-performance catalysts, carbon support materials, and water oxidation reactions. In addition, heat-resistant membranes also attract more attention for their ability to reduce water flooding in a cathode, but they would decrease the proton transfer rate and catalyst efficiency.

Obviously, there are still several problems in need of further study, such as (1) understanding the effects of different types of carbon materials' surface forms and structure; (2) developing a novel catalyst layer assembly including a high-performance catalyst, corrosion-resistant carbon material, a water oxidation catalyst, and hydrophobic agents; and (3) distinguishing the specific electrode reaction caused by the phenomenon of hydrogen pump and oxygen pump. Compared to a fuel cell system of water and heat management, this is simpler and more critical to catalyst layer design, and would make system management more efficient by diminishing several controlling limitations.

Acknowledgments: This paper was generously supported by the National Natural Science Foundation (No. 21206128).

Author Contributions: Daijun Yang, Bing Li, and Cunman Zhang collected experiment data, analyzed and synthesized fuel cell theory, Jue Wang and Bing Li revised the English expression. Congwei Qin analyzed references material, experiment data and wrote the paper.

Conflicts of Interest: The authors declare no conflict of interest.

References

1. Yi, B. *Fuel Cell-Principle, Technology and Application*; Chemical Industry: Beijing, China, 2003; p. 41.
2. Rajashekara, K.; Rathore, A.K. Power conversion and control for fuel cell systems in transportation and stationary power generation. *Electr. Power Compon. Syst.* **2015**, *43*, 1376–1387. [CrossRef]
3. Popovich, N. *US Department of Energy Hydrogen and Fuel Cells Program 2015 Annual Merit Review and Peer Evaluation Report: June 8–12, 2015, Arlington, Virginia*; NREL (National Renewable Energy Laboratory (NREL): Golden, CO, USA, 2015.
4. Zhou, F.; Andreasen, S.J.; Kær, S.K. Experimental study of cell reversal of a high temperature polymer electrolyte membrane fuel cell caused by H₂ starvation. *Int. J. Hydrogen Energy* **2015**, *40*, 6672–6680. [CrossRef]
5. Taniguchi, A.; Akita, T.; Yasuda, K.; Miyazaki, Y. Analysis of degradation in PEMFC caused by cell reversal during air starvation. *Int. J. Hydrogen Energy* **2008**, *33*, 2323–2329. [CrossRef]
6. Taniguchi, A.; Akita, T.; Yasuda, K.; Miyazaki, Y. Analysis of electrocatalyst degradation in PEMFC caused by cell reversal during fuel starvation. *J. Power Sources* **2004**, *130*, 42–49. [CrossRef]
7. Zhang, G.; Shen, S.; Guo, L.; Liu, H. Dynamic characteristics of local current densities and temperatures in proton exchange membrane fuel cells during reactant starvations. *Int. J. Hydrogen Energy* **2012**, *37*, 1884–1892. [CrossRef]
8. Yu, Y.; Li, H.; Wang, H.; Yuan, X.-Z.; Wang, G.; Pan, M. A review on performance degradation of proton exchange membrane fuel cells during startup and shutdown processes: Causes, consequences, and mitigation strategies. *J. Power Sources* **2012**, *205*, 10–23. [CrossRef]
9. Dou, M.; Hou, M.; Liang, D.; Shen, Q.; Zhang, H.; Lu, W.; Shao, Z.; Yi, B. Behaviors of proton exchange membrane fuel cells under oxidant starvation. *J. Power Sources* **2011**, *196*, 2759–2762. [CrossRef]
10. Castanheira, L.; Silva, W.O.; Lima, F.H.; Crisci, A.; Dubau, L.; Maillard, F.D.R. Carbon corrosion in proton-exchange membrane fuel cells: Effect of the carbon structure, the degradation protocol, and the gas atmosphere. *ACS Catal.* **2015**, *5*, 2184–2194. [CrossRef]
11. Knights, S.D.; Colbow, K.M.; St-Pierre, J.; Wilkinson, D.P. Aging mechanisms and lifetime of PEFC and DMFC. *J. Power Sources* **2004**, *127*, 127–134. [CrossRef]
12. Zhou, F.; Andreasen, S.J.; Kær, S.K.; Yu, D. Analysis of accelerated degradation of a HT-PEM fuel cell caused by cell reversal in fuel starvation condition. *Int. J. Hydrogen Energy* **2015**, *40*, 2833–2839. [CrossRef]
13. Zhang, S.; Yuan, X.-Z.; Hin, J.N.C.; Wang, H.; Friedrich, K.A.; Schulze, M. A review of platinum-based catalyst layer degradation in proton exchange membrane fuel cells. *J. Power Sources* **2009**, *194*, 588–600. [CrossRef]
14. Kang, J.; Jung, D.W.; Park, S.; Lee, J.-H.; Ko, J.; Kim, J. Accelerated test analysis of reversal potential caused by fuel starvation during PEMFCs operation. *Int. J. Hydrogen Energy* **2010**, *35*, 3727–3735. [CrossRef]
15. Yousfi-Steiner, N.; Moçotéguy, P.; Candusso, D.; Hissel, D. A review on polymer electrolyte membrane fuel cell catalyst degradation and starvation issues: Causes, consequences and diagnostic for mitigation. *J. Power Sources* **2009**, *194*, 130–145. [CrossRef]

16. Ralph, T.R.; Hudson, S.; Wilkinson, D.P. Electrocatalyst stability in PEMFCs and the role of fuel starvation and cell reversal tolerant anodes. *ECS Trans.* **2006**, *1*, 67–84.

17. Yang, X.-G.; Ye, Q.; Cheng, P. Hydrogen pumping effect induced by fuel starvation in a single cell of a PEM fuel cell stack at galvanostatic operation. *Int. J. Hydrogen Energy* **2012**, *37*, 14439–14453. [CrossRef]

18. Li, B.; Higgins, D.C.; Xiao, Q.; Yang, D.; Zhng, C.; Cai, M.; Chen, Z.; Ma, J. The durability of carbon supported Pt nanowire as novel cathode catalyst for a 1.5 kW PEMFC stack. *Appl. Catal. B* **2015**, *162*, 133–140. [CrossRef]

19. Sakamoto, S.; Karakane, M.; Maeda, H.; Miyake, Y.; Susai, T.; Isono, T. *Study of the Factors Affecting PEFC Life Characteristic*; Ecology and Energy Systems Research Center, SANYO Electric Co., Ltd.: Tochigi, Japan, 2000; pp. 326–8534.

20. Kim, S.; Shimpalee, S.; van Zee, J. The effect of stoichiometry on dynamic behavior of a proton exchange membrane fuel cell (PEMFC) during load change. *J. Power Sources* **2004**, *135*, 110–121. [CrossRef]

21. Rao, R.M.; Rengaswamy, R. A distributed dynamic model for chronoamperometry, chronopotentiometry and gas starvation studies in PEM fuel cell cathode. *Chem. Eng. Sci.* **2006**, *61*, 7393–7409.

22. Shen, Q.; Hou, M.; Yan, X.; Liang, D.; Zang, Z.; Hao, L.; Shao, Z.; Hou, Z.; Ming, P.; Yi, B. The voltage characteristics of proton exchange membrane fuel cell (PEMFC) under steady and transient states. *J. Power Sources* **2008**, *179*, 292–296. [CrossRef]

23. Liu, Z.; Yang, L.; Mao, Z.; Zhuge, W.; Zhang, Y.; Wang, L. Behavior of PEMFC in starvation. *J. Power Sources* **2006**, *157*, 166–176. [CrossRef]

24. Büchi, F.N.; Inaba, M.; Schmidt, T.J. *Polymer Electrolyte Fuel Cell Durability*; Springer: New York, NY, USA, 2009.

25. Ofstad, A.; Davey, J.; Sunde, S.; Borup, R.L. Carbon corrosion of a PEMFC during shut-down/start-up when using an air purge procedure. *ECS Trans.* **2008**, *16*, 1301–1311.

26. Hu, J.; Sui, P.; Kumar, S.; Djilali, N. Modelling and simulations of carbon corrosion during operation of a polymer electrolyte membrane fuel cell. *Electrochim. Acta* **2009**, *54*, 5583–5592. [CrossRef]

27. Patterson, T.W.; Darling, R.M. Damage to the cathode catalyst of a PEM fuel cell caused by localized fuel starvation. *Electrochem. Solid-State Lett.* **2006**, *9*, A183–A185. [CrossRef]

28. Lim, K.H.; Oh, H.-S.; Jang, S.-E.; Ko, Y.-J.; Kim, H.-J.; Kim, H. Effect of operating conditions on carbon corrosion in polymer electrolyte membrane fuel cells. *J. Power Sources* **2009**, *193*, 575–579. [CrossRef]

29. Ralph, T.; Hogarth, M. Catalysis for low temperature fuel cells. *Platin. Met. Rev.* **2002**, *46*, 117–135.

30. Liang, D.; Shen, Q.; Hou, M.; Shao, Z.; Yi, B. Study of the cell reversal process of large area proton exchange membrane fuel cells under fuel starvation. *J. Power Sources* **2009**, *194*, 847–853. [CrossRef]

31. Liang, D.; Dou, M.; Hou, M.; Shen, Q.; Shao, Z.; Yi, B. Behavior of a unit proton exchange membrane fuel cell in a stack under fuel starvation. *J. Power Sources* **2011**, *196*, 5595–5598. [CrossRef]

32. Tüber, K.; Pócza, D.; Hebling, C. Visualization of water buildup in the cathode of a transparent PEM fuel cell. *J. Power Sources* **2003**, *124*, 403–414. [CrossRef]

33. Natarajan, D.; van Nguyen, T. Current distribution in PEM fuel cells. Part 2: Air operation and temperature effect. *AIChE J.* **2005**, *51*, 2599–2608. [CrossRef]

34. Rangel, C.; Travassos, M.A.; Fernandes, V.R.; Silva, R.; Paiva, T. *Fuel Starvation: Irreversible Degradation Mechanisms in PEM Fuel Cells*; WHEC—World Hydrogen Energy Convention: Essen, Germany, 2010.

35. Zhang, J.; Yang, H.; Fang, J.; Zou, S. Synthesis and oxygen reduction activity of shape-controlled Pt_3Ni nanopolyhedra. *Nano Lett.* **2010**, *10*, 638–644. [CrossRef] [PubMed]

36. Yu, X.; Ye, S. Recent advances in activity and durability enhancement of Pt/C catalytic cathode in PEMFC: Part II: Degradation mechanism and durability enhancement of carbon supported platinum catalyst. *J. Power Sources* **2007**, *172*, 145–154. [CrossRef]

37. Wilkinson, D.; St-Pierre, J.; Vielstich, W.; Gasteiger, H.; Lamm, A. Handbook of fuel cells—Fundamentals, technology and applications. *Fuel Cell Technol. Appl.* **2003**, *3*, 611.

38. Maass, S.; Finsterwalder, F.; Frank, G.; Hartmann, R.; Merten, C. Carbon support oxidation in PEM fuel cell cathodes. *J. Power Sources* **2008**, *176*, 444–451. [CrossRef]

39. Barton, R.H. Preventing Voltage Reversal Conditions in Solid Polymer Electrolyte Cells; Heating Resistor and Rectifier in Series; Sensing Thermistor; Detecting a Low Voltage in a Fuel Cell. U.S. Patent 6,724,194, 20 April 2004.

40. Baumgartner, W.; Parz, P.; Fraser, S.; Wallnöfer, E.; Hacker, V. Polarization study of a PEMFC with four reference electrodes at hydrogen starvation conditions. *J. Power Sources* **2008**, *182*, 413–421. [CrossRef]

41. Perry, M.L.; Patterson, T.; Reiser, C. Systems strategies to mitigate carbon corrosion in fuel cells. *ECS Trans.* **2006**, *3*, 783–795.

42. Vahidi, A.; Stefanopoulou, A.; Peng, H. Model predictive control for starvation prevention in a hybrid fuel cell system. In Proceedings of the 2004 American Control Conference, 30 June–2 July 2004; IEEE: Boston, MA, USA; pp. 834–839.

43. Jung, N.; Chung, D.Y.; Ryu, J.; Yoo, S.J.; Sung, Y.-E. Pt-based nanoarchitecture and catalyst design for fuel cell applications. *Nano Today* **2014**, *9*, 433–456. [CrossRef]

44. Wang, J.; Yin, G.; Shao, Y.; Zhang, S.; Wang, Z.; Gao, Y. Effect of carbon black support corrosion on the durability of Pt/C catalyst. *J. Power Sources* **2007**, *171*, 331–339. [CrossRef]

45. Luo, Z.; Li, D.; Tang, H.; Pan, M.; Ruan, R. Degradation behavior of membrane–electrode-assembly materials in 10-cell PEMFC stack. *Int. J. Hydrogen Energy* **2006**, *31*, 1831–1837. [CrossRef]

46. Yoda, T.; Uchida, H.; Watanabe, M. Effects of operating potential and temperature on degradation of electrocatalyst layer for PEFCs. *Electrochim. Acta* **2007**, *52*, 5997–6005. [CrossRef]

47. Colón-Mercado, H.R.; Popov, B.N. Stability of platinum based alloy cathode catalysts in PEM fuel cells. *J. Power Sources* **2006**, *155*, 253–263. [CrossRef]

48. Gasteiger, H.A.; Kocha, S.S.; Sompalli, B.; Wagner, F.T. Activity benchmarks and requirements for Pt, Pt–alloy, and non-Pt oxygen reduction catalysts for PEMFCs. *Appl. Catal. B* **2005**, *56*, 9–35. [CrossRef]

49. Wan, L.J.; Moriyama, T.; Ito, M.; Uchida, H.; Watanabe, M. In situ STM imaging of surface dissolution and rearrangement of a Pt–Fe alloy electrocatalyst in electrolyte solution. *Chem. Commun.* **2002**. [CrossRef]

50. Paulus, U.; Wokaun, A.; Scherer, G.; Schmidt, T.; Stamenkovic, V.; Radmilovic, V.; Markovic, N.; Ross, P. Oxygen reduction on carbon-supported Pt–Ni and Pt–Co alloy catalysts. *J. Phys. Chem. B* **2002**, *106*, 4181–4191. [CrossRef]

51. Yu, P.; Pemberton, M.; Plasse, P. PtCo/C cathode catalyst for improved durability in PEMFCs. *J. Power Sources* **2005**, *144*, 11–20. [CrossRef]

52. Zhang, J.; Sasaki, K.; Sutter, E.; Adzic, R. Stabilization of platinum oxygen-reduction electrocatalysts using gold clusters. *Science* **2007**, *315*, 220–222. [CrossRef] [PubMed]

53. Stevens, D.; Hicks, M.; Haugen, G.; Dahn, J. Ex situ and in situ stability studies of PEMFC catalysts effect of carbon type and humidification on degradation of the carbon. *J. Electrochem. Soc.* **2005**, *152*, A2309–A2315. [CrossRef]

54. Kim, M.; Jung, N.; Eom, K.; Yoo, S.J.; Kim, J.Y.; Jang, J.H.; Kim, H.-J.; Hong, B.K.; Cho, E. Effects of anode flooding on the performance degradation of polymer electrolyte membrane fuel cells. *J. Power Sources* **2014**, *266*, 332–340. [CrossRef]

55. Antolini, E. Graphene as a new carbon support for low-temperature fuel cell catalysts. *Appl. Catal. B* **2012**, *123*, 52–68. [CrossRef]

56. Guo, S.; Sun, S. FePt nanoparticles assembled on graphene as enhanced catalyst for oxygen reduction reaction. *J. Am. Chem. Soc.* **2012**, *134*, 2492–2495. [CrossRef] [PubMed]

57. Artyushkova, K.; Atanassov, P.; Dutta, M.; Wessel, S.; Colbow, V. Structural correlations: Design levers for performance and durability of catalyst layers. *J. Power Sources* **2015**, *284*, 631–641. [CrossRef]

58. Artyushkova, K.; Pylypenko, S.; Dowlapalli, M.; Atanassov, P. Structure-to-property relationships in fuel cell catalyst supports: Correlation of surface chemistry and morphology with oxidation resistance of carbon blacks. *J. Power Sources* **2012**, *214*, 303–313. [CrossRef]

59. Wlodarczyk, R. Porous carbon materials for elements in low-temperature fuel cells. *Arch. Metall. Mater.* **2015**, *60*, 117–120. [CrossRef]

60. Oh, H.-S.; Lim, K.H.; Roh, B.; Hwang, I.; Kim, H. Corrosion resistance and sintering effect of carbon supports in polymer electrolyte membrane fuel cells. *Electrochim. Acta* **2009**, *54*, 6515–6521. [CrossRef]

61. Dicks, A.L. The role of carbon in fuel cells. *J. Power Sources* **2006**, *156*, 128–141.

62. Yang, W.; Wang, Y.; Li, J.; Yang, X. Polymer wrapping technique: An effective route to prepare Pt nanoflower/carbon nanotube hybrids and application in oxygen reduction. *Energy Environ. Sci.* **2010**, *3*, 144–149. [CrossRef]

63. Oh, H.S.; Kim, H. Efficient synthesis of Pt nanoparticles supported on hydrophobic graphitized carbon nanofibers for electrocatalysts using noncovalent functionalization. *Adv. Funct. Mater.* **2011**, *21*, 3954–3960. [CrossRef]

64. Lim, K.H.; Oh, H.-S.; Kim, H. Use of a carbon nanocage as a catalyst support in polymer electrolyte membrane fuel cells. *Electrochem. Commun.* **2009**, *11*, 1131–1134. [CrossRef]

65. Vinod Selvaganesh, S.; Selvarani, G.; Sridhar, P.; Pitchumani, S.; Shukla, A. Graphitic carbon as durable cathode-catalyst support for PEFCs. *Fuel Cells* **2011**, *11*, 372–384. [CrossRef]

66. Zhao, X.; Hayashi, A.; Noda, Z.; Kimijima, K.i.; Yagi, I.; Sasaki, K. Evaluation of change in nanostructure through the heat treatment of carbon materials and their durability for the start/stop operation of polymer electrolyte fuel cells. *Electrochim. Acta* **2013**, *97*, 33–41. [CrossRef]

67. Jang, S.-E.; Kim, H. Effect of water electrolysis catalysts on carbon corrosion in polymer electrolyte membrane fuel cells. *J. Am. Chem. Soc.* **2010**, *132*, 14700–14701. [CrossRef] [PubMed]

68. Jang, I.; Hwang, I.; Tak, Y. Attenuated degradation of a PEMFC cathode during fuel starvation by using carbon-supported IrO_2. *Electrochim. Acta* **2013**, *90*, 148–156. [CrossRef]

69. Reiser, C.A. Preventing Fuel Starvation of a Fuel Cell Stack. U.S. Patent 7,807,302, 5 October 2010.

70. Wilkinson, D.P.; Chow, C.Y.; Allan, D.E.; Johannes, E.P.; Roberts, J.A.; St-Pierre, J.; Longley, C.J.; Chan, J.K. Method and Apparatus for Operating an Electrochemical Fuel Cell with Periodic Fuel Starvation at the Anode. U.S. Patent 6,096,448, 1 August 2000.

71. Park, S.; Lee, J.-W.; Popov, B.N. A review of gas diffusion layer in PEM fuel cells: Materials and designs. *Int. J. Hydrogen Energy* **2012**, *37*, 5850–5865. [CrossRef]

72. Knights, S.D.; Taylor, J.L.; Wilkinson, D.P.; Wainwright, D.S. Fuel Cell Anode Structures for Voltage Reversal Tolerance. U.S. Patent 6,517,962, 11 February 2003.

73. Lim, C.; Wang, C. Effects of hydrophobic polymer content in GDL on power performance of a PEM fuel cell. *Electrochim. Acta* **2004**, *49*, 4149–4156. [CrossRef]

74. Park, S.; Popov, B.N. Effect of hydrophobicity and pore geometry in cathode GDL on PEM fuel cell performance. *Electrochim. Acta* **2009**, *54*, 3473–3479. [CrossRef]

75. Ismail, M.; Hughes, K.; Ingham, D.; Ma, L.; Pourkashanian, M. Effect of PtFe loading of gas diffusion layers on the performance of proton exchange membrane fuel cells running at high-efficiency operating conditions. *Int. J. Energy Res.* **2013**, *37*, 1592–1599. [CrossRef]

MDPI

Article

Effects of the Electrodeposition Time in the Synthesis of Carbon-Supported Pt(Cu) and Pt-Ru(Cu) Core-Shell Electrocatalysts for Polymer Electrolye Fuel Cells

Griselda Caballero-Manrique [1], Immad Muhammed Nadeem [2,3], Enric Brillas [1], Francesc Centellas [1], José Antonio Garrido [1], Rosa María Rodríguez [1] and Pere-Lluís Cabot [1,*]

1   Laboratori d'Electroquímica dels Materials i del Medi Ambient, Departament de Química Física, Facultat de Química, Universitat de Barcelona, Martí i Franquès 1-11, 08028 Barcelona, Spain; g_caballe@hotmail.com (G.C.-M.); brillas@ub.edu (E.B.); facentellas@ub.edu (F.C.); joseagarrido@ub.edu (J.A.G.); rosarodriguez@ub.edu (R.M.R.)
2   London Centre for Nanotechnology and Department of Chemistry, University College London, 20 Gordon Street, London, WC1H 0AJ, UK; immad.nadeem.14@ucl.ac.uk
3   Diamond Light Source Ltd, Harwell Science and Innovation Campus, Didcot, Oxfordshire, OX11 0DE, UK
*   Correspondence: p.cabot@ub.edu; Tel.: +34-93-4039236; Fax: +34-93-4021231

Academic Editors: Vincenzo Baglio and David Sebastián
Received: 22 June 2016; Accepted: 10 August 2016; Published: 18 August 2016

Abstract: Pt(Cu)/C and Pt-Ru(Cu)/C electrocatalysts with core-shell structure supported on Vulcan Carbon XC72R have been synthesized by potentiostatic deposition of Cu nanoparticles on the support, galvanic exchange with Pt and spontaneous deposition of Ru species. The duration of the electrodeposition time of the different species has been modified and the obtained electrocatalysts have been characterized using electrochemical and structural techniques. The High Resolution Transmission Electron Microscopy (HRTEM), Fast Fourier Transform (FFT) and Energy Dispersive X-ray (EDX) microanalyses allowed the determining of the effects of the electrodeposition time on the nanoparticle size and composition. The best conditions identified from Cyclic Voltammetry (CV) corresponded to onset potentials for CO and methanol oxidation on Pt-Ru(Cu)/C of 0.41 and 0.32 V vs. the Reversible Hydrogen Electrode (RHE), respectively, which were smaller by about 0.05 V than those determined for Ru-decorated commercial Pt/C. The CO oxidation peak potentials were about 0.1 V smaller when compared to commercial Pt/C and Pt-Ru/C. The positive effect of Cu was related to its electronic effect on the Pt shells and also to the generation of new active sites for CO oxidation. The synthesis conditions to obtain the best performance for CO and methanol oxidation on the core-shell Pt-Ru(Cu)/C electrocatalysts were identified. When compared to previous results in literature for methanol, ethanol and formic acid oxidation on Pt(Cu)/C catalysts, the present results suggest an additional positive effect of the deposited Ru species due to the introduction of the bifunctional mechanism for CO oxidation.

Keywords: core-shell Pt(Cu)/C electrocatalysts; core-shell Pt-Ru(Cu)/C electrocatalysts; potentiostatic deposition of Cu; Pt deposition by galvanic exchange; Ru spontaneous deposition; CO oxidation; methanol oxidation

1. Introduction

Direct Methanol Fuel Cells (DMFCs) operating under ambient temperature are gaining interest due to their safe and profitable use as portable power in the market for mobile phones, laptops and other portable electric devices [1–3]. However, the attractive low temperature DMFCs possess presents certain drawbacks. The anodic electrooxidation of methanol maintains unfavorable slow kinetics, which coupled with the tendency of methanol to migrate across the proton exchange membrane and oxidize at the cathode ultimately results in reduced operative capacity [4]. High surface area Pt-based nanostructured electrocatalysts are commonly employed. However, the methanol electrooxidation is a self-poisoning reaction because the electrogenerated CO intermediate strongly adsorbs onto the Pt surface, thus limiting the methanol adsorption [4–7]. CO interacts with the platinum surface in a linear, bridge and three fold configuration where each CO molecule strongly adsorbs onto one, two and three Pt atoms, respectively [8]. The carbon supported Pt becomes around 50% less efficient even on trace levels of CO [5]. Hence, widespread research has been conducted to prepare suitable nanostructured electrocatalysts to mitigate the problematic slow kinetics and self-poisoning.

CO poisoning can be addressed by alloying Pt with transition metals such as Ru, Mo, Re, W and others [5–10]. Ru appears between the most interesting ones [2,6,7,10,11]. The high tolerance to CO exhibited by the Pt-Ru alloys has been explained by two different mechanisms. The first one considers the favored oxidation of CO adsorbed on Pt by the Ru-OH species, which are formed by water dissociation on Ru at lower potentials than on Pt (bifunctional mechanism). The second one establishes the Pt-CO bond weakening via the electronic effect induced by Ru on Pt.

Currently, attention has been focused on synthesizing electrode materials with lower Pt content, thus allowing decreasing the cost of the catalyst and/or improving the catalyst performance [12–30]. They have been applied for very different reactions of interest including hydrogen, borohydride, methanol, CO, ethanol and formic acid oxidation and also oxygen reduction. Different nanoparticulate catalysts have been synthesized by deposition of a core of transition metal on carbon support which can be partially replaced by Pt through galvanic exchange [14–24,26–28,30]. This method is based on the displacement of the less noble metal by the nobler one under open circuit conditions. Particularly, Cu has been often used as the core metal in acidic solutions because the potential of the Cu^{2+}/Cu pair is sufficiently low but at the same time more positive than the potential of the standard hydrogen electrode (SHE) [14,16,18,20,24,26–28,30]. The displacement of Cu is not then complicated by hydrogen evolution. The galvanic exchange of Cu with $PtCl_6^{2-}$ can be represented as follows:

$$2Cu + PtCl_6^{2-} \rightarrow Pt + 2Cu^{2+} + 6Cl^- \tag{1}$$

which has a standard cell potential of $E° = 0.404$ V vs. SHE. The most part of the Pt-Cu/C catalysts in which the galvanic exchange has been used, have been prepared starting from Cu/C precursors obtained by chemical reduction of Cu^{2+} [24,26–28,30]. They have been mainly applied to methanol [24,28], ethanol [27] and formic acid oxidation [25,30] and oxygen reduction [26] The synthesis of nanoparticle catalysts starting from the electrochemical reduction of Cu^{2+} is scarcely found in the literature [18]. There is also an additional interest in studying whether there is a further positive effect in the introduction of Ru deposited species.

The suitable potential for the potentiostatic deposition of Cu core nanoparticles on carbon Vulcan XC72R to synthesize Pt(Cu)/C and Pt-Ru(Cu)/C core-shell electrocatalysts with a tentative constant Cu electrodeposition charge has been studied by us in a previous work [18]. The respective CO stripping peak potentials were about 0.1 and 0.2 V more negative than those corresponding to Pt/C and Ru-decorated Pt/C, thus indicating the higher activity of the Pt and the Pt-Ru shells towards the CO oxidation. This increased activity was explained by the structural effect induced by the Cu core. Moreover, the use of the Cu cores allowed the Pt economy in these catalysts. The main objective of this work is to study the effect of the electrodeposition times in the synthesis of these nanostructured electrocatalysts, i.e. time of Cu electrodeposition, galvanic exchange with Pt and spontaneous deposition of the Ru species. The morphology, particle size distribution and composition of the relevant specimens have been determined by means of Transmission Electron Microscopy (TEM), High Resolution (HR) TEM, Electron Diffraction, Fast Fourier Transform (FFT), and Energy Dispersive X-ray Spectroscopy (EDS). The electrochemical performance of the catalysts prepared for the CO and methanol oxidation has been tested using Cyclic Voltammetry (CV) and Linear Sweep Voltammetry on a Rotating Disk Electrode (RDE).

2. Results and Discussion

2.1. Charge of Cu Electrodeposition in the Synthesis Sequence i

Different catalysts were obtained starting from Cu electrodeposition at -0.1 V by changing the electrodeposition charge (q_{Cu}) between 20 and 50 mC. The measured deposition efficiencies were always close to 100%, in agreement with previous results of the authors [18]. The resulting Cu loads were in the range of 93–232 μg_{Cu} cm^{-2}. After each Cu electrodeposition, the different specimens were prepared following the sequence *i*. Some examples of the cyclic voltammograms recorded for the Pt(Cu)/C and the Pt-Ru(Cu)/C electrocatalysts in 0.5 M H$_2$SO$_4$ and for CO stripping are depicted in Figure 1a and b. The cyclic voltammograms depicted in Figure 1a present the same essential features as those previously reported for Pt-exposed surfaces [10,18,31]. Thus, the cyclic voltammogram for Pt(Cu)/C of Figure 1a shows the hydrogen adsorption/desorption region in the potential range from 0.0 to 0.3 V [31], the Pt oxidation formation about 0.65 V in the anodic sweep and the reduction of the Pt oxide around the cathodic peak potential of 0.8 V. Figure 1a also shows that the profile of the cyclic voltammogram for Pt-Ru(Cu)/C was quite similar, although the anodic and cathodic currents for the hydrogen adsorption/desorption region were smaller due to the partial blocking of the Pt sites by the Ru species.

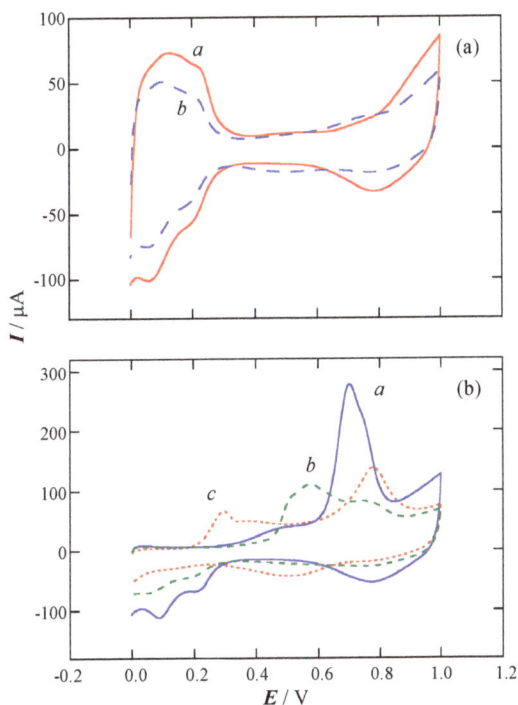

Figure 1. (a) Cyclic voltammograms of Pt(Cu)/C (curve *a*) and Pt-Ru(Cu)/C (curve *b*) in deaerated 0.5 M H$_2$SO$_4$ after a Cu electrodeposition charge (q_{Cu}) of 30 mC following the sequence *i*; (b) CO stripping experiments on Pt(Cu)/C for q_{Cu} of 45 (curve *a*) and 20 mC (curve *c*) and on Pt-Ru(Cu)/C for q_{Cu} = 45 mC (curve *b*). Sweep rate 20 mV s^{-1}.

The nature of the Ru species formed on Pt by spontaneous deposition was determined using commercial Pt/C [32] and only oxidized Ru species (RuO$_2$ and RuO$_x$H$_y$) were found. Considering that the atomic hydrogen is not adsorbed on the Ru species [7], the Pt coverage (θ) in the Pt-Ru(Cu)/C electrocatalyst can be estimated according to the following equation [33]:

$$\theta = (Q_{H,0} - Q_{H,1})/Q_{H,0} \tag{2}$$

where $Q_{H,0}$ and $Q_{H,1}$ are the mean charges involved in the hydrogen adsorption/desorption, after substracting the double layer charge, before and after the deposition of the Ru species, respectively. Note that the anodic and cathodic currents for Pt-Ru(Cu)/C in the potential region between 0.3 and 0.55 V, related to the double layer charge, were relatively greater than for Pt(Cu)/C. This can be explained by the pseudocapacitive behavior related to the hydroxylation of the Ru species [34,35]. The experimental results showed Ru coverages in the range 0.2–0.35 for q_{Cu} values between 20 and 40 mC. These values indicate that more than the half of the Pt surface was free from Ru species, falling in the suitable range for the best catalytic activity of the Pt-Ru/C catalysts in front of the CO, methanol and ethanol oxidation, which were about 0.25–0.3 for the Pt-Ru/C catalysts obtained by spontaneous deposition of Ru species on commercial Pt/C [36].

Figure 1b shows some examples of the CO stripping experiments both, for Pt(Cu)/C (curve *a*) and Pt-Ru(Cu)/C (curve *b*). The anodic sweep always starts at 0.0 V with negligible currents. Adsorbed hydrogen, if present, would be oxidized in the potential region between 0.0 and 0.3 V. However, the Pt sites for hydrogen adsorption are now blocked by strongly adsorbed CO, which is oxidized at potentials in the range of 0.4–0.8 V [36,37]. Pt can be oxidized in the Pt sites where CO has been removed, between 0.6 and 1.0 V. In the reverse sweep, Pt oxide is reduced to Pt (cathodic peak potential at about 0.8 V) and in the potential region between 0.3 and 0.0 V, hydrogen can be adsorbed on the Pt sites. Note that CO oxidation on Pt(Cu)/C (curve *a*) occurs at more positive potentials than on Pt-Ru(Cu)/C (curve *b*) because the Ru species favor the CO oxidation by means of the bifunctional mechanism [6,7,10]. We have included in this figure curve *c*, which corresponds to the catalyst preparation with a Cu electrodeposition charge of only 20 mC. This curve presents an anodic peak in the potential range 0.2–0.4 V, which is not found for electrodeposition charges of 30 mC or higher. According to previous work of the authors [18], it is due to the copper oxidation, thus indicating that there are Cu sites exposed to the electrolyte and therefore, Cu electrodeposition charges of 20 mC are not adequate to produce the core-shell electrocatalysts studied in this work. Previous work of Podlovchenko et al. [14,16] about the galvanic exchange between Cu and Pt reported that in some cases, depending on the amount of deposited Cu, on the solution stirring rate, and on the oxidation state of Pt in the complex, Pt was not able to completely cover the Cu surface. The reason for this is not clear. However, we may imagine that for very small Cu nanoparticles, the Pt complex is not able to sufficiently approach the Cu atoms which are partially occluded by the surface Pt metal atoms already deposited. Moreover, according to Equation 1, two Cu atoms per Pt complex are needed in the galvanic exchange. In addition we may suppose that Cu nuclei can be generated at relatively occluded points in the carbon. Probably, when they are too small, the Cu nuclei could not be easily reached by the Pt complex also by steric hindrance. Conversely, these partially occluded Cu atoms could probably be easily oxidized in the acidic aqueous environment.

Figure 2a collects the peak potential for CO oxidation on Pt(Cu)/C and on Pt-Ru(Cu)/C, curves *a* and *b* respectively, as a function of the Cu electrodeposition charge. It is shown here that the respective minimum peak potentials for CO oxidation on Pt(Cu)/C and Pt-Ru(Cu)/C are about 0.72 and 0.55 V. It is worth mentioning that these values are significantly more negative than those measured for the catalysts without the Cu core. Thus, the peak potential for CO oxidation on carbon-supported cubo-octahedral Pt nanoparticles was about 0.8 V [37], whereas on the carbon-supported Pt-Ru alloys they were in the range of 0.65–0.75 V [36]. This means that the Cu core-shell catalysts have higher CO tolerance than those without the Cu core. The ligand (electronic) effect of Cu on Pt would explain the shift of these potentials to more negative values, in agreement with previously reported results [26].

Note in addition to the CO stripping peaks in the core-shell, catalysts appear to be composed of at least two peaks (see curve *b* in Figure 1b). This indicates that there are Pt sites of different nature in the catalyst where CO can be adsorbed. The CO molecules are then oxidized at different potentials on the different sites, depending on their adsorption energy. As long as there is only one peak for CO oxidation for cubo-octahedral Pt and Pt-Ru alloy, we may conclude that such different Pt sites would result from the effect of the Cu core.

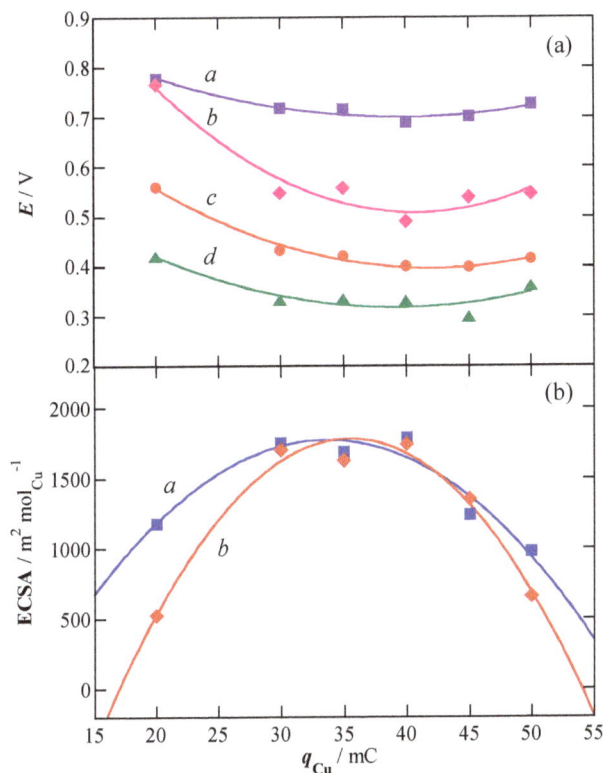

Figure 2. (a) Peak potentials for CO oxidation on Pt(Cu)/C (curve *a*) and on Pt-Ru(Cu)/C (curve *b*), and onset potentials for CO (curve *c*) and methanol oxidation (curve *d*) on Pt-Ru(Cu)/C, as a function of the electrodeposition charge q_{Cu}; (b) Dependence of the electrochemical surface area (ECSA) for CO oxidation on q_{Cu} for Pt(Cu)/C (curve *a*) and for Pt-Ru(Cu)/C (curve *b*). The electrocatalysts were prepared according to sequence *i*.

The onset potentials for CO and methanol oxidation on Pt-Ru(Cu)/C are also depicted in Figure 2a, curves *c* and *d*, respectively. They were identified as the potentials in which the current significantly increased from the base line. As shown in curve *d* of Figure 2a, the onset potential for the methanol oxidation on the Pt-Ru(Cu)/C electrocatalyst followed the same trend as for CO (curve *c*), with minimum values of 0.32 and 0.41 V at about 40 mC, also respectively. These values are about 0.05 V smaller than those obtained for Ru-decorated Pt/C catalysts [36], in agreement with the higher reactivity for the present core-shell structure.

Figure 2b depicts the electrochemical surface area (ECSA) values for CO oxidation after normalization by the actual amount of electrodeposited Cu (determined from the Cu electrodeposition charge after correcting by the Cu electrodeposition efficiency). They were calculated from dividing the ECSA by the amount of copper in order to determine the conditions in which the nanoparticles with the highest ECSAs for CO oxidation were obtained. It is apparent from this figure that the normalized ECSAs for Pt(Cu)/C (curve *a*) and Pt-Ru(Cu)/C (curve *b*) depended on the Cu electrodeposition charge. Note in addition that the ECSAs for the Pt(Cu)/C and Pt-Ru(Cu)/C catalysts prepared with Cu electrodeposition charges in the range of 30–45 mC are nearly coincident. It seems strange because in the Pt-Ru(Cu)/C nanoparticles the Pt surface must be partially covered by Ru species, all of them in oxidized form [32] and not able to adsorb CO [38]. Moreover, the successive cycling of Pt-Ru(Cu)/C in the same electrolyte was repetitive, thus indicating that such oxidized species remained on the Pt surface without being dissolved into the electrolyte [32]. However, they can be at least partially reduced to Ru metal during the potential cycling [36,39] and as long as CO can be adsorbed not only on the Pt sites but also on the Ru metal [10], CO adsorption is not only restricted to surface Pt, but also takes place on the Ru metal surface. The ECSA for CO oxidation on Pt-Ru(Cu)/C may then approach the value obtained for Pt(Cu)/C.

Figure 2b shows similar values for the ESCAs in the range of 30–40 mC but they are significantly smaller for 20, 45 and 50 mC. As a result, and within the experimental error (about 5%), the ECSA vs. q_{Cu} curves could be tentatively adjusted to a parabolic form. To gain a further insight into this behavior, the specimens prepared with q_{Cu} equal to 45 mC were examined by means of the TEM. The corresponding images together with the size distribution and the HRTEM and FFT analyses of Cu/C, Pt(Cu)/C, and Pt-Ru(Cu)/C are shown in Figures 3 and 4. According to Figure 3, the mean sizes of the nanoparticles were 4.3 ± 1.3, 4.9 ± 1.5, and 4.9 ± 1.4 nm, respectively. These values were somewhat higher than 3.9 and 3.6 nm for Cu/C and Pt-Ru(Cu)/C respectively, previously reported for q_{Cu} = 40 mC [18]. The parabolic form of the curves depicted in Figure 2a can then be explained by a size effect. The nanoparticle size increased with q_{Cu} and thus, when referring the ECSA per mol of Cu, the active area for the CO adsorption decreased. For the same reason, at q_{Cu} as low as 20 mC, the Cu nanoparticles were probably too small to allow for building up stable Pt(Cu)/C and Pt-Ru(Cu)/C electrocatalysts. The nanoparticles cannot be completely covered by Pt in these conditions and then, Cu dissolution in the potential cycling cannot be avoided, as discussed above in relation to the curve *c* depicted in Figure 1b.

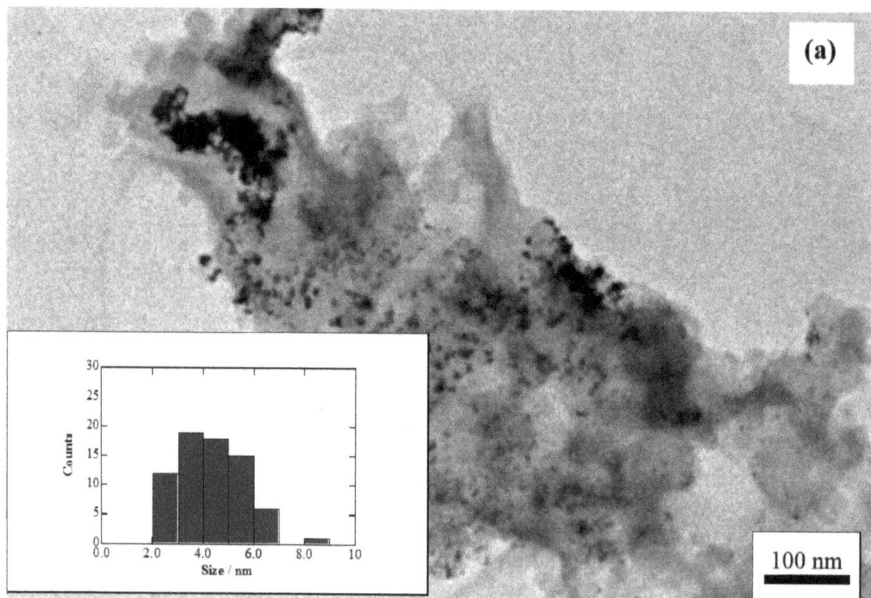

Figure 3. *Cont.*

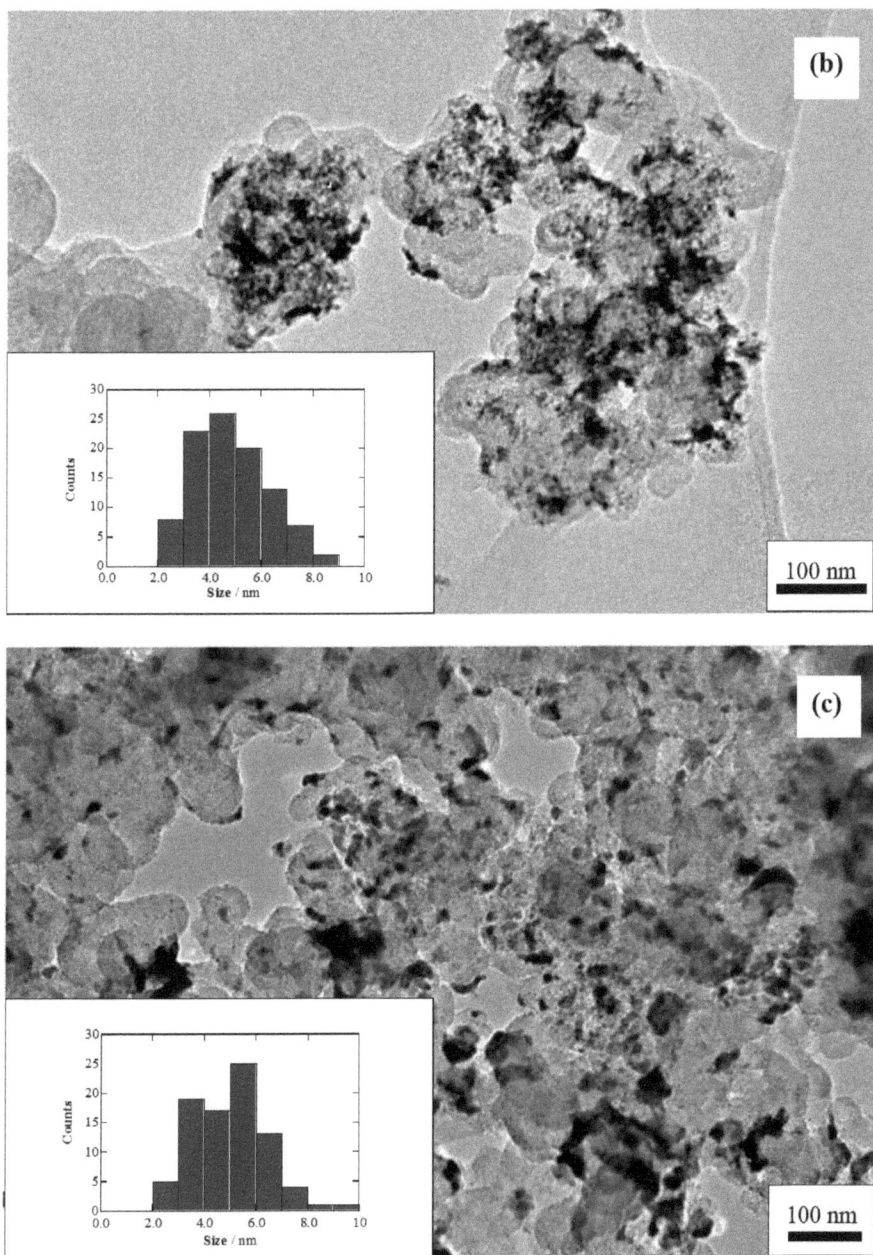

Figure 3. TEM analysis of the nanoparticles supported on Vulcan carbon XC72R prepared from a Cu electrodeposition charge of 45 mC and following the sequence *i*. (a) Cu/C; (b) Pt(Cu)/C; (c) Pt-Ru(Cu)/C. The insets show the size distribution of the nanoparticles.

Another interesting point resulting from Figures 3 and 4 is the composition of the electrocatalysts. The FFT analysis of the Cu nanoparticles exemplified in Figure 4a leads to an interplanar space of 0.215 nm, which clearly matches (relative error of about 2%) with the value of 0.219 nm corresponding to the Cu(111) crystallographic planes [40].

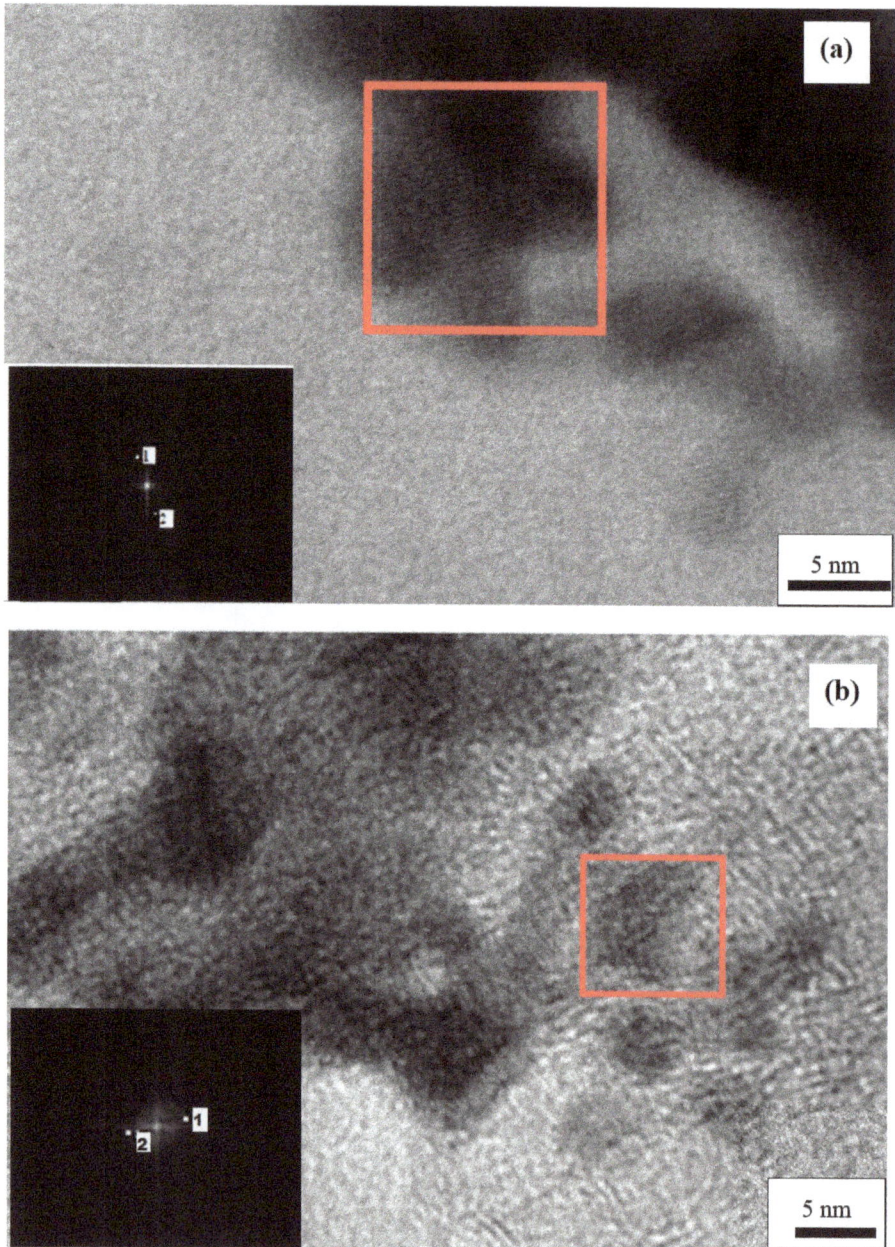

Figure 4. *Cont.*

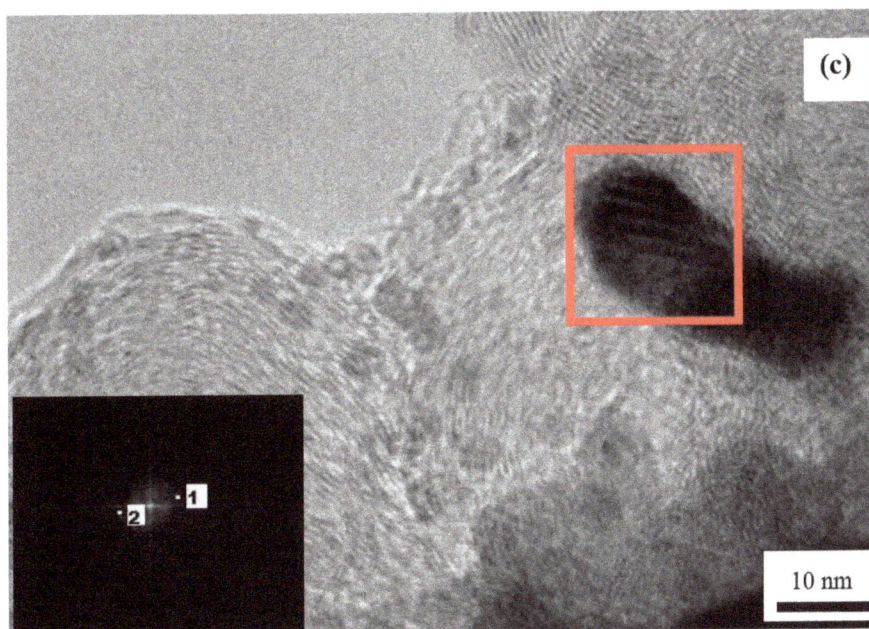

Figure 4. High Resolution Transmission Electron Microscopy (HRTEM) images of the: (**a**) Cu/C; (**b**) Pt(Cu)/C; (**c**) Pt-Ru(Cu)/C electrocatalysts of Figure 3. The insets depict the Fast Fourier Transform (FFT) analyses of the marked zones.

The interplanar spacings resulting from the FFT analysis of the marked region in Figure 4b,c are of 0.2244 and 0.2216 nm for Pt(cu) and Pt-Ru(Cu)/C, respectively, which can all be assigned to the Pt(111) planes [40] (relative errors of about 1 and 2%, also respectively). This indicates that Pt has essentially the same lattice structure as pure Pt, thus suggesting that the performance improvement in the CO and methanol oxidation discussed above is mainly due to the electronic effect of Cu on Pt. However, this is compatible with the generation of active Pt sites of different nature due to the Cu core effect. The corresponding EDS analyses of Pt(Cu)/C and Pt-Ru(Cu)/C gave Pt:Cu atomic ratios of about 1.4, that is somewhat smaller than the value of 1.6 obtained for the specimens prepared with q_{Cu} equal to 40 mC [18]. The smaller relative Pt content of the former can also be explained by the different nanoparticle size, because in the core-shell structure a higher relative amount of Cu is expected to remain under the Pt shell when increasing the nanoparticle size. These EDS results together with the absence of Cu oxidation peaks in the cyclic voltammograms of Figure 1a,b (curves *a* and *b*) clearly indicate the formation of the core-shell structures.

Figure 2 also allows concluding about the best preparation conditions of the Pt-Ru(Cu)/C catalysts for CO and methanol electrooxidation. These should be at least those leading to the most negative onset potentials for CO and methanol oxidation. According to this figure, this approximately corresponds to the most negative peak potential of CO stripping and also to the maximum ECSA for CO oxidation. The coincidence of the catalyst preparation conditions to obtain the lowest onset potentials for methanol oxidation together with the maximum ECSA values for CO oxidation is in agreement with the participation of CO as intermediate in the electrooxidation of methanol. Parallel results were also found for Pt-Ru/C catalysts obtained by spontaneous deposition of Ru species on commercial Pt/C [36], in which case the most negative potential for the CO stripping peak corresponded to the greatest ECSA for CO oxidation together with the highest currents for the methanol oxidation. According to the curves *b–d* in Figure 2a and curve *b* in Figure 2b, the most suitable Cu electrodeposition charge is estimated to be of 38 ± 2 mC, which corresponds to a Cu load of $5.4 (\pm 0.3) \times 10^2$ mC cm^{-2}. It has been determined in these conditions that 0.38 mol Pt is deposited per mol of electrodeposited Cu [18] and, therefore, the corresponding specific Pt load would be of 0.21 mg$_{Pt}$ cm^{-2}, which falls in the range of suitable Pt loads for catalytic purposes.

2.2. Pt Deposition Time by Galvanic Exchange in the Synthesis Sequence ii

After identifying the suitable Cu deposition charge, this was applied to obtain the Pt(Cu) core-shell structure with different times of galvanic exchange between Cu and Pt, following the synthesis sequence *ii*. Figure 5 depicts the CO stripping voltammograms for Pt(Cu)/C and Pt-Ru(Cu)/C for different times of galvanic exchange, showing the same essential features as those of Figure 2a, that is, CO stripping peaks at about 0.72 and 0.55 V for the former and for the latter respectively, in agreement with the better CO tolerance of Pt-Ru(Cu)/C. These stripping peak potentials are depicted in Figure 6, curves *a* and *c*, for both electrocatalysts at different times of galvanic exchange. As can be seen, these values did not significantly depend on the immersion time in the Pt(IV) solution. Figure 5 highlights that the highest stripping currents were obtained for 30 min. In fact, they increased from 10 to 30 min and then they remained almost constant or even decreased slightly. The same change takes place with the corresponding anodic charges. Note that the stripping voltammograms recorded for 30 (curves *b* and *e*) and 60 min (curves *c* and *f*) were very similar. As long as the charges of these anodic peaks approached to the ECSAs of CO oxidation, one can conclude that 30 min appeared to be the most adequate time for the Pt exchange with Cu. Increasing the immersion time in the Pt(IV) solution did not significantly affect the catalyst performance, most probably because no more Cu can be exchanged and the nanoparticles remain the same.

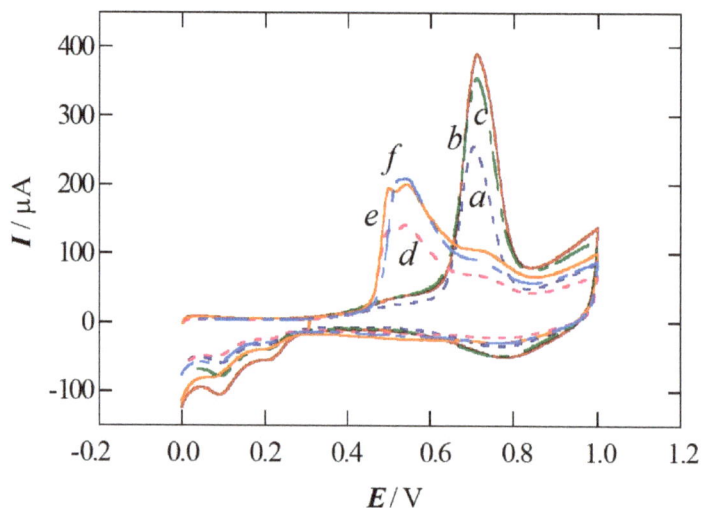

Figure 5. CO stripping curves for the different electrocatalysts recorded after different times of Pt exchange following the synthesis sequence *ii*: 20 (curves *a* and *d*), 30 (curves *b* and *e*) and 60 min (curves *c* and *f*). Curves *a*, *b* and *c* correspond to Pt(Cu)/C and curves *d*, *e* and *f*, to Pt-Ru(Cu)/C. Sweep rate 20 mV s^{-1}.

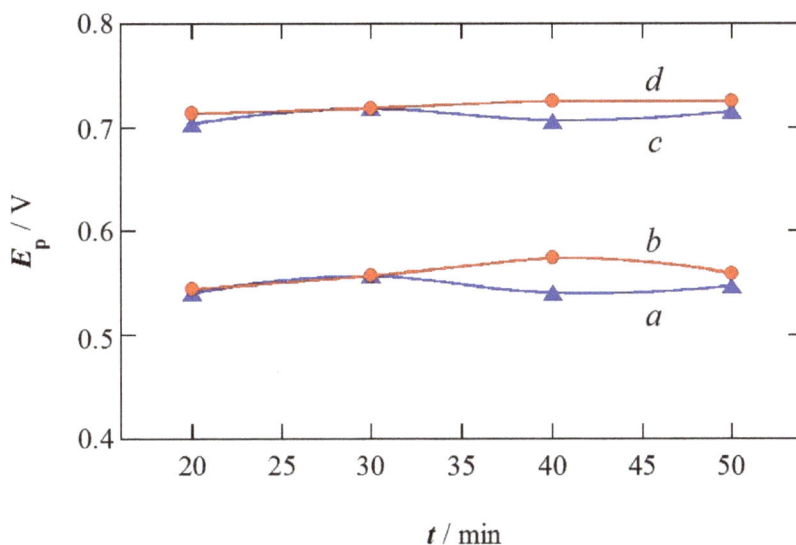

Figure 6. CO stripping peak potentials as a function of the immersion time in the Pt(IV) solution following the sequence *ii* (spontaneous deposition of the Ru species for 30 min, curves *a* and *c*) and in the Ru(III) solution following the sequence *iii* (galvanic exchange with Pt(IV) species for 30 min, curves *b* and *d*). Curves *a* and *b* refer to Pt-Ru(Cu)/C and curves *c* and *d*, to Pt(Cu)/C. In all cases, q_{Cu} was of 38 mC.

2.3. Spontaneous Deposition Time of Ru Species on Pt in the Synthesis Sequence iii

In this case, after 38 mC of Cu electrodeposition and 30 min of galvanic exchange with Pt, the spontaneous deposition of Ru species were allowed for different times in the quiescent solution. Representative CO stripping curves of the resulting Pt-Ru(Cu)/C catalysts are depicted in Figure 5. The CO stripping peak potentials for the Pt(Cu)/C and the Pt-Ru(Cu)/C catalysts as a function of the immersion time in the Ru(III) solution are represented in curves *b* and *d* of Figure 6, respectively. No important changes in these parameters were found when compared to the different immersion times in the Pt(IV) solution.

Figure 7 shows that the currents increased when passing from 10 to 20 min of spontaneous deposition of the Ru species and then, the stripping profile presented small changes. In fact, the ECSAs for CO oxidation are 1.11×10^3 m^2 mol$_{Cu}^{-1}$ for 10 min and increases to 1.70×10^3 m^2 mol$_{Cu}^{-1}$ for 20 min, being 1.71×10^3 m^2 mol$_{Cu}^{-1}$ for 50 min (see Figure 8a). In addition, the onset potentials for CO and methanol oxidation were minimal and about 0.41 and 0.32 V respectively, for about 22 ± 2 min of spontaneous deposition of the Ru species (see Figure 8b). At the same time, the coverage by the Ru species increased from about 0.2 after 10 min of spontaneous deposition to about 0.5 after 40 min. This variation of the onset potential for methanol oxidation on the Ru coverage is probably due to the deposition of the Ru species on active sites of the Pt shell structure. Again, suitable coverage of Ru species of about 0.3 were found for 20–30 min of spontaneous deposition of Ru(III) species, as discussed above.

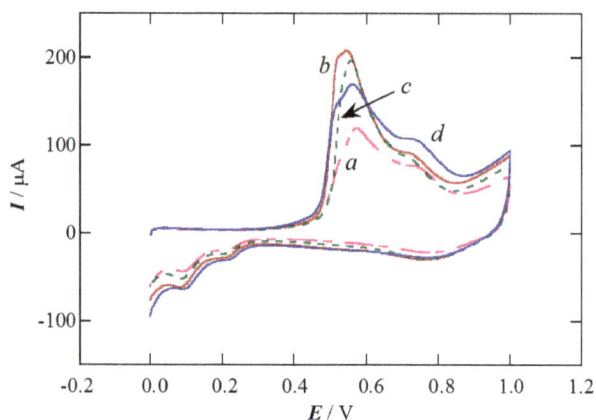

Figure 7. CO stripping curves for the different Pt-Ru(Cu)/C electrocatalysts obtained after different times of spontaneous deposition of Ru species following the synthesis sequence *iii*: 10 (curve *a*), 20 (*b*), 30 (*c*) and 50 (*d*) min. Sweep rate 20 mV s^{-1}.

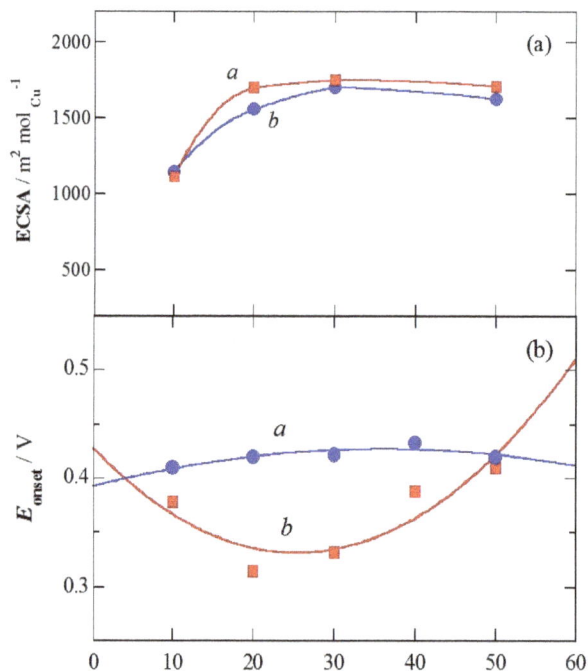

Figure 8. (a) Electrochemical surface areas for CO oxidation per mol of electrodeposited Cu on the Pt-Ru(Cu)/C (curve *a*) and the Pt(Cu)/C (curve *b*) electrocatalysts; (b) onset potentials for CO (curve *a*) and methanol oxidation (curve *b*) on the Pt-Ru(Cu)/C electrocatalyst in front of the spontaneous deposition time of Ru species from the quiescent solution.

Considering that 0.38 mol Pt are produced per mol of electrodeposited Cu [18], the voltammograms of the methanol oxidation were recorded and are shown in Figure 9, represented as specific currents per unit mass of Pt (mass activities). Note that the current density vs. potential curves referred to the electrode section (i.e. prior to the normalization by the Pt mass), presented the same sequence as that shown in this figure because the nanoparticles contained the same amount of Pt, only changing the amount of spontaneously deposited Ru species. It can be mentioned however, that the current density at 0.7 V corresponding to the curve *a* in Figure 9 was of 76 mA cm^{-2}. In agreement with the onset potentials for methanol oxidation, the best performance can be observed for 22 min of spontaneous deposition of the Ru(III) species. From these findings, the sequence *iii* with q_{Cu} of 38 mC with a rotation rate of 100 rpm, a galvanic exchange with the Pt(IV) solution for 30 min also at 100 rpm and finally a spontaneous deposition time of 22 min in the quiescent Ru(III) solution were identified to be the best conditions for the electrochemical preparation of the present core-shell electrocatalysts.

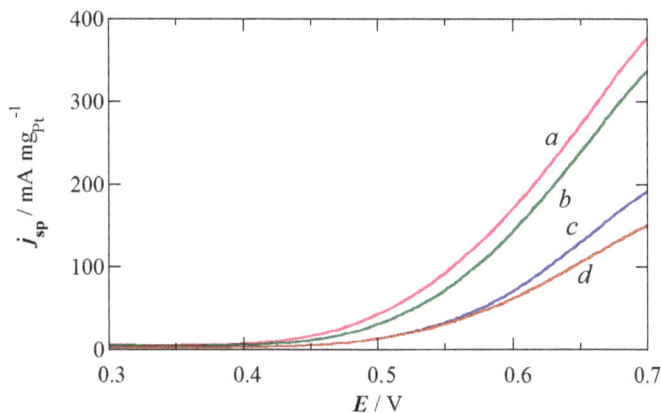

Figure 9. Linear sweep voltammograms corresponding to the methanol oxidation in 1.0 M CH$_3$OH + 0.5 M H$_2$SO$_4$ on different Pt-Ru(Cu)/C electrocatalysts synthesized from spontaneous deposition times of 22 (curve *a*), 30 (curve *b*), 10 (curve *c*) and 40 min (curve *d*), following the sequence *iii*. Sweep rate 20 mV s^{-1}.

Figure 9 also shows that the best mass activity achieved at 0.7 V was 375 mA mg$_{Pt}^{-1}$, corresponding to a spontaneous deposition time of Ru species of 22 min. The direct comparison with previous results in the literature is not possible because there are too many variables to fix. However, we have selected some values from literature in order to have a rough approach. Thus, for 2.0 M CH$_3$OH + 0.1 M H$_2$SO$_4$ on Pt(Cu)/C core-shell electrocatalysts, the best mass activity at 0.7 V and 50 mV s^{-1} was about 250 mA mg$_{Pt}^{-1}$, for a Pt:Cu molar ratio of 0.17:0.83 [24]. Note that the linear sweep voltammograms in Figure 9 have been obtained at 20 mV s^{-1} and 1.0 M CH$_3$OH. The mass activity reported for methanol oxidation at the same potential for Pt-Cu core-shell alloy with 9.5 wt % Pt at 5 mV s^{-1} in 0.5 M CH$_3$OH + 1 M HClO$_4$ was about 140 mA mg$_{Pt}^{-1}$ [28]. Therefore, the results reported in curve *a* of Figure 9 can be considered good. Other works report on different fuels. The best mass activity for ethanol oxidation at the same potential for Pt-Cu core-shell catalysts with similar content in Pt and Cu at 20 mV s^{-1} in 0.17 M CH$_3$CH$_2$OH + 0.5 M H$_2$SO$_4$ was about 4 mA mg$_{Pt}^{-1}$ [27]. In the case of formic acid oxidation, the mass activity also at 0.7 V for highly dispersed Pt-Cu/C catalysts via surface substitution and etching separation at 50 mV s^{-1} in 0.25 M HCOOH + 0.5 M H$_2$SO$_4$ was about 250 mA mg$_{Pt}^{-1}$ [30]. In a different synthesis process, novel excavated rhombic dodecahedral PtCu$_3$ nanocrystals with (110) facets prepared by a wet chemical route, gave about 600 mA mg$_{Pt}^{-1}$ in the same conditions of potential, sweep rate and electrolyte than the latter [25]. In all these cases, the mass activity in front of the fuel oxidation was always much better than that for commercial Pt/C and also with much better stability under cycling than the latter, assigned to the superior catalytic activity and selectivity of the Pt-Cu alloy in decreasing the generation of CO [30]. The ligand effect of Cu has also been argued to explain the superior activity of Pt-Cu for oxygen reduction [26]. In the present paper, the positive results can be assigned not only to the electronic effect of Cu but also to the positive effect of the deposited Ru species, which are easily hydroxylated and allow enhancing the CO oxidation due to the bifunctional mechanism [7,35,36,41]:

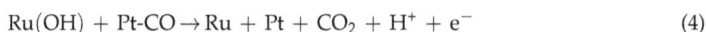

$$RuO_xH_y(OH) + Pt\text{-}CO \rightarrow RuO_xH_y + Pt + CO_2 + H^+ + e^- \tag{3}$$

$$Ru(OH) + Pt\text{-}CO \rightarrow Ru + Pt + CO_2 + H^+ + e^- \tag{4}$$

in which $RuO_xH_y(OH)$ and $Ru(OH)$ result from the hydroxylation of RuO_xH_y and Ru respectively.

3. Experimental Section

3.1. Materials and Reagents

The supporting carbon was Vulcan XC72R produced by Cabot Corporation, Boston, MA, USA (mean particle size and specific surface area of about 30 nm and 250 m^2 g^{-1} respectively [42]). It was deposited onto a Metrohm glassy carbon (GC) tip of 3 mm in diameter (section of 0.071 cm^2), previously polished with Micropolish II deagglomerated α-alumina (0.3 μm) and γ-alumina (0.05 μm) on a PSA-backed White Felt cloth from Buehler, Coventry, UK. The solutions were prepared using Millipore Milli Q high-purity water from Merck KGaA, Darmstadt, Germany (resistivity >18 MΩ cm at 25 °C), analytical grade 96 wt % H_2SO_4 from Acros Organics, $HClO_4$, hydrated $RuCl_3$, and H_2PtCl_6 from Merck, and $CuSO_4 \cdot 5H_2O$, and Na_2SO_4 from Panreac. N_2 and CO gases were Linde 3.0 (purity $\geq$ 99.9%).

3.2. Working Electrodes

The preparation and testing of the electrodes were performed by means of a PGSTAT100 potentiostat-galvanostat commanded by the NOVA 1.10 software both from Metrohm Autolab B.V., Utrecht, The Netherlands. The electrochemical cell was Metrohm 200 mL-volume with a double wall, which was connected to a MP-5 thermostat from JULABO GmbH, Seelbach, Germany to maintain the temperature to 25.0 $\pm$ 0.1 °C. A double junction Ag | AgCl | KCl(sat) (0.199 V vs. SHE at 25 °C) and a Pt rod were used as the reference and the auxiliary electrodes respectively. However, all the potentials given in this paper have been referred to the RHE. The electrocatalysts were prepared on the carbon support deposited on the GC tip, which was coupled to a RDE from Metrohm Autolab B.V., Utrecht, The Netherlands. N_2 was bubbled through the electrolyte before the deposition of Cu, Pt and the Ru species and also before the CV experiments. N_2 was passed over the electrolyte during the deposition processes and the electrochemical testing.

The Pt(Cu)/C and Pt-Ru(Cu)/C working electrodes were prepared as follows, based on the electrodeposition method previously described [18]. First, 20 μL of the carbon suspension (4 mg in 4 mL of water, sonicated for at least 45 min) were deposited onto the polished GC tip (0.28 mg_C cm^{-2}) and dried under the heat of a lamp. Afterwards, it was cleaned on the RDE in deaerated 0.5 M H_2SO_4 by CV scans within the limits of 0.0 and 1.0 V at 100, 50 and 20 mVs^{-1} for 10, 5 and 3 cycles, respectively (cleaning protocol). The electrodeposition of the catalysts consisted in the following consecutive steps: (a) potentiostatic Cu electrodeposition at −0.1 V and 100 rpm in 1.0 mM $CuSO_4$ + 0.1 M Na_2SO_4 + 0.01 M H_2SO_4 (Cu/C electrode); (b) Pt deposition on the Cu nuclei by galvanic exchange in 1 mM H_2PtCl_6 + 0.1 M $HClO_4$ at 100 rpm (Pt(Cu)/C); (c) spontaneous deposition of Ru species on the Pt(Cu)/C electrode in aged (for at least one week) and quiescent 8.0 mM $RuCl_3$ + 0.1 M $HClO_4$ (Pt-Ru(Cu)/C). The Cu electrodeposition efficiency was determined through the Cu oxidation charge in the same solution after sweeping the potential from 0.0 to 1.0 V at 10 mV s^{-1}. The variables studied were:

(*i*) Cu electrodeposition between 20 and 50 mC followed by the galvanic exchange with Pt for 30 min and the spontaneous deposition of Ru species for 30 min.

(*ii*) Galvanic exchange with Pt from 10 to 60 min after the best Cu electrodeposition charge (obtained from sequence *i*), followed by the spontaneous deposition of the Ru species for 30 min.

(*iii*) Spontaneous deposition of the Ru species between 10 and 60 min after the best Cu electrodeposition charge (from sequence *i*) and the best time of galvanic exchange with Pt (from sequence *ii*).

Immediately after preparation, the Cu/C electrode was carefully cleaned in water, whereas the Pt(Cu)/C and Pt-Ru(Cu)/C ones were submitted to the cleaning protocol described above. The CV profiles obtained from this protocol were always practically stationary after the second scan, confirming their stability and cleanliness.

3.3. Electrochemical Testing

The CO tolerance of the electrocatalysts was studied by CV in 0.5 M H_2SO_4. CO gas was first bubbled through the solution for 15 min while setting the electrode potential at 0.1 V. Afterwards, the dissolved CO was removed by N_2 bubbling through the solution for 30 min and then, the adsorbed CO was stripped by sweeping the potential between 0.0 and 1.0 V at 20 mV s^{-1} without stirring. The ECSA for the CO oxidation was estimated taking into account that the oxidation of a CO monolayer on polycrystalline Pt needs 420 μC cm^{-2} [4,9,43]. After CO stripping, the activity of the Pt(Cu)/C and Pt-Ru(Cu)/C catalysts was recovered as demonstrated by the consecutive cyclic voltammograms, which retraced those obtained before the CO adsorption.

The methanol oxidation performance for the different Pt-Ru(Cu)/C electrocatalysts was characterized by CV in a previously deaereated 1.0 M CH_3OH + 0.5 M H_2SO_4 solution between 0.0 and 0.7 V at 20 mVs^{-1}. The CV experiments before and after the methanol oxidation analyses in the deaerated 0.5 M H_2SO_4 confirmed that there was not loss of catalyst loading.

3.4. Structural Characterization

The morphological and structural characterization of the catalysts was performed by means of the TEM and HRTEM techniques using a 200 kV JEM 2100 LaB$_6$ transmission electron microscope from JEOL, Peabody, MA, USA, furnished with EDS facilities. For the microscopic examination, the electrocatalysts prepared on the GC tip were dispersed in 3 mL of *n*-hexane for 10 min by ultrasonication. A drop of the suspension was then placed on a holley-carbon Ni grid. The electrocatalyst was ready for examination after the solvent evaporation under the heat of a 40 W lamp for 5 min. The TEM and HRTEM images were taken using an Orius MultiScan 794 charge-coupled device (CCD) camera from Gatan, Pleasanton, CA, USA. More than one hundred nanoparticles were counted to depict their size distribution. The interplanar spacing of the nanoparticles was determined from the digital treatment of the HRTEM pictures by means of the Digital Micrograph software, version 3.7.0, from Gatan, Pleasanton, CA, USA. First, the Fourier diffractogram was obtained by the Fast Fourier Transform of the HRTEM images of selected areas performed with this software. Then, the reciprocal value of the interplanar spacing was given by the distance between each identified spot and the center of the diffractogram. The interplanar spacing thus calculated was contrasted with the results listed in the MinCryst Database of the Institute of Experimental Mineralogy, Chernogolovka, Moscow region, Russia, revision of August 2008 [40].

4. Conclusions

This work studied the performance of carbon-supported core-shell Pt(Cu)/C and Pt-Ru(Cu)/C electrocatalysts, obtained by Cu electrodeposition, galvanic exchange with Pt(IV) and spontaneous deposition of Ru species. The Cu electrodeposition potential was -0.1 V vs. RHE and the variables explored were Cu electrodeposition charge, time of galvanic exchange with Pt(IV) and time of spontaneous deposition of Ru species. The ECSA for CO oxidation was determined per mol of electrodeposited Cu in order to identify the preparation conditions to obtain the most suitable core-shell nanoparticles. It was found that these normalized ECSA values were practically the same for Cu electrodeposition charges q_{Cu} in the range of 30–40 mC. However, they decreased for q_{Cu} equal to or greater than 45 mC. The HRTEM analyses indicated that the nanoparticle size increased for q_{Cu} exceeding 40 mC, thus justifying the concomitant decrease in the ECSA. Electrodeposition charges about 20 mC were insufficient to create suitable core-shell nanoparticles probably because a significant number of Cu atoms were partially occluded in the Pt shells and in the carbon. In this case, the steric hindrance impeded the Pt complex approach. However, the further Cu oxidation in these points could be easier. Based on the peak potentials for CO oxidation and the onset potentials for CO and methanol oxidation together with ECSA values, the suitable Cu electrodeposition charge to obtain the Pt-Ru(Cu)/C electrocatalysts was 38 ± 2 mC. The most suitable time for galvanic exchange of Cu by Pt(IV) was 30 min, the Pt shells completely covering the Cu cores. Under these conditions, maximum ECSA values for CO oxidation, normalized per mol of electrodeposited Cu on the Pt-Ru(Cu)/C electrocatalyst were achieved. A spontaneous deposition time of the Ru species on Pt(Cu)/C of 22 ± 2 min led to the smallest onset potentials for CO and methanol oxidation, which were about 0.05 V smaller than those determined for Ru-decorated commercial Pt/C catalysts. These conditions yielded the highest normalized ECSA values for CO oxidation and the best specific anodic current for methanol oxidation. For these best conditions, surface coverage of Pt by the Ru species in the Pt-Ru(Cu)/C electrocatalyst were determined and their normalized ECSA values for CO oxidation were comparable to those of Pt(Cu)/C, thus suggesting that the deposited Ru species were partially reducible to Ru metal during the cyclic scans. The positive effect of Cu was related to the electronic effect of the Cu core on the Pt shells and also to the generation of new active sites for CO oxidation. The best mass activities obtained for the methanol oxidation were compared to those previously reported in literature for the same fuel and also for ethanol and formic acid oxidation on Pt(Cu)/C catalysts prepared by different procedures. The present results strongly indicate an additional positive effect of the Ru spontaneously deposited species due to the bifunctional mechanism for CO oxidation.

Acknowledgments: The authors thank the financial support of Secretaria Nacional de Ciencia, Tecnología e Innovación (SENACYT, Panama) fellowship received by Griselda Caballero-Manrique and the Erasmus Undergraduate Research Placement achieved by Immad Muhammed Nadeem. The authors also thank the CCiT-UB (Scientific and Technological Centers of the Universitat de Barcelona) for the microscopic analysis facilities.

Author Contributions: G.C.-M. and P.-L.C. conceived and designed the experiments; G.C.-M. and I.M.N. performed the experiments; G.C.-M., E.B., J.A.G. and P.-L.C. analyzed the data; F.C. and R.M.R. contributed reagents/materials/analysis tools; G.C.-M., P.-L.C and E.B. wrote the paper.

Conflicts of Interest: The authors declare no conflict of interest. The founding sponsors had no role in the design of the study; in the collection, analyses, or interpretation of data; in the writing of the manuscript, and in the decision to publish the results.

References

1. Costamagna, P.; Srinivasan, S. Quantum jumps in the PEMFC science and technology from the 1960s to the year 2000: Part 1. Fundamental scientific aspects. *J. Power Sources* **2001**, *102*, 242–252. [CrossRef]
2. Liu, H.; Song, Ch.; Zhang, L.; Zhang, J.; Wang, H.; Wilkinson, D. A review of anode catalysis in the direct methanol fuel cell. *J. Power Sources* **2006**, *155*, 95–110. [CrossRef]
3. Zainoodin, A.M.; Kamarudin, S.K.; Daub, W.R.W. Electrode in direct methanol fuel cells. Review. *Int. J. Hydrogen Energy* **2010**, *35*, 4606–4621. [CrossRef]
4. Dos Santos, L.; Colmati, F.; González, E.R. Preparation and characterization of supported Pt-Ru catalysts with a high Ru content. *J. Power Sources* **2006**, *159*, 869–877. [CrossRef]
5. Antolini, E. Platinum-based ternary catalysts for low temperature fuel cells: Part I. Preparation methods and structural characteristics. *Appl. Catal. B Environ.* **2007**, *74*, 324–336. [CrossRef]
6. Iwasita, T. Methanol and CO electrooxidation. In *Handbook of Fuel Cells-Fundamentals, Technology and Applications*; Vielstich, W., Gasteiger, H.A., Lamm, A., Eds.; John Wiley & Sons: New York, NY, USA, 2003; Volume 3, pp. 603–622.
7. Ruth, K.; Vogt, M.; Zuber, R. Development of CO-tolerant catalysts. In *Handbook of Fuel Cells-Fundamentals, Technology and Applications*; Vielstich, W., Gasteiger, H.A., Lamm, A., Eds.; John Wiley & Sons: New York, NY, USA, 2003; Volume 3, pp. 489–496.

8. Lamy, C.; Lima, A.; LeRhun, V.; Delime, F.; Coutanceau, C.; Léger, J.M. Recent advances in the development of direct alcohol fuel cells (DAFC). *J. Power Sources* **2002**, *105*, 283–296. [CrossRef]
9. Gasteiger, H.A.; Markovic, N.M.; Ross, N.P., Jr.; Cairns, E.J. CO electrooxidation on well-characterized Pt-Ru alloys. *J. Phys. Chem.* **1994**, *98*, 617–625. [CrossRef]
10. Gasteiger, H.A.; Markovic, N.M.; Ross, P.N., Jr. H_2 and CO electrooxidation on well-characterized Pt, Ru and Pt-Ru Rotating disk electrode studies of the pure gases including temperature effects. *J. Phys. Chem.* **1995**, *99*, 8290–8298. [CrossRef]
11. Aricó, A.S.; Baglio, V.; Di Blasi, A.; Modica, E.; Antonucci, P.L.; Antonucci, V. Analysis of the high-temperature methanol oxidation behavior at carbon-supported Pt-Ru catalysts. *J. Electroanal. Chem.* **2003**, *557*, 167–176. [CrossRef]
12. Tegou, A.; Papadimitriou, S.; Pavlidou, E.; Kokkinidis, G.; Sotiropoulos, S. Oxygen reduction at platinum- and gold-coated copper deposits on glassy carbon substrates. *J. Electroanal. Chem.* **2007**, *608*, 67–77. [CrossRef]
13. Papadimitriou, S.; Tegou, A.; Pavlidou, E.; Armyanov, S.; Valova, E.; Kokkinidis, G.; Sotiropoulos, S. Preparation and characterization of platinum- and gold-coated copper, iron, cobalt and nickel deposits on glassy carbon substrates. *Electrochim. Acta* **2008**, *53*, 6559–6567. [CrossRef]
14. Podlovchenko, B.I.; Gladysheva, T.D.; Filatov, Y.; Yashina, L.V. The use of galvanic displacement in synthesizing Pt(Cu) catalysts with the core-sell structure. *Russ. J. Electrochem.* **2010**, *46*, 1189–1197. [CrossRef]
15. Kuznetsov, V.V.; Podlovchenko, B.I.; Kavyrshina, K.V.; Maksimov, Y.M. Oxidation of methanol on Pt(Mo) electrodes obtained using galvanic displacement method. *Russ. J. Electrochem.* **2010**, *46*, 1353–1359. [CrossRef]
16. Podlovchenko, B.I.; Krivchenko, V.A.; Maksimov, Y.M.; Gladysheva, T.D.; Yashina, L.V.; Evlashin, S.A.; Pilevsky, A.A. Specific features of the formation on Pt(Cu) catalysts by galvanic displacement with carbon nanowalls used as support. *Electrochim. Acta* **2012**, *76*, 137–144. [CrossRef]
17. Tegou, A.; Armyanov, S.; Valova, E.; Steenhaut, O.; Hubin, A.; Kokkinidis, G.; Sotiropoulus, S. Mixed platinum-gold electrocatalysts for borohydride oxidation prepared by the galvanic replacement of nickel deposits. *J. Electroanal. Chem.* **2009**, *634*, 104–110. [CrossRef]
18. Caballero-Manrique, G.; Velázquez-Palenzuela, A.; Centellas, F.; Garrido, J.A.; Arias, C.; Rodríguez, R.M.; Brillas, E.; Cabot, P.L. Electrochemical synthesis and characterization of carbon-supported Pt and Pt-Ru nanoparticles with Cu cores for CO and methanol oxidation in polymer electrolyte fuel cells. *Int. J. Hydrogen Energy* **2014**, *39*, 12859–12869. [CrossRef]
19. Mohl, M.; Kumar, A.; Reddy, A.L.M.; Kukovecz, A.; Konya, Z.; Kiricsi, I.; Vajtai, R.; Ajayan, P.M. Synthesis of catalytic porous metallic nanorods by galvanic exchange reaction. *J. Phys. Chem. C* **2010**, *114*, 389–393. [CrossRef]
20. Mohl, M.; Dobo, D.; Kukovecz, A.; Konya, Z.; Kordas, K.; Wei, J.; Vajtai, R.; Ajayan, P.M. Formation of CuPd and CuPt bimetallic nanotubes by galvanic replacement reaction. *J. Phys. Chem. C* **2011**, *115*, 9403–9409. [CrossRef]
21. Cheng, B.; Cheng, D.; Zhu, J. Synthesis of PtCu nanowires in nonoaqueous solvent with enhanced activity and stability for oxygen reduction reaction. *J. Power Sources* **2014**, *267*, 380–387. [CrossRef]
22. Kokkinidis, G.; Papoutsis, A.; Stoychev, D.; Milchev, A. Electroless deposition of Pt on Ti—Catalytic activity for the hydrogen evolution reaction. *J. Electroanal. Chem.* **2000**, *486*, 48–55. [CrossRef]
23. Kokkinidis, G.; Lazarov, V.; Papoutsis, A.; Stoychev, D.; Milchev, A. Electroless deposition of Pt on Ti: Part II. Catalytic activity for oxygen reduction. *J. Electroanal. Chem.* **2001**, *511*, 20–30. [CrossRef]
24. Liu, Y.; Huang, Y.; Xie, Y.; Yang, Z.; Huang, H.; Zhou, Q. Preparation of highly dispersed CuPt nanoparticles on ionic-liquid-assisted grapheme sheets for direct methanol fuel cell. *Chem. Eng. J.* **2012**, *197*, 80–87. [CrossRef]
25. Jia, Y.; Jiang, Y.; Zhang, J.; Zhang, L.; Chen, Q.; Xie, Z.; Zheng, L. Unique excavated rhombic dodecahedral $PtCu_3$ alloy nanocrystals constructed with ultrathin nanosheets of high-energy (110) facets. *J. Am. Chem. Soc.* **2014**, *136*, 3748–3751. [CrossRef] [PubMed]
26. Geboes, B.; Mintsouli, I.; Wouters, B.; Georgieva, J.; Kakaroglou, A.; Sotiropoulos, S.; Valova, E.; Armyanov, S.; Hubin, A.; Breugelmans, T. Surface and electrochemical characterization of a Pt-Cu/C nano-structured electrocatalysts, prepared by galvanic displacement. *Appl. Catal. B: Environ.* **2014**, *150–151*, 249–256. [CrossRef]
27. Ammam, M.; Easton, E.B. PtCu/C and Pt(Cu)/C catalysts: Synthesis, characterization and catalytic activity towards ethanol electrooxidation. *J. Power Sources* **2013**, *222*, 79–87. [CrossRef]

28. Mintsouli, I.; Georgieva, J.; Armyanov, S.; Valova, E.; Avdeev, G.; Hubin, A.; Steenhaut, O.; Dille, J.; Tsiplakides, D.; Balomenou, S.; et al. Pt-Cu electrocatalysts for methanol oxidation prepared by partial galvanic replacement of Cu/carbon powder precursors. *Appl. Catal. B: Environ.* **2013**, *136–137*, 160–167. [CrossRef]

29. Qiu, H.J.; Xu, H.T.; Li, X.; Wang, J.Q.; Wang, Y. Core-shell-structured nanoporous PtCu with high C content and enhanced catalytic performance. *J. Mater. Chem. A* **2015**, *3*, 7939–7944. [CrossRef]

30. Huang, Y.; Zhao, T.; Zeng, L.; Tan, P.; Xu, J. A facile approach for preparation of highly dispersed platinum-copper/carbon nanocatalyst toward formic acid electro-oxidation. *Electrochim. Acta* **2016**, *190*, 956–963. [CrossRef]

31. Solla-Gullón, J.; Vidal-Iglesias, F.J.; Herrero, E.; Feliu, J.M.; Aldaz, A. CO monolayer oxidation on semi-spherical and preferentially oriented (100) and (111) platinum nanoparticles. *Electrochem. Commun.* **2006**, *8*, 189–194. [CrossRef]

32. Velázquez-Palenzuela, A.; Brillas, E.; Arias, C.; Centellas, F.; Garrido, J.A.; Rodríguez, R.M.; Cabot, P.L. Structural characterization of Ru-modified carbon-supported Pt nanoparticles using spontaneous deposition with CO oxidation activity. *J. Phys. Chem. C* **2012**, *116*, 18469–18478.

33. MacDonald, J.P.; Gualtieri, B.; Runga, N.; Teliz, E.; Zinola, C.F. Modification of platinum surfaces by spontanoeous deposition: methanol oxidation electrocatalysis. *Int. J. Hydrogen Energy* **2008**, *33*, 7048–7061. [CrossRef]

34. Gasteiger, H.A.; Markovic, N.; Ross, P.N., Jr.; Cairns, E.J. Temperature-dependent methanol electro-oxidation on well-characterized Pt-Ru alloys. *J. Electrochem. Soc.* **1994**, *141*, 1795–1803. [CrossRef]

35. Sugimoto, W.; Yokoshima, K.; Murakami, Y.; Takasu, Y. Charge storage mechanism of nanostructured anhydrous and hydrous ruthenium-based oxides. *Electrochim. Acta* **2006**, *52*, 1742–1748. [CrossRef]

36. Velázquez-Palenzuela, A.; Brillas, E.; Arias, C.; Centellas, F.; Garrido, J.A.; Rodríguez, R.M.; Cabot, P.L. Carbon monoxide, methanol and ethanol electro-oxidation on Ru decorated carbon-supported Pt nanoparticles prepared by spontaneous deposition. *J. Power Sources* **2013**, *225*, 163–171. [CrossRef]

37. Schmidt, T.J.; Gasteiger, H.A.; Stäb, G.D.; Urban, P.M.; Kolb, D.M.; Behm, R.J. Characterization of high-surface-area electrocatalysts using a rotating disk electrode configuration. *J. Electrochem. Soc.* **1998**, *145*, 2354–2358. [CrossRef]

38. Gómez de la Fuente, J.L.; Martínez-Huerta, M.V.; Rojas, S.; Hernández-Fernández, P.; Terreros, P.; Fierrro, J.L.G.; Peña, M.A. Tailoring and structure of PtRu nanoparticles supported on functionalized carbon for DMFC applications: New evidence of the hydrous ruthenium oxide phase. *Appl. Catal. B Environ.* **2009**, *88*, 505–514. [CrossRef]

39. Chrzanowski, W.; Wieckowski, A. Ultrathin films of ruthenium on low index platinum single crystal surfaces: An electrochemical study. *Langmuir* **1997**, *13*, 5974–5978. [CrossRef]

40. WWW-MINCRYST. Crystallographic and Crystallochemical Database for Minerals and their Structural Analogues. Institute of Experimental Mineralogy. Russian Academy of Sciences. Available online: http://database.iem.ac.ru/mincryst/index.php (accessed on 7 January 2016).

41. Rolison, D.R.; Hagans, P.L.; Swider, K.E.; Long, J.W. Role of hydrous ruthenium oxide in Pt-Ru direct methanol fuel cell anode electrocatalysts. The importance of mixed electron/proton conductivity. *Langmuir* **1999**, *15*, 774–779. [CrossRef]

42. Cabot Corporation. Specialty Chemicals and Performance Materials. Available online: http://www.cabotcorp.com (accessed on 10 October 2015).

43. Maillard, F.; Savinova, E.; Stimming, U. CO monolayer oxidation on Pt nanoparticles: Further insights into the particle size effects. *J. Electroanal. Chem.* **2007**, *599*, 221–232. [CrossRef]

catalysts

MDPI

Article

One Pot Synthesis of Pt/Graphene Composite Using Polyamidoamine/Chitosan as a Template and Its Electrocatalysis for Methanol Oxidation

Yanli Wang [1], Zhongshui Li [1,2], Shuhong Xu [1], Fengling Lei [1] and Shen Lin [1,2,]*

[1] College of Chemistry & Chemical Engineering, Fujian Normal University, Fuzhou 350007, China; yanliwang408@gmail.com (Y.W.); zsli@fjnu.edu.cn (Z.L.); xushuhongxu@gmail.com (S.X.); fenglinglei408@gmail.com (F.L.)

[2] Fujian Key Laboratory of Polymer Materials, Fujian Normal University, Fuzhou 350007, China

* Correspondence: shenlin@fjnu.edu.cn; Tel./Fax: +86-591-2286-7399

Academic Editors: Vincenzo Baglio and David Sebastián
Received: 17 August 2016; Accepted: 19 October 2016; Published: 24 October 2016

Abstract: A one-pot hydrothermal strategy was used to synthesize Pt/GNs (PAMAM) & Pt/GNs (CS) composites. Pt nanoparticles are deposited onto graphene sheets (GNs) via synchronous reduction of K_2PtCl_4 and graphene oxide (GO) under hydrothermal conditons without additional reducing agent. During the synthesis process, polyamidoamine (PAMAM) or chitosan (CS) was used as a template respectively to obtain shape controlled Pt particles on the surface of GNs, leading to the formation of flower-like Pt nanoclusters for Pt/GNs (PAMAM) and uniform spherical Pt nanoparticles for Pt/GNs (CS). PAMAM and CS are simultaneously served as intrinsic reducing agents to accelerate reduction process; ensuring excellent electrical conductivity of the composites. Electrochemical tests show that Pt/GNs (PAMAM) and Pt/GNs (CS) have much higher electrocatalytic activity and better stability toward methanol oxidation reaction (MOR) in comparison with counterpart Pt/GNs and the commercially available 20% Pt/C catalyst (Pt/C) due to their better dispersion of Pt particles, stronger interaction between Pt and substrate materials, and better electron transfer capability.

Keywords: platinum; polyamidoamine; chitosan; graphene; methanol oxidation

---

1. Introduction

Hydrothermal synthesis method is widely used to synthesize composite materials with excellent properties. Boppella et al. [1] reported the formation of oriented ZnO structures with tunable percentage of exposed polar facets via a simple hydrothermal route in aqueous base environment. Yu et al. [2] demonstrated unique hollow Pt-ZnO nanocomposite microspheres with hierarchical structure under mild solvothermal conditions. Especially, this method has an advantage for the synthesis of graphene-based nanomaterials due to the possible synchronous reduction of graphene oxide and the corresponding metal precursors [3,4]. Lei et al. [5] have reported one-pot hydrothermal synthesis of an efficient anodic electro-photo catalyst $Pt/SnO_2/GNs$ (EDTA) for direct methanol fuel cell (DMFCs) applications. Li et al. [6] developed a facile hydrothermal approach to efficiently synthesize Pt-NCs/rGO composites with high shape selectivity and enhanced catalytic activity for MOR. Yun et al. [7] reported the in situ hydrothermal synthesis of 3D macroporous rGO aerogels/palladium nanoparticle hybrids for electrocatalytic applications. In these reports, they emphasized that the shapes, assembly or dispersion of as-prepared nanoparticles by hydrothermal synthesis method can significantly impact their performance. However, hydrothermal reaction is a "covert operation" and unable to intervene in the reaction process. Therefore, in order to effectively control the micro structure of the products, the raw materials must be strictly selected and the reasonable chemical reactions process needs to be predetermined.

Polyamidoamine (PAMAM) is one of the most popular dendrimers with a near-spherical structure [8,9]. It contains rich cavities within molecules and has a large number of active functional groups on the surface of molecules, which makes it able to graft other ions, molecules or particles on the core, peripheral or embedded cavity by electrostatic interaction and covalent coordination [10,11]. Maiyalagan, T. [9] has anchored PAMAM on functionalized carbon nanofibers (CNF) to facilitate the controlled dispersion of Pt–Ru nanoparticles which are used to catalyze methanol oxidation. Chitosan (CS) is an abundant linear polymer, in which there exist a large number of hydroxyls and aminos in polymer chain segments [12–14]. Especially, its –NH$_2$ groups can be protonated in acidic medium, causing them to have a positive charge. The positive protonated amines and the rich hydroxyl in chitosan can act as the mooring point and steric hindrance agent of metal nanoparticles [15,16]. Li et al. [17] have assembled worm-like Pt nanoparticles on PW$_{12}$-GNs support using CS as a shape directing agent for catalyzing methanol oxidation. Thus, it can be reasonable to deduce that the unique structures of dendritic and linear macromolecule above can be used as the agent for controlling Pt nanoparticles shape and dispersion under hydrothermal conditions.

In this case, a one-pot hydrothermal method was used to synthesize Pt/graphene composites Pt/GNs (PAMAM) and Pt/GNs (CS) without additional reducing agent. The fourth generation amine-terminated polyamidoamine dendrimers (PAMAM) and chitosan (CS) not only act as templates to control the morphology of Pt particles but also serve as internal reductants to accelerate reduction reaction, which effectively makes up the "covert operation" shortcomings of hydrothermal reaction to present a new paradigm for the synthesis of graphene based nano-materials with desired morphology. As-synthesized composites show remarkably higher electrocatalytic activity and enhanced resistance to CO poisoning compared with counterpart Pt/GNs and commercially available 20% Pt/C catalyst (Pt/C). Especially, as-synthesized composites show better stability toward methanol oxidation reaction (MOR), which is helpful to overcome the fast decay in catalytic reaction process for Pt-based catalysts.

2. Results and Discussion

2.1. Characterization

X-ray photoelectron spectroscopy (XPS) of Pt/GNs, Pt/GNs (CS) and Pt/GNs (PAMAM) are shown in Figure 1. The C1s spectra show that carbon elements in complex exist in three kinds of chemical bonds which are mainly in the form of C–C, C–O/C–N and –CONH [18]. It is calculated that the percentages of the above three kinds of carbon functional groups are 43.24%, 15.82%, 40.94% for Pt/GNs (PAMAM) and 43.21%, 18.81%, 37.98% for Pt/GNs (CS), respectively. Correspondingly, they are 60.32%, 17.39%, 22.29% for Pt/GNs, respectively. Evidently, the C–O/C–N components in the three complexes are drastically decreased in comparison with pristine GO [19], which indicates that GO was converted into graphene sheets (GNs) at high temperature and pressure [20–22]. Furthermore, the proportion of –CONH bond components for Pt/GNs (PAMAM) and Pt/GNs (CS) are 40.94% and 37.98%, respectively, both are much higher than that for Pt/GNs, which may be due to that some of PAMAM and CS chains have been bonded to the graphene surface during the GO reduction process [21]. The Pt 4f spectrum of Pt/GNs is displayed in Figure 1D. Based on curve fitting with a mixed Gaussian-Lorentzian line shape, Pt seems to exist in various states. The most intense doublet (71.19 eV and 74.66 eV) is ascribed to metallic Pt (0) [23]. The doublet (71.92 eV and 75.51 eV) can be assigned to Pt (II) chemical state (PtO and Pt(OH)$_2$) [24]. Correspondingly, Pt 4f of Pt/GNs (CS) exhibits the clear peaks which can be deconvoluted to reveal Pt (0) species at 71.39 eV and 74.85 eV (Figure 1E). The peaks assigned to Pt (II) species are also observed at 72.17 eV & 76.14 eV. Similarly, Pt (0) species is the main species for Pt in Pt/GNs (CS). As shown in Figure 1F, Pt 4f spectra of Pt/GNs (PAMAM) show the presence of metallic Pt (0) (71.38 eV, 74.84 eV), PtO/Pt(OH)$_2$ (72.4 eV, 77.71 eV), respectively. It is calculated that the percentages of Pt (0) for Pt/GNs, Pt/GNs (CS), and Pt/GNs (PAMAM) are 73.49%, 68.36%, 62.21%, respectively, and the percentages of Pt (II) for Pt/GNs, Pt/GNs (CS), and Pt/GNs (PAMAM) are 26.51%, 31.64%, 37.79%, respectively. It is reported that the existence of appropriate amount of oxide species may promote methanol oxidation reaction in acid medium [25,26]. Furthermore, it can be seen that the binding energies of Pt 4f7/2 and Pt 4f5/2 for Pt (0) in Pt/GNs (PAMAM) and Pt/GNs (CS) shift to a higher value in comparison with that in Pt/GNs, indicating that Pt/GNs (PAMAM) and Pt/GNs (CS) have stronger interaction between Pt and GNs (support) due to the introduction of PAMAM and CS [27].

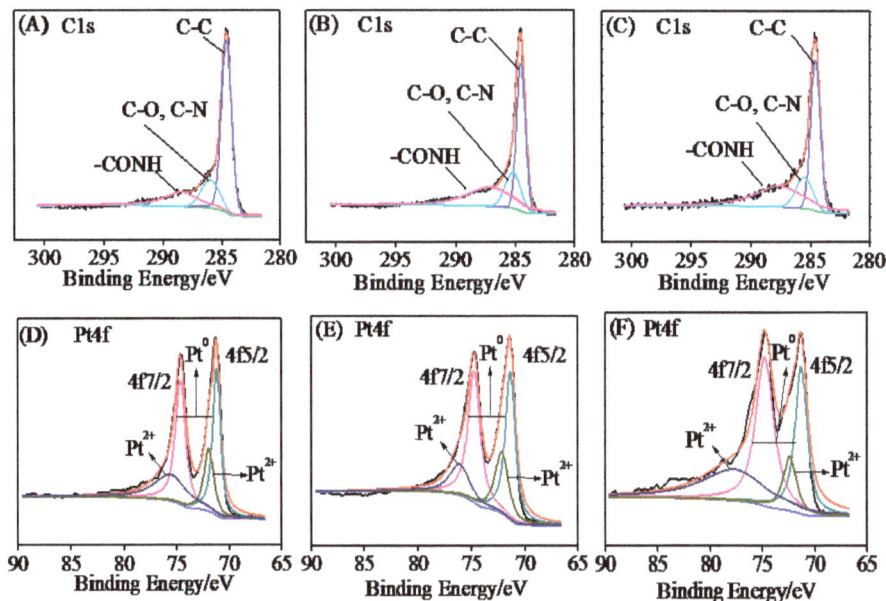

Figure 1. C1s XPS spectra of Pt/GNs (**A**); Pt/GNs (CS) (**B**); Pt/GNs (PAMAM-polyamidoamine) (**C**); Pt 4f XPS spectra of Pt/GNs (**D**); Pt/GNs (CS) (**E**); Pt/GNs (PAMAM) (**F**).

Transmission electron microscope (TEM) and high-resolution transmission electron microscope (HRTEM) images of Pt/GNs (PAMAM), Pt/GNs (CS), and Pt/GNs are shown in Figure 2. It is found that the Pt nanoparticles in Pt/GNs (PAMAM) were mainly in the form of flower-like clusters consisting of primary nanoparticles (Figure 2A,B), and the size of the primary particles were uniform, with the average diameter of about 5 nm as seen in the inset of Figure 2A. ThenHRTEM image in Figure 2C presents the details fringes of Pt for Pt/GNs (PAMAM). The lattice fringes with a d-spacing of 0.226 nm is attributed to the spacing of the (111) planes in face-centered cubic (fcc) Pt [28,29], which further suggests the zero-valent state of Pt and high crystallinity of Pt nanoparticles. Figure 2D,E show the Pt nanoparticles of Pt/GNs (CS) are highly dispersed on the surface of GNs, mainly in the form of spherical particles. As depicted in the inset of Figure 2D, the particle size of Pt nanoparticles was very uniform, with the average diameter of about 3 nm. It should be noted that uniform and small spherical Pt nanoparticles have a significant advantage of large surface areas. Similarly, the HRTEM image in Figure 2F presents the details fringes of Pt for Pt/GNs (CS), and the lattice fringes with a d-spacing of 0.225 nm correspond to the (111) plane of fcc Pt. In contrast, Pt particles in Pt/GNs (Figure 2G–H) are big and non-uniform, with the average diameter about 10–34 nm (the inset of Figure 2G). Furthermore, the lattice spacing for dark Pt particles is 0.227 nm (Figure 2I). Obviously, the dispersity of Pt nanoparticles in Pt/GNs (PAMAM) and Pt/GNs (CS) is drastically improved compared with Pt/GNs, and their morphologies can be effectively controlled by PAMAM and CS.

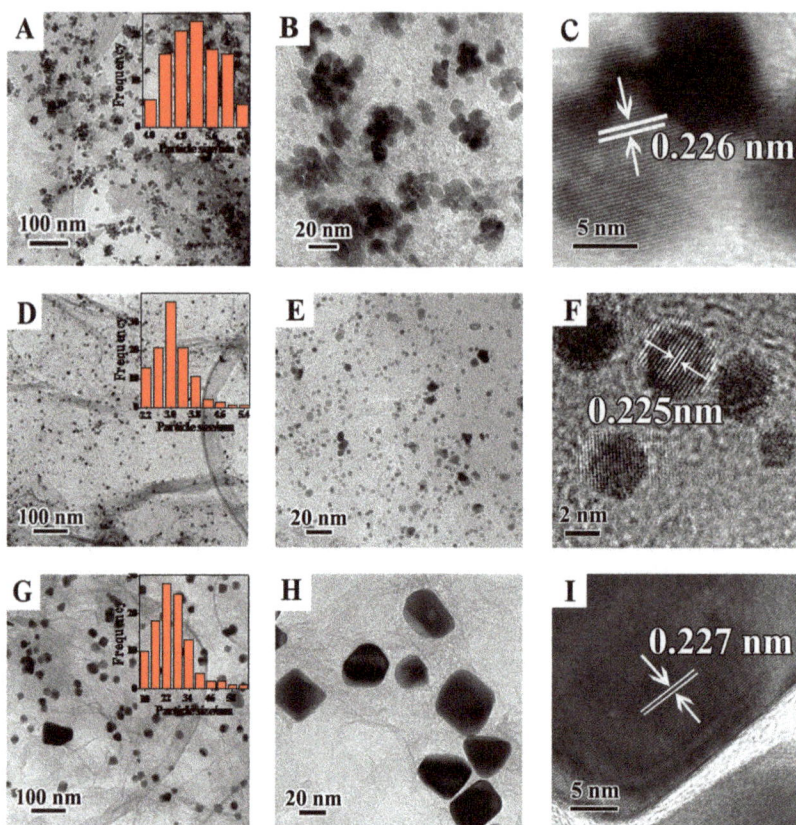

Figure 2. TEM images of Pt/GNs (PAMAM) (**A,B**); Pt/GNs (CS) (**D,E**); Pt/GNs (**G,H**). Size-distribution of Pt particles in Pt/GNs (PAMAM) (the inset of (**A**)); Pt/GNs (CS) (the inset of (**D**)); Pt/GNs (the inset of (**G**)). HRTEM images of Pt particles in Pt/GNs (PAMAM) (**C**); Pt/GNs (CS) (**F**); Pt/GNs (**I**).

The shape of Pt nanoparticles on Pt/GNs (PAMAM) is different from that of the Pt nanoparticles on Pt/GNs (CS). As for Pt/GNs (PAMAM), the primary Pt nanoparticles are formed with uniform size due to the mooring role of the protonated amines in PAMAM. Meanwhile, with the template effect of the dendrimer stucture, these primary nanoparticles are connected to each other to form flower-like nanoclusters. Similarly, CS also can be protonated under acidic conditions to greatly enhance the solubility of CS and promote polymer chain segment stretching in aqueous solution [16], which brings about the electrostatic interaction between $PtCl_4^{2-}$ and protonated $-NH^{3+}$ in the polymer chain. The interaction contributes to the rapid nucleation of Pt particles after the reduction of $PtCl_4^{2-}$. The rich hydroxyls in CS act as a steric hindrance agent to protect Pt particles from further aggregation [30], resulting in the formation of uniform Pt nanoparticles. Saidi and Fichthorn have pointed out that both thermodynamics and kinetics likely played a role in the formation of these nanostructures [31,32]. On one hand, PAMAM (or CS) can induce kinetic Pt particles shapes by regulating the relative Pt fluxes to desired facets (Pt (111) in this work) [32]. On the other hand, the freshly formed primary Pt particles are thermodynamically unstable because of their high surface energy, and they tend to aggregate driven by the minimization of interfacial energy [33,34]. With the template of the dendrimer structure, Pt primary particles on Pt/GNs (PAMAM) are inclined to aggregate into nanoclusters; on the contrary, the rich hydroxyls in CS block Pt primary particles surface diffusion, leading to uniform spherical particles. Additionally, it is reported that organic amines can serve as a reducing agent to reduce GO into GNs [35,36]. Therefore, PAMAM and CS should promote to reduce GO and $PtCl_4^{2-}$ into GNs and Pt under hydrothermal conditions, accelerating the nucleation rate of Pt particles and favoring the formation of Pt nanoparticles of smaller size. Thus, PAMAM and CS not only act as a template to control the morphology of Pt nanoparticles but also serve as an intrinsic reducing agent to increase the dispersity of Pt nanoparticles.

Powder X-ray diffraction (XRD) patterns of Pt/GNs, Pt/GNs (CS), and Pt/GNs (PAMAM) are shown in Figure 3A. The diffraction peaks at 40.0°, 46.5°, 67.8°, 81.6°, and 86.2° (2θ value) can be assigned to (111), (200), (220), (311), and (222) crystalline planes of Pt (0) fcc lattice, respectively [37], which further indicates good crystallinity and zero-valent state of Pt in the obtained three composites. The diffraction peak for Pt (220) is used to estimate the size of Pt nanoparticles with the Scherrer Equation (1) [30,37].

$$d = K\lambda/\beta\cos\theta \tag{1}$$

It is calculated that the average size of Pt nanoparticles in Pt/GN, Pt/GNs (CS) and Pt/GNs (PAMAM) is about 21.5, 4.0 and 5.5 nm, respectively, which are consistent with the TEM analysis.

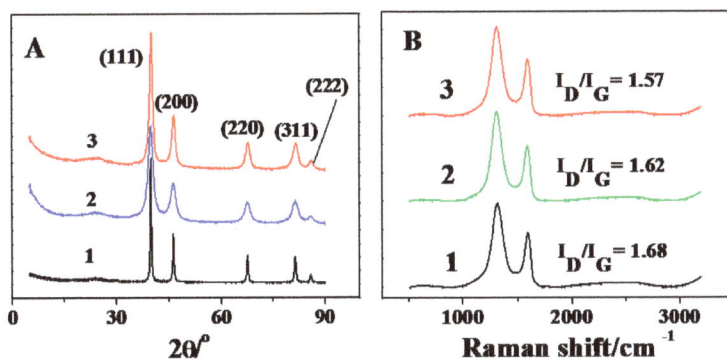

Figure 3. XRD patterns (**A**) and Raman spectra (**B**) of Pt/GNs (1), Pt/GNs (CS) (2), and Pt/GNs(PAMAM) (3).

Figure 3B displays Raman spectra of Pt/GNs, Pt/GNs (CS), and Pt/GNs (PAMAM). The D band (about 1310 cm^{-1}) originates from the defects in the curved graphene sheet and staging disorder, while the G band (about 1597 cm^{-1}) was associated with the graphitic hexagonpinch mode [38,39]. The I_D/I_G intensity ratio can be used to measure the crystalline quality of graphite or graphene via different kinds of treatment, increasing with the amount of disorder for grapheme-based materials [40–43]. It is found that the values of I_D/I_G for Pt/GNs (PAMAM) and Pt/GNs (CS) are 1.57 and 1.62, both of which are lower than the value of I_D/I_G for the Pt/GNs value (1.68), indicating that the disorder degree of GNs is decreased in Pt/GNs (PAMAM) and Pt/GNs (CS). A possible reason for this is that PAMAM and CS promote GO to be reduced into GNs, in favor of the recovery of the original structure for GNs. GNs with lower degrees of disorder can maintain their good electron transfer capability and endow graphene-based composites with better catalytic performance [42,44].

2.2. Electrocatalysis

Figure 4 displays the cyclic voltammetry (CV) curves of the different composites conducted at room temperature in 0.5 M H_2SO_4 solution at 100 mV·s^{-1}. For all composites, typical hydrogen adsorption/desorption peaks and Pt oxide formation/reduction peaks can be observed. Their electrochemical surface area (ECSA) is evaluated by the integrated charge (Q_H) in the hydrogen adsorption region, with Equation (2) [42].

$$\text{ECSA} = Q_\text{H}/(210 \text{ uC}\cdot\text{cm}^{-2} \times \text{Pt loading}) \tag{2}$$

By calculation, it is found that the specific ECSA of Pt/GNs (PAMAM) and Pt/GNs (CS) is 117.8 $\text{m}^2\cdot\text{g}^{-1}$ and 103.7 $\text{m}^2\cdot\text{g}^{-1}$, respectively, which is higher than that of Pt/GNs (47.7 $\text{m}^2\cdot\text{g}^{-1}$) and commercial catalyst Pt/C (75.1 $\text{m}^2\cdot\text{g}^{-1}$) (Table 1). The higher ECSA of Pt/GNs (PAMAM) and Pt/GNs (CS) are obviously related to the unique morphology and uniform distribution of Pt nanoparticles, which is helpful to improve their catalytic performance [37].

Table 1. Electrochemical parameters of as-synthesized different composites [a].

Composites	ECSA/$\text{m}^2\cdot\text{g}^{-1}$ (Electrochemical Surface Area)	Onset Potential [b]/V	Massactivity/$\text{mA}\cdot\text{mg}^{-1}$
Pt/C-JM	75.1	0.45	455
Pt/GNs	47.7	0.40	556
Pt/GNs (CS)	117.8	0.36	1031
Pt/GNs (PAMAM-polyamidoamine)	103.7	0.35	1203

[a] The reproducibility of electrochemical experiment data in our work was ensured by repeated experiments, and their standard deviation is less than 3%; [b] The onset potential is defined as the potential at which 10% of the current value at the peak potential was reached in this work [34].

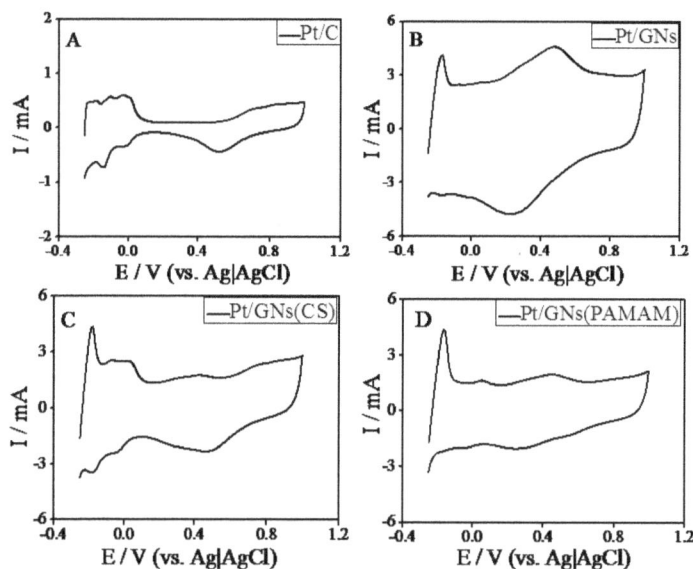

Figure 4. CV curves in 0.5 M H_2SO_4 at 100 $\text{mV}\cdot\text{s}^{-1}$: Pt/C (**A**); Pt/GNs (**B**); Pt/GNs (CS) (**C**); Pt/GNs (PAMAM) (**D**).

The electrocatalytic activities of as-synthesized composites for MOR were discussed by analyzing their CV curves carried out in 0.5 M H_2SO_4 containing 1 M CH_3OH solution at a scan rate of 100 $\text{mV}\cdot\text{s}^{-1}$ (Figure 5A,B). It can be seen from Figure 5B that the Pt/GNs (CS) and Pt/GNs (PAMAM) exhibit a mass activity (forward peak current density) of 1031 $\text{mA}\cdot\text{mg}^{-1}$ and 1203 $\text{mA}\cdot\text{mg}^{-1}$ (listed in Table 1), which is 2.3 and 2.6 times higher than that of Pt/C (455 $\text{mA}\cdot\text{mg}^{-1}$), respectively. It is observed that the onset potential value for Pt/GNs (CS) and Pt/GNs (PAMAM) is 0.36 and 0.35 V, respectively, which is lower than that for Pt/GNs (0.40 V) and Pt/C (0.45 V) (Table 1). Lower onset potential may contribute to superior electro-catalytic activity for methanol oxidation [37,45]. It is reported that the functional-modified graphene is in favor of inhibiting the irreversible aggregation process of Pt/graphene composites [46]. In this case, the introduction CS and PAMAM into the graphene layer may play a functional-modified role to prevent the aggregation of graphene composites, resulting in remarkably improved electrocatalytic activity.

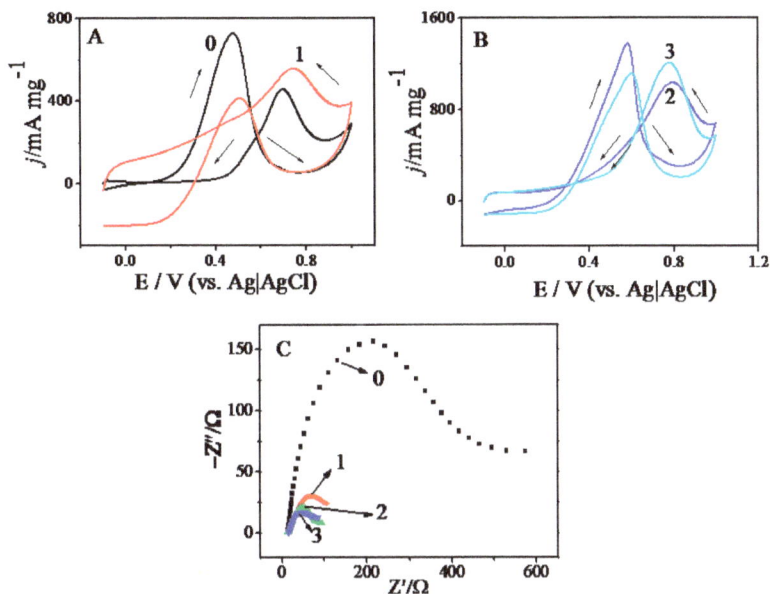

Figure 5. (**A**) CV curves in 1 M CH_3OH + 0.5 M H_2SO_4 at 100 mV·s^{-1}: Pt/C (**0**), Pt/GNs (**1**); (**B**) CV curves in 1 M CH_3OH + 0.5 M H_2SO_4 at 100 mV·s^{-1}: Pt/GNs (CS) (**2**), Pt/GNs (PAMAM) (**3**); (**C**) Nyquist plots of EIS for methanol electrooxidation in 1 M CH_3OH + 0.5 M H_2SO_4 solution: Pt/C (**0**), Pt/GNs (**1**), Pt/GNs (CS) (**2**), Pt/GNs (PAMAM) (**3**).

In addition, the enhanced catalytic activity may be ascribed to the superior electric conductivities of the as-synthesized composites, which can be proven by EIS analysis. The Nyquist plots of EIS for Pt/GNs (PAMAM), Pt/GNs (CS), Pt/GNs and Pt/C in 1 M CH_3OH + 0.5 M H_2SO_4 solution at 0.5 V are shown in Figure 5C. The diameter of semicircle at high frequencies is a measure of charge transfer resistance related to the charge transfer reaction kinetics [47–49]. Obviously, the diameters of semicircle for Pt/GNs (PAMAM) and Pt/GNs (CS) are much lower than those of Pt/GNs and Pt/C, which demonstrates that Pt/GNs (PAMAM) and Pt/GNs (CS) have lower charge transfer resistance and faster reaction rate for MOR [50]. This suggests that the introduction of PAMAM and CS can keep the good electron transfer capability of graphene-based composites to endow them with better catalytic performance.

In order to directly observe their anti-poisoning ability to CO-like intermediate species, CO stripping curves of Pt/GNs (PAMAM), Pt/GNs (CS), Pt/GNs and Pt/C were measured by oxidation of pre-adsorbed and saturated CO in the 0.5 M H_2SO_4 solution at 100 mV·s^{-1} (Figure 6A). The peak potential of CO oxidation for Pt/GNs (PAMAM) (curve 3) and Pt/GNs (CS) (curve 2) is 0.59 and 0.64 V, respectively, which is lower than that for Pt/GNs (0.67 V) and Pt/C (0.72 V). Their more negative CO oxidation peak demonstrates that CO on Pt/GNs (PAMAM) and Pt/GNs (CS) surface is easier to oxidize and remove from the Pt surface [27], suggesting an enhanced poisoning tolerance. The possible reasons for the greater tolerance to CO-like species are as follows: firstly, there are abundant cavities in the core of PAMAM and a large number of nitrogen-containing groups on the surface of PAMAM [51], and CS also possesses abundant $-NH_2$ and $-OH$ groups on its polymer chains [52], which results in a complex interaction between PAMAM (CS) mooring groups and Pt particles [30,53]. The strong interaction can induce modulation in the electronic structure of Pt particles and decrease the Pt–CO binding energy to reduce the CO adsorption on Pt active sites [9,12]. This point can be proved by Pt 4f XPS spectra analysis. As shown in Figure 1D, the binding energy of Pt 4f for Pt/GNs (PAMAM) and Pt/GNs (CS) presents a positive shift due to the introduction of PAMAM and CS, which means that the interaction between Pt and support materials is much stronger than that of their corresponding counterpart Pt/GNs [27,29]. A higher binding energy would increase metal bond strength and reduce the potential of metal to form strong bonds with absorbed reactants, in favor of removing the strongly adsorbed CO-like species during MOR [29]. Secondly, the protonated NH_3^+ groups or $-OH$ groups can enhance the hydrophilic properties of RGO to promote water activation [16], and as a result, the adsorbed $OH-$ species at the surface of Pt particles promote the oxidation of CO [30].

Chronoamperometry tests were carried out at 0.70 V for 3600 s to assess the electrocatalytic stability of different composites. As shown in Figure 6B, their current densities decay quickly during the initial minutes, which may be due to CO-like intermediate species poisoning on the Pt surface during the early stage of MOR [54]. Following this, the current densities decreased slowly and reach a quasi-stationary value after 3600 s. It was found that Pt/GNs (PAMAM) (curve 3) and Pt/GNs (CS) (curve 2) present the lower declining rate and the higher quasi-stationary current density in contrast with Pt/GNs (curve 1) and Pt/C (curve 0), suggesting an enhanced catalytic stability. The results are in agreement with CO stripping curves analysis. The superior catalytic activity and stability of Pt/GNs (PAMAM) and Pt/GNs (CS) may be related to the fact that the abundant protonated NH_3^+ groups or $-OH$ groups can effectively stabilize Pt particles against gathering, endowing Pt active sites with a stronger ability to refresh [16,30,51,53].

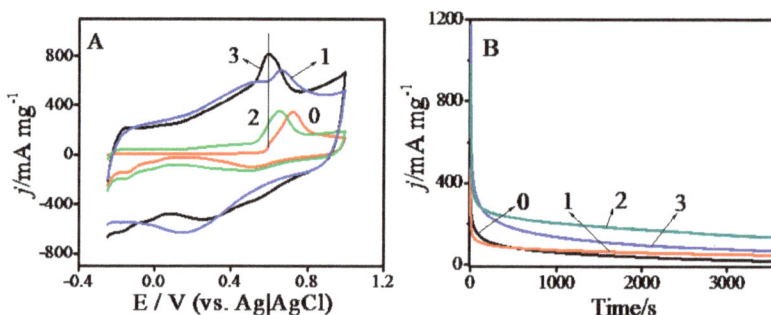

Figure 6. (**A**) CO-stripping voltammograms in 0.5 M H_2SO_4: Pt/C (**0**), Pt/GNs (**1**), Pt/GNs (CS) (**2**), Pt/GNs (PAMAM) (**3**); (**B**) Chronoamperometric curves in 1 M CH_3OH + 0.5 M H_2SO_4: Pt/C (**0**), Pt/GNs (**1**), Pt/GNs (CS) (**2**), Pt/GNs (PAMAM) (**3**).

3. Materials and Methods

3.1. Materials

Natural graphite powder (about 325 mesh) was purchased from Alfa Aesar (Ward Hill, MA, USA). Commercial Pt/C catalyst (Hispec 3000, wt % 20%) was purchased from Johnson-Matthey (London, UK). Fourth generation amine-terminated polyamidoamine dendrimers (PAMAM) with the highest available purity (10 wt % in methanol) were purchased from Sigma Aldrich (Darmstadt, Germany). Chitosan (CS) was purchased from Sinopharm Chemical Reagent Co. Ltd. (Shanghai, China), and then completely deacetylated as follows [55]: 80% deacetylation CS was added to 50% NaOH solution and stirred for 6 h at 95 °C under N_2 protection. After that, the product was purified and lyophilized. Potassium chloride (99%), hydrogen peroxide (30%), potassium permanganate, potassium sulfate, concentrated sulfuric acid, phosphorus pentoxide, methanol, and ethanol were purchased from Sinopharm Chemical Reagent Co. Ltd. (Shanghai, China), and all the chemicals were of analytical grade. The water used in the experiments was distilled water.

3.2. Synthesis of Pt/GNs (PAMAM) and Pt/GNs (CS)

Graphene oxide (GO) was synthesized using a modified Hummers' method [56,57], and the detailed steps were described previously [5]. The overall synthetic route of Pt/GNs (PAMAM) and Pt/GNs (CS) was illustrated in Scheme 1. In a typical procedure of Pt/GNs (PAMAM), 50 mg of GO were dispersed in 25 mL of distilled water by sonication for 1 h. Simultaneously, after removing methanol from 0.45 mL 10% PAMAM methanol solution, PAMAM was dissolved into 20 mL water and its pH was adjusted to 3 with dilute hydrochloric acid. Then 5 mL of K_2PtCl_4 (1 g/100 mL) were added and stirring for 30 min. After the completion of the stirring, the mixture was put into the GO solution and further sonicated for 1 h to ensure sufficient mixing, and its pH was adjusted to 3 again. Afterwards, the mixed suspension was transferred into a Teflon-lined stainless steel autoclave and reacted at 180 °C for 12 h under autogenous pressure. When hydrothermal reaction was completed, autoclave was allowed to cool naturally to room temperature. Finally, the product was purified through repeated washing and centrifugation (at 10,000 rpm for 20 min), then the black precipitate was lyophilized and Pt/GNs (PAMAM) were obtained. The Pt/GNs (CS) composite was synthesized by the same process, adjusting the pH of mixed solution to 3. For comparison, Pt/GNs was synthesized by similar procedure without addition of PAMAM or CS. Pt actual contents in the composites were determined by ICP-AES (Thermo Scientific, Pleasanton, CA, USA) with the values of 31.42% for Pt/GNs, 32.50% for Pt/GNs (CS) and 35.71% for Pt/GNs (PAMAM), respectively.

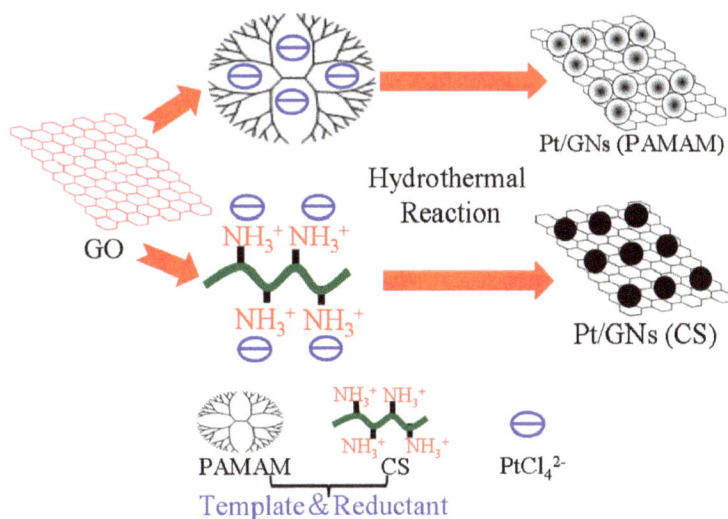

Scheme 1. The synthetic route of Pt/GNs (PAMAM) and Pt/GNs (CS) composites.

3.3. Characterization

The microscopic feature and morphology of the composites were characterized by a high-resolution transmission electron microscope (HRTEM, TECNAI G2, FEI, Hillsboro, OR, USA) operating at 200 K. XPS was recorded on monochromatic Al Ka radiation (1486.6 eV) using a Thermo Scientific VG ESCALAB 250 spectrometer (Thermo Scientific). XRD patterns were determined at a scanning rate of $5° \cdot min^{-1}$ on a Philips X' pert Pro diffractometer (PANalytical B.V., Holland, The Netherlands), using Cu Ka radiation. Raman spectra were measured through a Renishaw-invia Raman micro-spectrometer equipped with a 514 nm diode laser excitation on a 300 lines·mm^{-1} grating. The amount of actual Pt loading was determined using inductively coupled plasma atomic emission spectroscopy (ICP-AES, ICAP6300).

3.4. Electrochemical Measurements

All electrochemical tests were performed with a standard three-electrode system on a CHI 660C electrochemical workstation (ChenHua, Shanghai, China) at room temperature. The three electrodes cell included an Ag/AgCl (saturated KCl) electrode as a reference electrode, a platinum column as a counter-electrode and a modified glassy carbon electrode as a working electrode. Its working electrode was modified as follows: 5 mg of catalyst were ultrasonically dispersed into 1 mL ethanol to form homogeneous ink, then 5 µL of ink were dropped onto the surface of pre-polished glass carbon electrode (3 mm in diameter). Subsequently, 7.5 µL diluted 0.5% Nafion solution was dropped to fix the samples. Cyclic voltammetry (CV) was tested in 0.5 M H_2SO_4 or 0.5 M H_2SO_4 + 1.0 M CH_3OH solution at room temperature with a scan rate of 100 mV·s^{-1}. The CO stripping voltammograms were recorded by oxidation of preadsorbed CO (COad) in 0.5 M H_2SO_4 solution at 100 mV·s^{-1}. CO gas was purged into 0.5 M H_2SO_4 solution at a constant potential of 0.1 V for 1800 s to ensure the complete adsorption of CO onto the samples. The excess CO in the electrolyte was driven out by purging N_2 for 15 min. Chronoamperometry was conducted at 0.70 V in a solution of 0.5 M H_2SO_4 + 1.0 M CH_3OH for a period of 3600 s. Electrochemical impedance spectra (EIS) were performed in a solution containing 0.5 M H_2SO_4 and 1.0 M CH_3OH at 25 °C. Its perturbation potential was 5 mV, and the frequency ranged from 0.01 Hz to 100 kHz.

4. Conclusions

In summary, a one-pot hydrothermal method was used to synthesize Pt/GNs (PAMAM) and Pt/GNs (CS) composites. Under hydrothermal conditions, PAMAM and CS serve as templates to control the morphology of Pt particles, resulting in the formation of flower-like Pt nanoclusters for Pt/GNs (PAMAM) and uniform spherical Pt nanoparticles for Pt/GNs (CS). Meanwhile, PAMAM and CS play a promotion role in the reduction process to accelerate the nucleation rate of Pt particles and lead to improved recovery of the sp^2 bond for GNs. The controlled morphology of Pt nanoparticles on Pt/GNs (PAMAM) and Pt/GNs (CS) is an important driving force to improve the performance of the catalysts. The introduction of PAMAM and CS results in stronger interaction between Pt and support materials and better electron transfer capability of grapheme-based composites, which synergistically contributes to the significantly improved catalytic activity (with values of 1031 mA·mg^{-1} and 1203 mA·mg^{-1} for Pt/GNs (CS) and Pt/GNs (PAMAM), respectively), stability and CO poisoning tolerance. The study above provides a green, simple and low-cost way to improve electrocatalytic performance and enhance the effective utilization of Pt catalysts in DMFC anodic reaction.

Acknowledgments: This work was financially supported by the National Natural Science Foundation of China (No. 21171037) and Research Foundation of the Education Department of Fujian Province (No. JA13082 & No. JB13010).

Author Contributions: Zhongshui Li planned and designed the experiments. Yanli Wang performed the experimental works. Zhongshui Li, Shuhong Xu, and Fengling Lei contributed to the data analysis. Yanli Wang wrote the manuscript; Shen Lin supervised the project and revised manuscript; all authors discussed the results and approved of the final version of the manuscript.

Conflicts of Interest: The authors declare no conflict of interest.

References

1.   Boppella, R.; Anjaneyulu, K.; Basak, P.; Manorama, S.V. Facile synthesis of face oriented ZnO crystals: Tunable polar facets and shape induced enhanced photocatalytic performance. *J. Phys. Chem. C* **2013**, *117*, 4597–4605. [CrossRef]

2. Yu, C.L.; Yang, K.; Xie, Y.; Fan, Q.Z.; Yu, J.C.; Shu, Q.; Wang, C.Y. Novel hollow Pt-ZnO nanocomposite microspheres with hierarchical structure and enhanced photocatalytic activity and stability. *Nanoscale* **2013**, *5*, 2142–2151. [CrossRef] [PubMed]

3. Guo, W.; Zhang, F.; Lin, C.; Wang, Z.L. Direct growth of TiO_2 nanosheet arrays on carbon fibers for highly efficient photocatalytic degradation of methyl orange. *Adv. Mater.* **2012**, *24*, 4761–4764. [CrossRef] [PubMed]

4. Yu, X.; Kuai, L.; Geng, B.Y. CeO_2/rGO/Pt sandwich nanostructure: rGO-enhanced electron transmission between metal oxide and metal nanoparticles for anodic methanol oxidation of direct methanol fuel cells. *Nanoscale* **2012**, *4*, 5738–5743. [CrossRef] [PubMed]

5. Lei, F.L.; Li, Z.S.; Ye, L.T.; Wang, Y.L.; Lin, S. One-pot synthesis of Pt/SnO_2/GhTs and its electro-photo-synergistic catalysis for methanol oxidation. *Int. J. Hydrog. Energy* **2016**, *41*, 255–264. [CrossRef]

6. Li, F.M.; Gao, X.Q.; Xue, Q.; Li, S.N.; Chen, Y.; Lee, J.M. Reduced graphene oxide supported platinum nanocubes composites: One-pot hydrothermal synthesis and enhanced catalytic activity. *Nanotechnology* **2015**, *26*, 065603. [CrossRef] [PubMed]

7. Yun, S.; Lee, S.; Shin, C.; Park, S.; Kwon, S.J.; Park, H.S. One-pot self-assembled, reduced graphene oxide/palladium nanoparticle hybrid aerogels for electrocatalytic applications. *Electrochim. Acta* **2015**, *180*, 902–908. [CrossRef]

8. Alvarez, A.; Guzman, C.; Rivas, S.; Godinez, L.A.; Sacca, A.; Carbone, A.; Passalacqua, E.; Arriaga, L.G.; Ledesma-Garcia, J. Composites membranes based on nafion and PAMAM dendrimers for PEMFC applications. *Int. J. Hydrog. Energy* **2014**, *39*, 16686–16693. [CrossRef]

9. Maiyalagan, T. Pt-Ru nanoparticles supported PAMAM dendrimer functionalized carbon nanofiber composite catalysts and their application to methanol oxidation. *J. Solid State Electrochem.* **2009**, *13*, 1561–1566. [CrossRef]

10. Crespilho, F.N.; Huguenin, F.; Zucolotto, V.; Olivi, P.; Nart, F.C.; Oliveira, O.N. Dendrimers as nanoreactors to produce platinum nanoparticles embedded in layer-by-layer films for methanol-tolerant cathodes. *Electrochem. Commun.* **2006**, *8*, 348–352. [CrossRef]

11. Sun, L.S.; Ca, D.V.; Cox, J.A. Electrocatalysis of the hydrogen evolution reaction by nanocomposites of poly(amidoamine)-encapsulated platinum nanoparticles and phosphotungstic acid. *J. Solid State Electrochem.* **2005**, *9*, 816–822. [CrossRef]

12. Noroozifar, M.; Khorasani-Motlagh, M.; Ekrami-Kakhki, M.-S.; Khaleghian-Moghadam, R. Enhanced electrocatalytic properties of Pt-chitosan nanocomposite for direct methanol fuel cell by $LaFeO_3$ and carbon nanotube. *J. Power Sources* **2014**, *248*, 130–139. [CrossRef]

13. Cogo, L.C.; Batisti, M.V.; Pereira-da-Silva, M.A.; Oliveira, O.N., Jr.; Nart, F.C.; Huguenin, F. Layer-by-layer films of chitosan, poly(vinyl sulfonic acid), and platinum for methanol electrooxidation and oxygen electroreduction. *J. Power Sources* **2006**, *158*, 160–163. [CrossRef]

14. Osifo, P.O.; Masala, A. The influence of chitosan membrane properties for direct methanol fuel cell applications. *J. Fuel Cell Sci. Technol.* **2011**, *9*, 011003. [CrossRef]

15. Pan, Y.; Bao, H.; Li, L. Noncovalently functionalized multiwalled carbon nanotubes by chitosan-grafted reduced graphene oxide and their synergistic reinforcing effects in chitosan films. *ACS Appl. Mater. Interfaces* **2011**, *3*, 4819–4830. [CrossRef] [PubMed]

16. Bao, H.; Pan, Y.; Ping, Y.; Sahoo, N.G.; Wu, T.; Li, L.; Li, J.; Gan, L.H. Chitosan-functionalized graphene oxide as a nanocarrier for drug and gene delivery. *Small* **2011**, *7*, 1569–1578. [CrossRef] [PubMed]

17. Li, Z.S.; Lei, F.L.; Ye, L.T.; Zhang, X.F.; Lin, S. Controlled synthesis of Pt/CS/PW12-GNs composite as an anodic electrocatalyst for direct methanol fuel cells. *J. Nanopart. Res.* **2015**, *17*, 192. [CrossRef]

18. Yu, D.; Dai, L. Self-assembled graphene/carbon nanotube hybrid films for supercapacitors. *J. Phys. Chem. Lett.* **2010**, *1*, 467–470. [CrossRef]

19. Wang, X.; Zhang, X.; He, X.; Ma, A.; Le, L.; Lin, S. Facile electrodeposition of flower-like PMo12-Pt/rGO composite with enhanced electrocatalytic activity towards methanol oxidation. *Catalysts* **2015**, *5*, 1275–1288. [CrossRef]

20. Huang, H.; Liu, Y.; Gao, Q.; Ruan, W.; Lin, X.; Li, X. Rational construction of strongly coupled metal-metal oxide-graphene nanostructure with excellent electrocatalytic activity and durability. *ACS Appl. Mater. Interfaces* **2014**, *6*, 10258–10264. [CrossRef] [PubMed]

21. Li, Z.S.; Huang, X.M.; Zhang, X.F.; Zhang, L.; Lin, S. The synergistic effect of graphene and polyoxometalates enhanced electrocatalytic activities of Pt-{PEI-GNs/[PMo$_{12}$O^{40}]$^{3-}$}$_n$ composite films regarding methanol oxidation. *J. Mater. Chem.* **2012**, *22*, 23602–23607. [CrossRef]

22. Liu, X.; Du, H.; Sun, X.W. High-performance photoresponse of carbon-doped ZnO/reduced graphene oxide hybrid nanocomposites under UV and visible illumination. *RSC Adv.* **2014**, *4*, 5136–5140. [CrossRef]

23. You, H.J.; Zhang, F.L.; Liu, Z.; Fang, J.X. Free-standing Pt-Au hollow nanourchins with enhanced activity and stability for catalytic methanol oxidation. *ACS Catal.* **2014**, *4*, 2829–2835. [CrossRef]

24. Wang, H.; Wang, R.F.; Li, H.; Wang, Q.F.; Kang, J.A.; Lei, Z.Q. Facile synthesis of carbon-supported pseudo-core@shell PdCu@Pt nanoparticles for direct methanol fuel cells. *Int. J. Hydrog. Energy* **2011**, *36*, 839–848. [CrossRef]

25. Park, K.W.; Choi, J.H.; Kwon, B.K.; Lee, S.A.; Sung, Y.E.; Ha, H.Y.; Hong, S.A.; Kim, H.; Wieckowski, A. Chemical and electronic effects of Ni in Pt/Ni and Pt/Ru/Ni alloy nanoparticles in methanol electrooxidation. *J. Phys. Chem. B* **2002**, *106*, 1869–1877. [CrossRef]

26. Biesinger, M.C.; Payne, B.P.; Grosvenor, A.P.; Lau, L.W.M.; Gerson, A.R.; Smart, R.S. Resolving surface chemical states in XPS analysis of first row transition metals, oxides and hydroxides: Cr, Mn, Fe, Co and Ni. *Appl. Surf. Sci.* **2011**, *257*, 2717–2730. [CrossRef]

27. Cui, Z.M.; Jiang, S.P.; Li, C.M. Highly dispersed MoOx on carbon nanotube as support for high performance Pt catalyst towards methanol oxidation. *Chem. Commun.* **2011**, *47*, 8418–8420. [CrossRef] [PubMed]

28. Huang, H.; Chen, Q.; He, M.; Sun, X.; Wang, X. A ternary Pt/MnO$_2$/graphene nanohybrid with an ultrahigh electrocatalytic activity toward methanol oxidation. *J. Power Sources* **2013**, *239*, 189–195. [CrossRef]

29. Yu, S.; Liu, Q.; Yang, W.; Han, K.; Wang, Z.; Zhu, H. Graphene-CeO$_2$ hybrid support for Pt nanoparticles as potential electrocatalyst for direct methanol fuel cells. *Electrochim. Acta* **2013**, *94*, 245–251. [CrossRef]

30. Wietecha, M.S.; Zhu, J.; Gao, G.; Wang, N.; Feng, H.; Gorring, M.L.; Kasner, M.L.; Hou, S. Platinum nanoparticles anchored on chelating group-modified graphene for methanol oxidation. *J. Power Sources* **2012**, *198*, 30–35. [CrossRef]

31. Saidi, W.A.; Feng, H.J.; Fichthorn, K.A. Binding of polyvinylpyrrolidone to Ag surfaces: Insight into a structure-directing agent from dispersion-corrected density functional theory. *J. Phys. Chem. C* **2013**, *117*, 1163–1171. [CrossRef]

32. Al-Saidi, W.A.; Feng, H.J.; Fichthorn, K.A. Adsorption of polyvinylpyrrolidone on Ag surfaces: Insight into a structure-directing agent. *Nano Lett.* **2012**, *12*, 997–1001. [CrossRef] [PubMed]

33. Qi, X.; Balankura, T.; Zhou, Y.; Fichthorn, K.A. How structure-directing agents control nanocrystal shape: Polyvinylpyrrolidone-mediated growth of Ag nanocubes. *Nano Lett.* **2015**, *15*, 7711–7717. [CrossRef] [PubMed]

34. Li, Z.S.; Zhang, L.; Huang, X.M.; Ye, L.T.; Lin, S. Shape-controlled synthesis of Pt nanoparticles via integration of graphene and beta-cyclodextrin and using as a noval electrocatalyst for methanol oxidation. *Electrochim. Acta* **2014**, *121*, 215–222. [CrossRef]

35. Wu, T.; Wang, X.; Qiu, H.; Gao, J.; Wang, W.; Liu, Y. Graphene oxide reduced and modified by soft nanoparticles and its catalysis of the Knoevenagel condensation. *J. Mater. Chem.* **2012**, *22*, 4772–4779. [CrossRef]

36. Kim, N.H.; Kuila, T.; Lee, J.H. Simultaneous reduction, functionalization and stitching of graphene oxide with ethylenediamine for composites application. *J. Mater. Chem. A* **2013**, *1*, 1349–1358. [CrossRef]

37. Qiu, J.-D.; Wang, G.-C.; Liang, R.-P.; Xia, X.-H.; Yu, H.-W. Controllable deposition of platinum nanoparticles on graphene as an electrocatalyst for direct methanol fuel cells. *J. Phys. Chem. C* **2011**, *115*, 15639–15645. [CrossRef]

38. Guo, Q.; Zheng, Z.; Gao, H.L.; Ma, J.; Qin, X. SnO$_2$/graphene composite as highly reversible anode materials for lithium ion batteries. *J. Power Sources* **2013**, *240*, 149–154. [CrossRef]

39. Estudillo-Wong, L.A.; Vargas-Gomez, A.M.; Arce-Estrada, E.M.; Manzo-Robledo, A. TiO$_2$/c composite as a support for Pd-nanoparticles toward the electrocatalytic oxidation of methanol in alkaline media. *Electrochim. Acta* **2013**, *112*, 164–170. [CrossRef]

40. Zhang, M.; Xie, J.; Sun, Q.; Yan, Z.; Chen, M.; Jing, J. Enhanced electrocatalytic activity of high Pt-loadings on surface functionalized graphene nanosheets for methanol oxidation. *Int. J. Hydrog. Energy* **2013**, *38*, 16402–16409. [CrossRef]

41. Zhong, X.; Wang, Z.; Huang, Y.; Yu, Y.; Feng, Q.; Li, Q. Fabrication of Pt nanoparticles on ethylene diamine functionalized graphene for formic acid electrooxidation. *Int. J. Hydrog. Energy* **2014**, *39*, 15920–15927. [CrossRef]

42. Luo, Z.; Yuwen, L.; Bao, B.; Tian, J.; Zhu, X.; Weng, L.; Wang, L. One-pot, low-temperature synthesis of branched platinum nanowires/reduced graphene oxide (BPtNW/RGO) hybrids for fuel cells. *J. Mater. Chem.* **2012**, *22*, 7791–7796. [CrossRef]

43. Liu, R.; Li, S.; Yu, X.; Zhang, G.; Zhang, S.; Yao, J.; Keita, B.; Nadjo, L.; Zhi, L. Facile synthesis of Au-nanoparticle/polyoxometalate/graphene tricomponent nanohybrids: An enzyme-free electrochemical biosensor for hydrogen peroxide. *Small* **2012**, *8*, 1398–1406. [CrossRef] [PubMed]

44. Shi, J.-J.; Yang, G.-H.; Zhu, J.-J. Sonoelectrochemical fabrication of PDDA-RGO-PdPt nanocomposites as electrocatalyst for DAFCs. *J. Mater. Chem.* **2011**, *21*, 7343–7349. [CrossRef]

45. Luo, B.; Yan, X.; Xu, S.; Xue, Q. Polyelectrolyte functionalization of graphene nanosheets as support for platinum nanoparticles and their applications to methanol oxidation. *Electrochim. Acta* **2012**, *59*, 429–434. [CrossRef]

46. Mayavan, S.; Jang, H.-S.; Lee, M.-J.; Choi, S.H.; Choi, S.-M. Enhancing the catalytic activity of Pt nanoparticles using poly sodium styrene sulfonate stabilized graphene supports for methanol oxidation. *J. Mater. Chem. A* **2013**, *1*, 3489–3494. [CrossRef]

47. Cui, X.; Wu, S.N.; Jungwirth, S.; Chen, Z.B.; Wang, Z.H.; Wang, L.; Li, Y.X. The deposition of Au-Pt core-shell nanoparticles on reduced graphene oxide and their catalytic activity. *Nanotechnology* **2013**, *24*, 295402. [CrossRef] [PubMed]

48. Xiang, D.; Yin, L. Well-dispersed and size-tuned bimetallic PtFe$_x$ nanoparticle catalysts supported on ordered mesoporous carbon for enhanced electrocatalytic activity in direct methanol fuel cells. *J. Mater. Chem.* **2012**, *22*, 9584–9593. [CrossRef]

49. Yang, S.D.; Shen, C.M.; Lu, X.J.; Tong, H.; Zhu, J.J.; Zhang, X.G.; Gao, H.J. Preparation and electrochemistry of graphene nanosheets-multiwalled carbon nanotubes hybrid nanomaterials as Pd electrocatalyst support for formic acid oxidation. *Electrochim. Acta* **2012**, *62*, 242–249. [CrossRef]

50. Huang, H.; Chen, H.; Sun, D.; Wang, X. Graphene nanoplate-Pt composite as a high performance electrocatalyst for direct methanol fuel cells. *J. Power Sources* **2012**, *204*, 46–52. [CrossRef]

51. Qian, L.; Yang, X. Dendrimer films as matrices for electrochemical fabrication of novel gold/palladium bimetallic nanostructures. *Talanta* **2008**, *74*, 1649–1653. [CrossRef] [PubMed]

52. Guo, W.; Xu, L.; Li, F.; Xu, B.; Yang, Y.; Liu, S.; Sun, Z. Chitosan-assisted fabrication and electrocatalytic activity of the composite film electrode of heteropolytungstate/carbon nanotubes. *Electrochim. Acta* **2010**, *55*, 1523–1527. [CrossRef]

53. Qian, L.; Yang, X. Polyamidoamine dendrimers-assisted electrodeposition of gold-platinum bimetallic nanoflowers. *J. Phys. Chem. B* **2006**, *110*, 16672–16678. [CrossRef] [PubMed]

54. Guo, S.; Dong, S.; Wang, E. Three-dimensional Pt-on-Pd bimetallic nanodendrites supported on graphene nanosheet: Facile synthesis and used as an advanced nanoelectrocatalyst for methanol oxidation. *ACS Nano* **2010**, *4*, 547–555. [CrossRef] [PubMed]

55. Wu, Z.; Feng, W.; Feng, Y.; Liu, Q.; Xu, X.; Sekino, T.; Fujii, A.; Ozaki, M. Preparation and characterization of chitosan-grafted multiwalled carbon nanotubes and their electrochemical properties. *Carbon* **2007**, *45*, 1212–1218. [CrossRef]

56. Kovtyukhova, N.I.; Ollivier, P.J.; Martin, B.R.; Mallouk, T.E.; Chizhik, S.A.; Buzaneva, E.V.; Gorchinskiy, A.D. Layer-by-layer assembly of ultrathin composite films from micron-sized graphite oxide sheets and polycations. *Chem. Mater.* **1999**, *11*, 771–778. [CrossRef]

57. Zeng, Q.O.; Cheng, J.S.; Tang, L.H.; Liu, X.F.; Liu, Y.Z.; Li, J.H.; Jiang, J.H. Self-assembled graphene-enzyme hierarchical nanostructures for electrochemical biosensing. *Adv. Funct. Mater.* **2010**, *20*, 3366–3372. [CrossRef]

MDPI

Article

Facile Synthesis of Bimetallic Pt-Ag/Graphene Composite and Its Electro-Photo-Synergistic Catalytic Properties for Methanol Oxidation

Shuhong Xu [1,†], Lingting Ye [2,†], Zhongshui Li [1], Yanli Wang [1], Fengling Lei [1] and Shen Lin [1,*]

[1] College of Chemistry & Chemical Engineering, Fujian Normal University, Fuzhou 350007, China; xushuhongxu@gmail.com (S.X.); zsli@fjnu.edu.cn (Z.L.); yanliwang408@gmail.com (Y.W.); fenglinglei408@gmail.com (F.L.)

[2] Key Lab of Design & Assembly of Functional Nanostructure, Fujian Institute of Research on the Structure of Matter, Chinese Academy of Sciences, Fuzhou 350002, Fujian, China; ltye@fjirsm.ac.cn

* Correspondence: shenlin@fjnu.edu.cn; Tel.: +86-591-2286-7399; Fax: +86-591-2286-7399

† These authors contributed equally to this work.

Academic Editors: Vincenzo Baglio and David Sebastián
Received: 25 July 2016; Accepted: 9 September 2016; Published: 16 September 2016

Abstract: A Pt-Ag/graphene composite (Pt-Ag/GNs) was synthesized by the facile aqueous solution method, in which Ag^+ was first transformed into Ag_2O under UV light irradiation, and then Ag_2O, Pt^{2+}, and graphene oxide (GO) were simultaneously reduced by formic acid. It was found that Pt-Ag bimetallic nanoparticles were highly dispersed on the surface of graphene, and their size distribution was narrow with an average diameter of 3.3 nm. Electrocatalytic properties of the Pt-Ag/GNs composite were investigated by cyclic voltammograms (CVs), chronoamperometry (CA), CO-stripping voltammograms, and electrochemical impedance spectrum (EIS) techniques. It was shown that the Pt-Ag/GNs composite has much higher catalytic activity and stability for the methanol oxidation reaction (MOR) and better tolerance toward CO poisoning when compared with Pt/GNs and the commercially available Johnson Matthey 20% Pt/C catalyst (Pt/C-JM). Furthermore, the Pt-Ag/GNs composite showed efficient electro-photo-synergistic catalysis for MOR under UV or visible light irradiation. Particularly in the presence of UV irradiation, the Pt-Ag/GNs composite exhibited an ultrahigh mass activity of 1842.4 mA·mg^{-1}, nearly 2.0 times higher than that without light irradiation (838.3 mA·mg^{-1}).

Keywords: platinum; silver; graphene; methanol oxidation; electro-photo catalysis

---

1. Introduction

Platinum has attracted significant attention as a catalyst for the methanol oxidation reaction (MOR) [1]. However, the monometal Pt catalyst is not only expensive, but also easily poisoned by the adsorption of CO-like intermediate species, reducing the electrocatalytic activity of the catalyst during methanol oxidation [2]. Binary or ternary Pt-based catalysts containing other transition metals, such as NiAuPt [3], PtRu [4], PtCu [5], FePt [6], PtSn [7], PtCo [8], PtRhNi [9], or PtPd [10] can evidently improve the performance in catalyzing MOR compared with the monometallic Pt catalysts. Particularly, they can alleviate CO_{ads} poisoning and expose more Pt active sites, resulting in lower noble metal dosages and lower costs. Among the transition metals, Ag can provide rich surface oxygen-containing species to promote the oxidation of CO-like species and prevent the catalyst from poisoning [11]. Ag has a high degree of free electron mobility and exhibits molecular-like excited-state properties with well-defined absorption and emission features, and its surface plasmon resonance (SPR) effect has a prominent contribution to the enhanced photocatalytic activity [12,13]. Therefore, Pt-Ag bimetallic nano-structures are expected to provide an efficient synergistic effect between the electro-catalytic and photo-catalytic properties for MOR.

The unique properties of graphene nanosheets (GNs), such as larger specific surface areas, excellent optical transmittance, high chemical and thermal stability, and outstanding electrical conductibility, make it a superior support material for Pt-based catalysts [14,15]. Referring to our previously reported catalysts, such as $Pt/SnO_2/GNs$ [16] and $Pt/TiO_2/GNs$ [17], it is known that GNs can not only improve the dispersion of Pt nanoparticles, but also can act as the acceptor and the transfer body of the photo-electron, giving rise to charge transfer and separation efficiency of the photo-generated electron-hole pairs which leads to a better photocatalytic activity.

Herein, we explored a facile method to synthesize Pt-Ag/graphene composite (Pt-Ag/GNs) with the use of formic acid as a reducing agent (Scheme 1). In the synthetic process, HCOOH has the advantages of simple operation, environmental friendliness, and mild reaction conditions, and the Ag_2O intermediate species formed under UV light irradiation contribute to the good dispersion of the bimetallic Pt-Ag nanoparticles on GNs with narrow size distribution. In comparison with commercially available Johnson Matthey 20% Pt/C (denoted as Pt/C-JM) and Pt/GNs, the as-synthesized Pt-Ag/GNs exhibit much higher catalytic activity and stability for MOR and better resistance to CO poisoning. More importantly, with the aid of the integrated photo-response of Ag and GNs, methanol electro-catalytic oxidation and photo-catalytic oxidation reactions are synergistically coupled on Pt-Ag/GNs under UV irradiation or visible light irradiation, bringing about increased catalytic activity and stability.

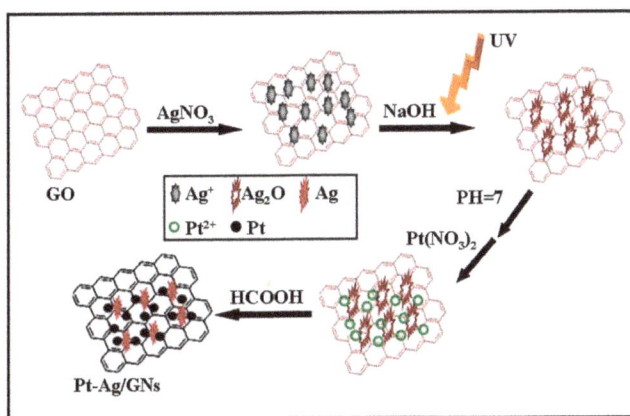

Scheme 1. Schematic of the preparation procedure of the Pt-Ag/GNs composite.

2. Results and Discussion

Figure 1 shows the XRD patterns for the following samples: (a) Ag/GNs, (b) Pt/GNs and (c) Pt-Ag/GNs. The diffraction pattern for the Pt/GNs catalyst shows the typical peaks of the face centered-cubic (fcc) structure of Pt (0). Peaks are located at 39.8°, 46.3°, 67.6° and 81.4° corresponding to the (111), (200), (220) and (311) planes of Pt (0) [18,19]. This indicates that the reduction of the Pt^{2+} precursor to Pt (0) by formic acid has been effective. Pt-Ag/GNs (curve c) shows broader and asymmetrical peaks in comparison to both Ag/GNs and Pt/GNs. The reason for this is that these peaks are the combination of the peaks for both Pt and Ag, which indicates that both elements are not alloyed. This is also supported by the the Powder Diffraction File of International Centre for Diffraction Data (ICDD PDF) of Pt (No. 04-0802) and Ag (No. 04-0783) in Figure 1. Furthermore, the representative diffraction peaks assigned to the (111) and (220) planes of Ag_2O located at 32.8° and 54.9°, respectively, are not observed (Figure 1) [20], suggesting that the intermediate species Ag_2O was completely transformed into Ag after the reduction with formic acid.

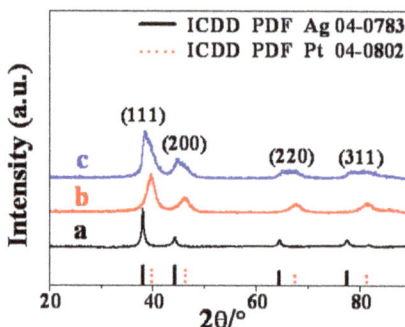

Figure 1. XRD patterns of Ag/GNs (a), Pt/GNs (b) and Pt-Ag/GNs (c).

The morphologies of the Pt-Ag/GNs and Pt/GNs composites were characterized by transmission electron microscopy (TEM) and high-resolution TEM (HRTEM). The TEM image of Pt/GNs shows apparent aggregation of the Pt particles on the surface of GNs (Figure 2A). By contrast, the spherical bimetallic nanoparticles of the Pt-Ag/GNs composite are uniformly dispersed on the surface of the GNs with an average size of about 3.3 nm (Figure 2D,E), and a relatively narrow size distribution peak (the inset of Figure 2E). HRTEM images of Pt/GNs and Pt-Ag/GNs (Figure 2C,F) show lattice fringes with spacing of 0.225 nm, which is in good agreement with the (111) planes of fcc Pt (0) [21]. In addition, the lattice spacing of 0.235 nm shown in Figure 2F can be assigned to the (111) planes of metallic Ag [22], further proving the formation of separate Ag nanoparticles in Pt-Ag/GNs. This also indicates that the Pt and Ag nanoparticles are separately deposited on the GNs, instead of forming a Pt-Ag alloy. The reason for this may be that Ag^+ was first transformed into Ag_2O under UV irradiation [23], and then the heterogeneous phase between the precursor Pt^{2+} and Ag_2O can prevent the formation of a binary alloy structure, which results in the formation of the heterostructured Pt-Ag nanoparticles. It is worth mentioning that (111) Pt planes are adjacent to (111) Ag planes, suggesting that Ag_2O provides an anchor for the Pt^{2+} precursor during the reduction process, resulting in a proper dispersion of Ag and Pt nanoparticles on the GNs after reduction by formic acid. The morphologies of Pt and Ag nanoparticles on GNs were further analyzed by high-angle annular dark-field scanning transmission electron microscopy (HAADF-STEM) and elemental mapping images. As shown in Figure 3A, Pt and Ag nanoparticles are highly dispersed onto the GNs surface, and it is difficult to discriminate Pt from Ag because of their full integration. Elemental mapping (Figure 3B,C) of the Pt-Ag/GNs composite show that the Pt and Ag nanoparticles are in good co-existence and in intimate contact, which further verifies that Ag_2O provides an anchor for the precursor of Pt^{2+}. The selected area electron diffraction (SAED) pattern demonstrates that the Pt-Ag/GNs composite has a typical polycrystalline structure (the inset of Figure 2F).

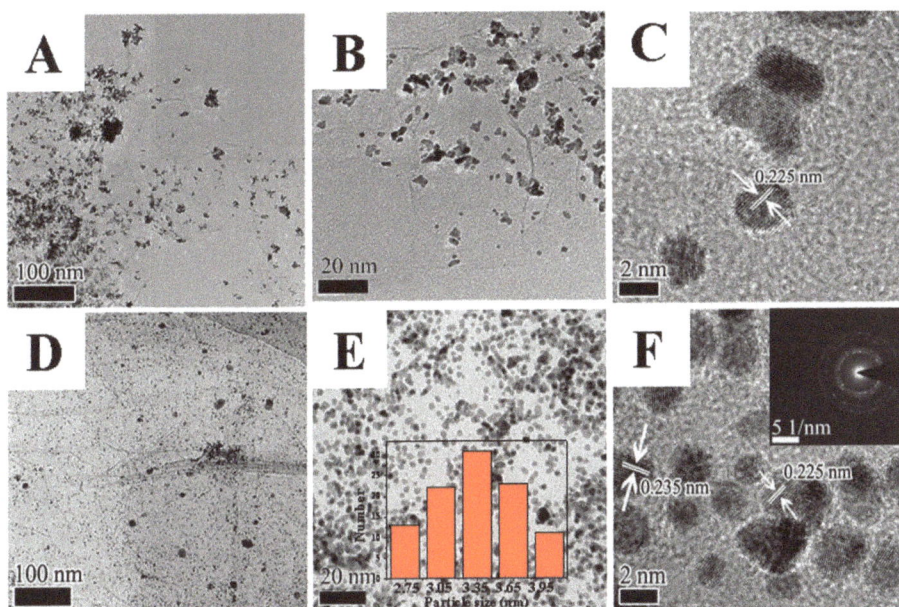

Figure 2. Transmission electron microscopy (TEM) and high-resolution TEM (HRTEM) images of Pt/GNs (**A**, **B**, **C**) and Pt-Ag/GNs (**D**, **E**, **F**); particle size distributions of Pt-Ag/GNs in the inset of (**E**); SAED of Pt-Ag/GNs in the inset of (**F**).

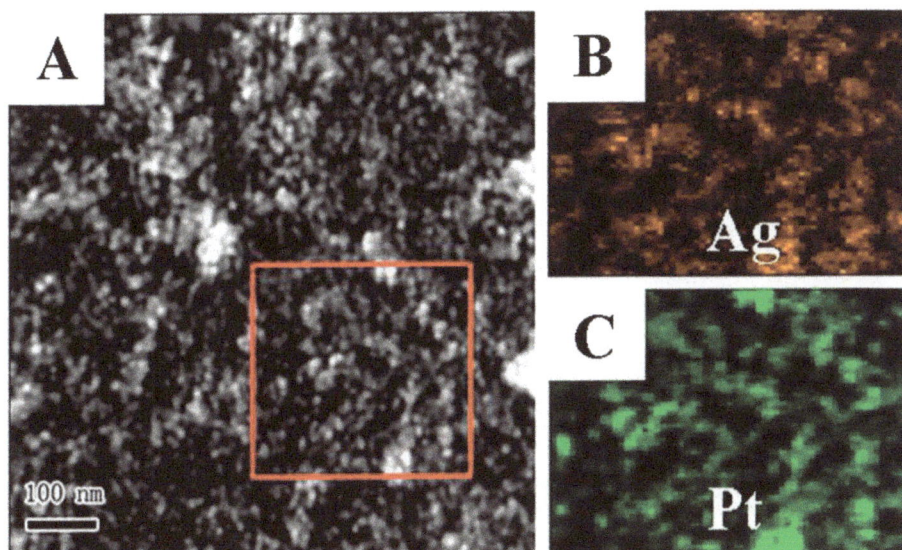

Figure 3. High-angle annular dark-field scanning transmission electron microscopy (HAADF-STEM) images for Pt-Ag/GNs (**A**); elemental mapping images for Ag (**B**) and Pt (**C**).

X-ray photoelectron spectra (XPS) were used to further analyze the valence states and surface composition of the as-synthesized composites, displayed in Figure 4. The binding energies of Ag $3d_{5/2}$

and Ag $3d_{3/2}$ for Pt-Ag/GNs are 368.1 eV and 374.0 eV (Figure 4A), respectively, demonstrating that the deposited Ag nanoparticles are in a zero-valent state [22,24]. Figure 4B shows Pt 4f spectra of Pt/GNs (curve a) and Pt-Ag/GNs (curve b). The peaks located at 71.08 and 74.38 eV for Pt-Ag/GNs and at 71.38 and 74.79 eV for Pt/GNs can be assigned to the binding energies of Pt $4f_{7/2}$ and Pt $4f_{5/2}$, respectively. Their values are in line with that of platinum in the zero-valent state, which further verifies the formation of Pt (0) nanoparticles in Pt/GNs and Pt-Ag/GNs composites [14]. However, the binding energy of Pt 4f for Pt-Ag/GNs is significantly lower than that for Pt/GNs. The reason may be that Pt has a higher electro-negativity than Ag, so Pt can withdraw electrons from the neighboring Ag atoms [25]. Furthermore, there exists an electrostatic interaction between the metal nanoparticles and the graphene sheets because of the remaining oxygen containing functional groups on the surface of the GNs, which allows a high dispersion of Pt-Ag nanoparticles on GNs to produce stronger electronic interactions between metal nanoparticles and supports [26,27], inducing their binding energy shift.

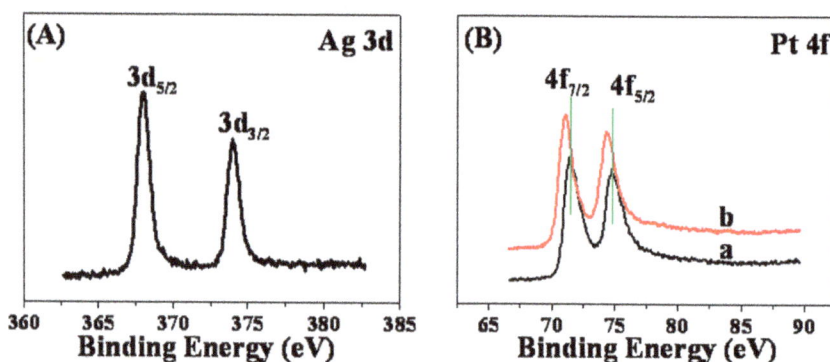

Figure 4. (**A**) Ag 3d X-ray photoelectron spectra (XPS) profile of Pt-Ag/GNs; (**B**) Pt 4f XPS profiles for Pt/GNs (a) and Pt-Ag/GNs (b).

Raman spectra were used to discuss the structural changes of the graphene based composites during the preparation process. As shown in Figure 5, the D band (associated with a breathing mode of k-point phonons of A_{1g} symmetry) and the G band (corresponding to the E_{2g} phonon of the C sp^2 atom) are observed [28]. I_D/I_G, namely the intensity ratio of the D and G bands, can be used to predict the extent of defects in carbonaceous materials [29]. It is found that the I_D/I_G value of the Pt-Ag/GNs composite (curve b) is obviously smaller than that of Pt/GNs (curve a), which indicates that there is a lower degree of disorder for the graphene structure.

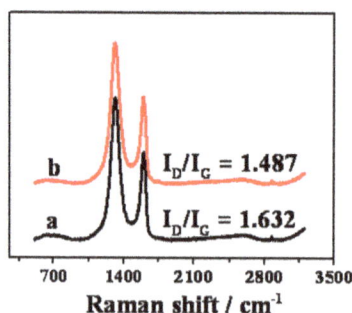

Figure 5. Raman spectra of Pt/GNs (a) and Pt-Ag/GNs (b).

Cyclic voltammograms (CVs) of Pt/C-JM (curve a), Pt/GNs (curve b) and Pt-Ag/GNs (curve c) in a 0.5 M H_2SO_4 solution were collected to evaluate the electrochemical properties of the different composites, shown in Figure 6. In general, the hydrogen adsorption/desorption peaks of the CV curves can be utilized to measure the electrochemically active surface area (ECSA) of Pt-based catalysts on the basis of Equation (1) [30].

$$ECSA = Q_H/(210 \,\mu C \cdot cm^{-2} \times Pt \,\, loading) \tag{1}$$

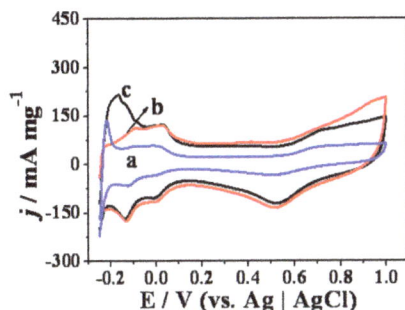

Figure 6. Cyclic voltammograms (CVs) of Pt/C-JM (a), Pt/GNs (b) and Pt-Ag/GNs (c) in 0.5 M H_2SO_4; scan rate: 100 mV·s^{-1}.

The ECSA values for the Pt-Ag/GNs (curve c) is 120.3 m^2·g^{-1}, which is about 1.5 times and 2.5 times higher than that of the Pt/GNs (curve b) (80.5 m^2·g^{-1}) and the Pt/C-JM (curve a) (48.9 m^2·g^{-1}), respectively. The larger ECSA mainly arises from better dispersion of the Ag and Pt nanoparticles on the GNs (as shown in TEM analysis), which can offer more active sites.

In order to further investigate their catalytic activities for methanol oxidation reaction (MOR), CVs in 1 M CH_3OH + 0.5 M H_2SO_4 solution were conducted (Figure 7). As shown in Figure 7, all of the curves have two significant oxidation peaks: the forward scan oxidation peak is assigned to the peak current of methanol oxidation, and the reverse scan oxidation peak is attributed to the peak current of the CO intermediate species oxidation. As presented in Figure 7, the forward scan current density (mass activity) increased up to 838.3 mA·mg^{-1} on Pt-Ag/GNs without light irradiation (*t* = 0 min), which is significantly higher than that of Pt/GNs (653.8 mA·mg^{-1}) and Pt/C-JM (402.6 mA·mg^{-1}). The improved catalytic activity may originate from the role of Ag. First, the introduction of Ag into Pt-Ag/GNs brings about the good distribution of the Ag and Pt nanoparticles on the GNs, which generates more active sites for MOR (as shown in Figure 6). Secondly, Ag has a lower electro-negativity than Pt, so Pt can withdraw electrons from the neighboring Ag atoms, which produces stronger electronic interactions between the Pt and Ag nanoparticles (as analyzed in the Pt 4f spectra) [25]. The interaction may effectively weaken the capacity to bind to adsorbed intermediates of Pt [31]. Thirdly, the coexistence of Ag and Pt nanoparticles on the surface of GNs can increase the electron transfer efficiency via an interfacial interaction among the Ag, Pt and GNs, which was verified by Electrochemical Impedance Spectra (EIS) analysis. As shown in Figure 8, the Nyquist plots of EIS for Pt-Ag/GNs, Pt/GNs and Pt/C-JM were obtained in 0.5 M H_2SO_4 + 1.0 M CH_3OH solution at 0.5 V. The diameter of the primary semicircle can be used to analyze the charge transfer resistance of the catalyst, which describes the rate of charge transfer during the methanol oxidation reaction [32]. The semicircle radius of Pt-Ag/GNs (curve c) is much smaller than that of Pt/GNs (curve b) and Pt/C-JM (curve a), which illustrates that Pt-Ag/GNs have better charge transfer and a faster methanol oxidation reaction rate [33,34].

Most interestingly, it was found that the peak current value of the Pt-Ag/GNs composite was significantly increased under UV or visible light irradiation. Particularly, the Pt-Ag/GNs composite exhibited a remarkably increased value (about 1842.4 mA·mg^{-1}) after UV irradiation for 30 min, which is nearly 2.0 times higher than that without light irradiation (838.3 mA·mg^{-1}) (Figure 7A). The value is also far more than that of reported commercial PtRu-C (E-TEK, Secaucus, New Jersey, NJ, USA) [35] or PtRu-C prepared by the ethanol reduction method [36]. However, the current density of Pt/GNs and Pt/C-JM are only slightly increased (Figure 7B,C). The obvious improvement in the catalytic activity of Pt-Ag/GNs under external light irradiation apparently stem from the introduction of Ag, which may endow Pt-Ag/GNs with an efficient electro-photo synergistic catalysis for MOR under UV or visible light irradiation. In order to further check whether the Pt-Ag/GNs composite can effectively respond to ultraviolet light or visible light irradiation, chronoamperometric curves of Pt-Ag/GNs were measured by turning on/off UV or visible light every 10 s in 1 M CH$_3$OH + 0.5 M H$_2$SO$_4$ solution (presented in Figure 9). When UV or visible light was turned on, the current density increased rapidly and temporarily up to a steady state; as the light was turned off, the current density decreased instantly. In the first illumination cycle, the current intensities were boosted from 0.325 to 2.171 (mA·cm^{-2}) (568.0% increase) under UV irradiation (curve c) and from 0.325 to 1.522 (mA·cm^{-2}) (368.3% increase) under visible irradiation, respectively (curve b). The increase of current density in the first illumination cycle is apparently higher than that of the reported silicon-based Pt-Ag nano-forests catalyst under visible irradiation [37]. Therefore, the as-synthesized Pt-Ag/GNs should be a more efficient Pt-Ag bimetal electro-photo synergistic catalyst for MOR under solar irradiation.

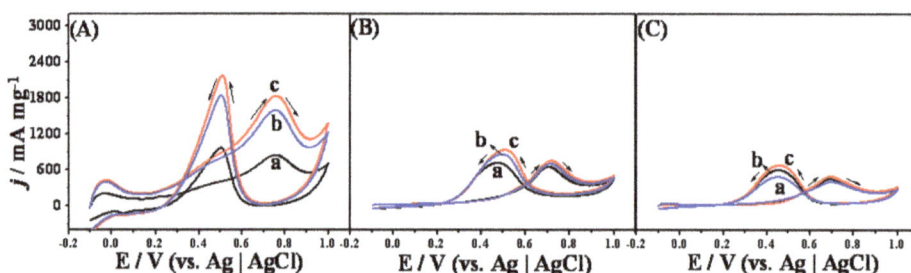

Figure 7. CVs of (**A**) Pt-Ag/GNs, (**B**) Pt/GNs and (**C**) Pt/C-JM in 0.5 M H$_2$SO$_4$ + 1.0 M CH$_3$OH; curves a, without irradiation; curves b, after visible irradiation for 30 min; curves c, after UV irradiation for 30 min.

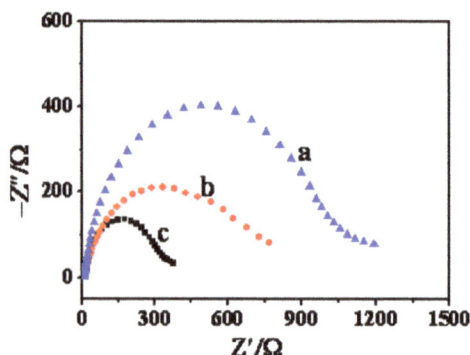

Figure 8. Nyquist plots of Electrochemical Impedance Spectra (EIS) in 0.5 M H$_2$SO$_4$ + 1.0 M CH$_3$OH: Pt/C-JM (a), Pt/GNs (b), and Pt-Ag/GNs (c).

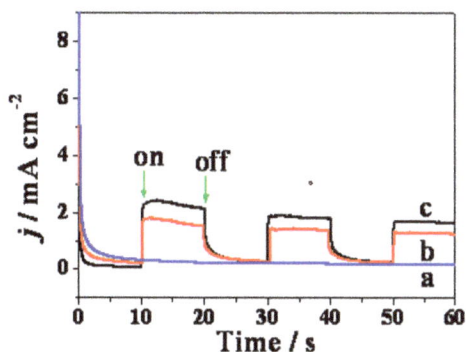

Figure 9. Photocurrent responses of Pt-Ag/GNs: (a) under dark; (b) under intermittent visible irradiation (every 10 s); (c) under intermittent UV irradiation (every 10 s).

The catalytic stability of Pt-Ag/GNs, Pt/GNs, and Pt/C-JM for methanol oxidation was further tested by chronoamperometry in 0.5 M H_2SO_4 containing 1.0 M CH_3OH solution at 0.70 V for 2000 s (Figure 10). As shown in Figure 10, rapid initial current density decay is observed because of CO-like intermediate species (mainly CO_{ads}) poisoning during the methanol electrooxidation reaction [38]. The slower decay slope and the significantly higher final steady-state current density of Pt-Ag/GNs without irradiation (curve c), compared with that of Pt/GNs (curve b) and Pt/C-JM (curve a), indicate that the introduction of Ag into Pt-Ag/GNs improves its electro-catalytic stability for MOR. The enhanced tolerance ability to CO can be reflected in the CO-stripping voltammograms test. As shown in Figure 11, CO-stripping voltammograms of Pt/C-JM (curve a), Pt/GNs (curve b), and Pt-Ag/GNs (curve c) were measured by the oxidation of pre-adsorbed and saturated CO in 0.5 M H_2SO_4 solution at a scan rate of 100 mV·s^{-1}. The peak potential of CO oxidation for Pt-Ag/GNs, Pt/GNs, and Pt/C-JM was 0.67, 0.70 and 0.71 V, respectively. Obviously, Pt-Ag/GNs exhibited a more negative peak potential. The negative shift of the CO oxidation peak demonstrates easier oxidation of CO on the surface of the Pt-Ag/GNs catalyst, indicative of a better anti-poisoning ability. Ag nanoparticles are oxophilic species and can remove the CO-like adsorbing species on Pt active sites at a lower potential through facile formation of Ag(OH)ads [39], contributing to better tolerance towards CO poisoning for Pt-Ag/GNs.

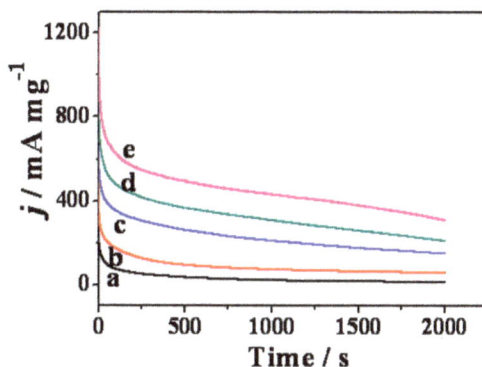

Figure 10. Chronoamperograms in 0.5 M H_2SO_4 + 1.0 M CH_3OH: Pt/C-JM (a), Pt/GNs (b), Pt-Ag/GNs (c), Pt-Ag/GNs under durative visible irradiation (d) and Pt-Ag/GNs under durative UV irradiation (e).

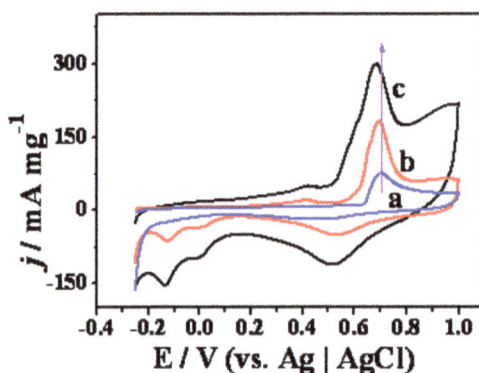

Figure 11. CO-stripping voltammograms of Pt/C-JM (a), Pt/GNs (b), and Pt-Ag/GNs (c) in 0.5 M H_2SO_4; scan rate: 100 mV·s^{-1}.

Particularly, the current density of Pt-Ag/GNs decay more slowly in the presence of light irradiation, and the steady-state current density is significantly enhanced under durative UV (curve e) or visible light irradiation (curve d), which are respectively nearly 2.1 times and 1.4 times higher than that without light irradiation (Figure 10). The results further confirm that the incorporation of Ag brings about the positive effects of light irradiation on the catalytic stability and tolerance to CO poisoning. This is also supported by the ratio of the forward anodic peak current (I_f) to the reverse anodic peak current (I_b) under irradiation. It is well known that the I_f/I_b ratio can also be used to describe the catalyst tolerance to the accumulation of CO-like intermediate species [40–42]. The I_f/I_b of Pt-Ag/GNs is 0.85 after UV irradiation for 30 min, which is higher than that of Pt/GNs (0.80) and Pt/C-JM (0.72) (Figure 7), respectively, implying more valid oxidation of methanol to carbon dioxide during the anodic scan and the removal of poisoning species from the surface of Pt-Ag/GNs under UV irradiation. It should be noted that Pt-Ag/GNs not only exhibited better catalytic activity, stability and tolerance to CO under UV irradiation, but its visible light assisted electro-photo synergistic catalysis performance is also more effective than that of the reported silicon-based Pt-Ag nano-forests catalyst [37]. The efficient synergistic catalysis under UV and visible light irradiation is in favor of harvesting the natural sunlight (UV and visible light) to produce clean energy and provides very promising practical applications. It is well known that noble metal Ag nanoparticles (NPs) show strong visible light absorption due to their surface plasmon resonance (SPR), which is produced by the collective oscillations of surface electrons [43,44]. The unique characteristics of Ag contribute to the beneficial effects of electro-photo-synergistic catalysis under visible light irradiation. Furthermore, GNs can effectively extend the light absorption edge to the visible light region [45]. By integrating with Pt-Ag nanoparticles, GNs are helpful for endowing Pt-Ag/graphene composites with the visible light response for photocatalytic methanol oxidation. Due to the local electromagnetic field and the excellent UV-Visible light absorption capabilities of Ag NPs, the excited electrons can be transferred quickly to GNs, and GNs can be used as the photocatalytic carriers to extract the photo-generated electrons and immediately transfer them to Pt particles, to limit self-photoreduction processes and more easily release Pt active sites [17]. Therefore, the efficient synergistic effects of photo-catalysis and electro-catalysis for MOR are achieved under external light irradiation.

3. Experimental Section

3.1. Materials

Graphite powder (~325 mesh, 99.9995%) was purchased from Alfa Aesar (Ward Hill, Massachusetts, MA, USA). 5% Nafion 117 solution and commercial Pt/C (20%) catalyst (denoted as Pt/C-JM) were purchased from Johnson Matthey (London, UK). AgNO$_3$, Pt(NO$_3$)$_2$, NaOH, KMnO$_4$, H$_2$O$_2$ (30%), K$_2$S$_2$O$_8$, P$_2$O$_5$, H$_2$SO$_4$, acetone, methanol, and ethanol were all analytical reagent and purchased from Sinopharm Chemical Reagent Co., Ltd. (Shanghai, China) and used without further purification. The resistance of deionized water (DI) was 18.2 MΩ.

3.2. Preparation of Pt-Ag/GNs

Graphene oxide (GO) was prepared from purified natural graphite using Hummers' method with some minor modification [46,47]. The product was collected by centrifugation after thoroughly washing with dilute hydrochloric acid and deionized water. After that, the obtained precipitate was dried in vacuum for further use. The overall synthetic route of Pt-Ag/GNs is illustrated in Scheme 1. In a typical synthetic process, 50 mg of GO was dispersed in 50 mL of distilled water by sonication for 30 min. Then, 30 mg of AgNO$_3$ was added to the mixture under stirring for 30 min. Subsequently, 50 mg of solid sodium hydroxide was added under UV irradiation and the mixture was stirred for another 30 min, which caused that Ag$^+$ was transformed into Ag$_2$O [23]. After removing the ultraviolet lamp and adjusting the pH to 7, 3 mL of Pt(NO$_3$)$_2$ (1g/100 mL) were added. After stirring for 30 min, 3 mL of HCOOH (0.5 M) were added. Subsequently, the mixture was stirred for 48 h under ambient condition. After repeated centrifugation (10,000 rpm, 20 min) and thorough washing, the final Pt-Ag/GNs precipitate was lyophilized for 48 h. The counterpart Pt/GNs was prepared by a similar method in the absence of AgNO$_3$. The Pt content was determined by inductively coupled plasma atomic emission spectrometry (ICP-AES). The actual loading amount of Pt on the GNs for Pt-Ag/GNs and Pt/GNs was about 20.7 wt % and 25.9 wt %, respectively.

3.3. Characterization Method

The microscopic features and morphology of the samples were investigated through TEM and HRTEM with a TECNAI G2 high-resolution transmission electron microscope (FEI, Hillsboro, Oregon, OR, USA) operating at 200 K. Elemental distribution mapping was performed in a HAADF-STEM with an Oxford INCA energy dispersive X-ray detector (FEI). XPS was conducted on a VG ESCALAB 250 spectrometer (Thermo Scientic, Pleasanton, California, CA, USA) with Al Kα radiation (1486.6 eV). XRD patterns were obtained with an X'pert Pro X-ray diffractometer (PANalytical B.V., Holland, The Netherlands), equipped with a Cu Kα radiation source. Raman spectra were recorded on a Renishaw-in-Via Raman (Renishaw, London, UK) micro-spectrometer equipped with a 514 nm wavelength laser line. The actual amount of Pt loading in the catalysts was determined by ICP-AES (ICAP6300, Thermo Scientic).

3.4. Electrochemical Measurements

A conventional three electrode cell was used for the electrochemical measurement on a CHI 660C electrochemical workstation (ChenHua, Shanghai, China), with a platinum column as the counter electrode, a standard Ag|AgCl (3 M KCl) electrode as the reference, and a modified glassy carbon electrode (GCE, 3.0 mm in diameter) as the working electrode, respectively. To prepare the working electrode, 5 mg of catalyst sample were first dispersed into 1 mL of ethanol by ultrasonication, and then 5 μL of the previously prepared ethanol suspension were dropped onto the pre-cleaned surface of the GCE. Subsequently, an aliquot of Nafion solution (7.5 μL, 0.5%) was dropped on the coated GCE. CVs were recorded in 0.5 M H$_2$SO$_4$ or 0.5 M H$_2$SO$_4$ solution containing 1.0 M CH$_3$OH at room temperature with a scan rate of 100 mV·s^{-1}. CO-stripping voltammograms were performed in a 0.5 M N$_2$-saturated H$_2$SO$_4$ solution, and then CO was bubbled for 30 min to allow the complete adsorption of CO onto the composites while the working electrode was kept at 0.1 V vs. Ag|AgCl. Excess CO in the electrolyte was then purged out with N$_2$ for 15 min. Chronoamperometric curves were obtained for 2000 s in 0.5 M H$_2$SO$_4$ solution containing 1.0 M CH$_3$OH at 0.70 V. EIS was carried out in 0.5 M H$_2$SO$_4$ in the presence of 1.0 M CH$_3$OH with a potential of 0.5 V, and the Nyquist plots were recorded between 0.1 and 100,000 Hz.

3.5. Photo-Electrochemical Measurements

A 15 W UV lamp (ZW3D15WZ105, Cnlight, Guangdong, China) was used to provide UV light (200 < λ < 400 nm) or visible light (λ > 400 nm) via different optical filters. It was placed at the bottom of the electrochemical cell, facing the working electrode at a distance of 15 cm. The irradiating power density was measured as 0.3 mW·cm^{-2}. Cyclic voltammograms and chronoamperometric curve analyses were carried out after external light irradiation for 30 min.

4. Conclusions

In summary, Pt-Ag bimetallic nanoparticles with small size and uniform size distribution were highly dispersed on the surface of GNs in a Pt-Ag/GNs composite with the aid of synchronous reduction by HCOOH. The good dispersion of the Pt-Ag nanoparticles, the strong interaction between the Pt and the Ag nanoparticles, and the good electron transfer capacity of the Pt-Ag/GNs composite contributed to its higher electrocatalytic activity, better tolerance of CO, and better stability for MOR. Under UV or visible light irradiation, Pt-Ag/GNs showed an efficient synergistic photoelectrocatalysis for MOR, which provides an effective strategy to utilize natural sunlight (UV and visible light) to produce clean energy, and favors potential applications of electro-photo catalysts in direct methanol fuel cell anodic reactions.

Acknowledgments: This work was financially supported by the National Natural Science Foundation of China (No. 21171037), and the Research Foundation of the Education Department of Fujian Province (No. JA13082 & No. JB13010).

Author Contributions: Zhongshui Li planned and designed the experiments; Shuhong Xu and Lingting Ye performed the experimental work; Zhongshui Li, Yanli Wang, and Fengling Lei contributed to the data analysis; Shuhong Xu and Lingting Ye wrote the manuscript; Shen Lin supervised the project and the revised manuscript; all authors discussed the results and approved of the final version of the manuscript.

Conflicts of Interest: The authors declare no conflict of interest.

References

1. Jiang, S.P.; Liu, Z.; Hao, L.T.; Mu, P. Synthesis and characterization of PDDA-stabilized Pt nanoparticles for direct methanol fuel cells. *Electrochim. Acta* **2006**, *51*, 5721–5730. [CrossRef]

2. Yu, L.H.; Xi, J.Y. TiO$_2$ nanoparticles promoted Pt/C catalyst for ethanol electro-oxidation. *Electrochim. Acta* **2012**, *67*, 166–171. [CrossRef]

3. Dutta, A.; Ouyang, J.Y. Ternary NiAuPt nanoparticles on reduced graphene oxide as catalysts toward the electrochemical oxidation reaction of ethanol. *ACS Catal.* **2015**, *16*, 1371–1380. [CrossRef]

4. Bo, Z.; Hu, D.; Kong, J.; Yan, J.H.; Cen, K.F. Performance of vertically oriented graphene supported platinum–ruthenium bimetallic catalyst for methanol oxidation. *J. Power Sources* **2015**, *273*, 530–537. [CrossRef]

5. Ammam, M.; Easton, E.B. PtCu/C and Pt(Cu)/C catalysts: Synthesis, characterization and catalytic activity towards ethanol electrooxidation. *J. Power Sources* **2013**, *222*, 79–87. [CrossRef]

6. Guo, J.X.; Sun, Y.F.; Zhang, X.; Tang, L.; Liu, H.T. FePt nanoalloys anchored reduced graphene oxide as high-performance electrocatalysts for formic acid and methanol oxidation. *J. Alloys Compd.* **2014**, *604*, 286–291. [CrossRef]

7. Hoseini, S.J.; Barzegar, Z.; Bahrami, M.; Roushani, M.; Rashidi, M. Organometallic precursor route for the fabrication of PtSn bimetallic nanotubes and Pt$_3$Sn/reduced-graphene oxide nanohybrid thin films at oil–water interface and study of their electrocatalytic activity in methanol oxidation. *J. Organomet. Chem.* **2014**, *769*, 1–6. [CrossRef]

8. Kepeniene, V.; Jablonskiene, J.; Vaiciuniene, J.; Kondrotas, R.; Juskenas, R. Investigation of graphene supported platinum-cobalt nanocomposites as electrocatalysts for ethanol oxidation. *J. Electrochem. Soc.* **2014**, *59*, 135–136.

9. Shen, Y.; Zhang, M.Z.; Xiao, K.J.; Xi, Z.Y. Synthesis of Pt, PtRh, and PtRhNi alloys supported by pristine graphene nanosheets for ethanol electrooxidation. *ChemCatChem* **2014**, *6*, 3254–3261. [CrossRef]

10. Lu, Y.Z.; Jiang, Y.T.; Wu, H.B.; Wei, C. Nano-PtPd cubes on graphene exhibit enhanced activity and durability in methanol electrooxidation after CO stripping-cleaning. *J. Phys. Chem. C* **2013**, *117*, 2926–2938. [CrossRef]

11. Li, L.Z.; Chen, M.X.; Huang, G.B.; Yang, N.; Zhang, L.; Wang, H.; Liu, Y.; Wang, W.; Gao, J.P. A green method to prepare Pd–Ag nanoparticles supported on reduced graphene oxide and their electrochemical catalysis of methanol and ethanol oxidation. *J. Power Sources* **2014**, *263*, 13–21. [CrossRef]

12. Kale, M.J.; Avanesian, T.; Christopher, P. Direct photocatalysis by plasmonic nanostructures. *ACS Catal.* **2013**, *4*, 116–128. [CrossRef]

13. Hou, W.B.; Cronin, S.B. A review of surface plasmon resonance-enhanced photocatalysis. *Adv. Funct. Mater.* **2013**, *23*, 1612–1619. [CrossRef]

14. Gao, L.N.; Yue, W.B.; Tao, S.S.; Fan, L.Z. Novel strategy for preparation of graphene-Pd, Pt composite, and its enhanced electrocatalytic activity for alcohol oxidation. *Langmuir* **2012**, *29*, 957–964. [CrossRef] [PubMed]

15. Sun, Y.Q.; Wu, Q.; Shi, G.Q. Graphene based new energy materials. *Energy Environ. Sci.* **2011**, *4*, 1113–1132. [CrossRef]

16. Lei, F.L.; Li, Z.S.; Ye, L.T.; Wang, Y.L.; Lin, S. One-pot synthesis of Pt/SnO$_2$/GNs and its electro-photo-synergistic catalysis for methanol oxidation. *Int. J. Hydrogen Energy* **2015**, *41*, 255–264. [CrossRef]

17. Ye, L.T.; Li, Z.S.; Zhang, L.; Lei, F.L.; Lin, S. A green one-pot synthesis of Pt/TiO$_2$/Graphene composites and its electro-photo-synergistic catalytic properties for methanol oxidation. *J. Colloid Interface Sci.* **2014**, *433*, 156–162. [CrossRef] [PubMed]

18. Zhang, H.L.; Hu, C.G.; He, X.S.; Liu, H.; Du, G.J.; Zhang, Y. Pt support of multidimensional active sites and radial channels formed by SnO$_2$ flower-like crystals for methanol and ethanol oxidation. *J. Power Sources* **2011**, *196*, 4499–4505. [CrossRef]

19. Wen, Z.L.; Yang, S.D.; Liang, Y.Y.; Wei, H.; Hao, T.; Liang, H.; Zhang, X.G.; Song, Q.J. The improved electrocatalytic activity of palladium/graphene nanosheets towards ethanol oxidation by tin oxide. *Electrochim. Acta* **2010**, *56*, 139–144. [CrossRef]

20. Ma, S.S.; Xue, J.J.; Zhou, Y.M.; Zhang, Z.W. Photochemical synthesis of ZnO/Ag$_2$O heterostructures with enhanced ultraviolet and visible photocatalytic activity. *J. Mater. Chem. A* **2014**, *2*, 7272–7280. [CrossRef]

21. Li, S.S.; Lv, J.J.; Hu, Y.Y.; Zheng, J.N.; Chen, J.R.; Wang, A.J.; Feng, J.J. Facile synthesis of porous Pt–Pd nanospheres supported on reduced graphene oxide nanosheets for enhanced methanol electrooxidation. *J. Power Sources* **2014**, *247*, 213–218. [CrossRef]

22. Zhang, L. A facile synthesis of flower-shaped TiO$_2$/Ag microspheres and their application in photocatalysts. *RSC Adv.* **2014**, *4*, 54463–54468. [CrossRef]

23. Sarkar, D.; Ghosh, C.K.; Mukherjee, S.; Chattopadhyay, K.K. Three dimensional Ag$_2$O/TiO$_2$ type-II (p–n) nanoheterojunctions for superior photocatalytic activity. *ACS Appl. Mater. Interfaces* **2013**, *5*, 331–337. [CrossRef] [PubMed]

24. Wang, S.L.; Qian, H.H.; Hu, Y.; Dai, W.; Zhong, Y.J.; Chen, J.F.; Hu, X. Facile one-pot synthesis of uniform TiO$_2$-Ag hybrid hollow spheres with enhanced photocatalytic activity. *Dalton Trans.* **2013**, *42*, 1122–1128. [CrossRef] [PubMed]

25. Cao, J.Y.; Guo, M.W.; Wu, J.Y.; Xu, J.; Wang, W.C.; Chen, Z.D. Carbon-supported Ag@Pt core–shell nanoparticles with enhanced electrochemical activity for methanol oxidation and oxygen reduction reaction. *J. Power Sources* **2015**, *277*, 155–160. [CrossRef]

26. Ganesan, P.; Prabu, M.; Sanetuntikul, J.; Shanmugam, S. Cobalt sulfide nanoparticles grown on nitrogen and sulfur codoped graphene oxide: An efficient electrocatalyst for oxygen reduction and rvolution reactions. *ACS Catal.* **2015**, *5*, 3625–3637. [CrossRef]

27. Kim, Y.; Shanmugam, S. Polyoxometalate-reduced graphene oxide hybrid catalyst: Synthesis, structure, and electrochemical properties. *ACS Appl. Mater. Interfaces* **2013**, *5*, 12197–12204. [CrossRef] [PubMed]

28. Graf, D.; Molitor, F.; Ensslin, K.; Stampfer, C.; Jungen, A.; Hierold, C.; Wirtz, L. Spatially resolved raman spectroscopy of single- and few-layer graphene. *Nano Lett.* **2007**, *7*, 238–242. [CrossRef] [PubMed]

29. Routh, P.; Das, S.; Shit, A.; Bairi, P.; Das, P.; Nandi, A.K. Graphene quantum dots from a facile sono-fenton reaction and its hybrid with a polythiophene graft copolymer toward photovoltaic application. *ACS Appl. Mater. Interfaces* **2013**, *5*, 12672–12680. [CrossRef] [PubMed]

30. Liu, X.W.; Duan, J.L.; Chen, H.L.; Zhang, Y.F.; Zhang, X.L. A carbon riveted Pt/graphene catalyst with high stability for direct methanol fuel cell. *Microelectron. Eng.* **2013**, *110*, 354–357. [CrossRef]

31. Yan, Y.; Wei, W.; Liu, Y.Q.; Wang, F.X.; Zhe, Z.; Lei, Z.Q. Carbon supported heterostructured Pd–Ag nanoparticle: Highly active electrocatalyst for ethylene glycol oxidation. *Int. J. Hydrogen Energy* **2015**, *40*, 2225–2230.

32. Wang, J.J.; Yin, G.P.; Zhang, J.; Wang, Z.B.; Gao, Y.Z. High utilization platinum deposition on single-walled carbon nanotubes as catalysts for direct methanol fuel cell. *Electrochim. Acta* **2007**, *52*, 7042–7050. [CrossRef]

33. Yang, C.Z.; Laak, N.K.v.d.; Chan, K.Y.; Zhang, X. Microwave-assisted microemulsion synthesis of carbon supported Pt-WO$_3$ nanoparticles as an electrocatalyst for methanol oxidation. *Electrochim. Acta* **2012**, *75*, 262–272. [CrossRef]

34. Huang, H.L.; Liu, Y.J.; Gao, Q.Z.; Ruan, W.S.; Lin, X.M.; Xin, L. Rational construction of strongly coupled metal–metal oxide–graphene nanostructure with excellent electrocatalytic activity and durability. *ACS Appl. Mater. Interfaces* **2014**, *6*, 10258–10264. [CrossRef] [PubMed]

35. Maiyalagan, T.; Alaje, T.O.; Scott, K. Highly stable Pt–Ru nanoparticles supported on three-dimensional cubic ordered mesoporous carbon (Pt-Ru/CMK-8) as promising electrocatalysts for methanol oxidation. *J. Phys. Chem. C* **2012**, *116*, 2630–2638. [CrossRef]

36. Kashyout, A.B.; Bakr, A.; Nassr, A.B.A.A.; Giorgi, L.; Maiyalagan, T.; Youssef, B.A.B. Electrooxidation of methanol on carbon supported Pt-Ru nanocatalysts prepared by ethanol reduction method. *Int. J. Electrochem. Sci.* **2011**, *6*, 379–393.

37. Lin, C.T.; Shiao, M.H.; Chang, M.N.; Chu, N.; Chen, Y.W.; Peng, Y.H.; Liao, B.H.; Huang, H.J.; Hsiao, C.N.; Tseng, F.G. A facile approach to prepare silicon-based Pt-Ag tubular dendritic nano-forests (tDNFs) for solar-light-enhanced methanol oxidation reaction. *Nanoscale Res. Lett.* **2015**, *10*, 1–8. [CrossRef] [PubMed]

38. Li, Z.S.; Huang, X.M.; Zhang, X.F.; Zhang, L.; Lin, S. The synergistic effect of graphene and polyoxometalates enhanced electrocatalytic activities of Pt-{PEI-GNs/[PMo$_{12}$O$_{40}$]$^{3-}$}$_n$ composite films regarding methanol oxidation. *J. Mater. Chem.* **2012**, *22*, 23602–23607. [CrossRef]

39. Zheng, J.N.; Lv, J.J.; Li, S.S.; Xue, M.W.; Wang, A.J.; Feng, J.J. One-pot synthesis of reduced graphene oxide supported hollow Ag@Pt core-shell nanospheres with enhanced electrocatalytic activity for ethylene glycol oxidation. *J. Mater. Chem.* **2014**, *2*, 3445–3451. [CrossRef]

40. Maiyalagan, T.; Dong, X.C.; Chen, P.; Wang, X. Electrodeposited Pt on three-dimensional interconnected graphene as a free-standing electrode for fuel cell application. *J. Mater. Chem.* **2012**, *22*, 5286–5290. [CrossRef]

41. Maiyalagan, T. Pt–Ru nanoparticles supported PAMAM dendrimer functionalized carbon nanofiber composite catalysts and their application to methanol oxidation. *J. Solid State Electrochem.* **2009**, *13*, 1561–1566. [CrossRef]

42. Sanetuntikul, J.; Ketpang, K.; Shanmugam, S. Hierarchical nanostructured Pt$_8$Ti-TiO$_2$/C as an efficient and durable anode catalyst for direct methanol fuel cells. *ACS Catal.* **2015**, *5*, 7321–7327. [CrossRef]

43. Guo, R.; Zhang, G.K.; Liu, J. Preparation of Ag/AgCl/BiMg$_2$VO$_6$ composite and its visible-light photocatalytic activity. *Mater. Res. Bull.* **2013**, *48*, 1857–1863. [CrossRef]

44. Jin, R.; Cao, Y.; Mirkin, C.A.; Kelly, K.L.; Schatz, G.C.; Zheng, J.G. Photoinduced conversion of silver nanospheres to nanoprisms. *Science* **2001**, *294*, 1901–1903. [CrossRef] [PubMed]

45. Liu, Y.P.; Fang, L.; Lu, H.D.; Li, Y.W.; Hu, C.Z.; Yu, H.G. One-pot pyridine-assisted synthesis of visible-light-driven photocatalyst Ag/Ag$_3$PO$_4$. *Appl. Catal. B* **2012**, *115–116*, 245–252. [CrossRef]

46. Kovtyukhova, N.I.; Ollivier, P.J.; Martin, B.R.; Mallouk, T.E.; Chizhik, S.A.; Buzaneva, E.V.; Gorchinskiy, A.D. Layer-by-layer assembly of ultrathin composite films from micron-sized graphite oxide sheets and polycations. *Chem. Mater.* **1999**, *11*, 771–778. [CrossRef]

47. Zeng, Q.; Cheng, J.S.; Tang, L.H.; Liu, X.F.; Liu, Y.Z.; Li, J.H.; Jiang, J.H. Self-assembled graphene–enzyme hierarchical nanostructures for electrochemical biosensing. *Adv. Funct. Mater.* **2010**, *20*, 3366–3372. [CrossRef]

catalysts

Article

Pulsed Laser Deposition of Platinum Nanoparticles as a Catalyst for High-Performance PEM Fuel Cells

Hamza Qayyum [1,2,3], Chung-Jen Tseng [4], Ting-Wei Huang [4] and Szu-yuan Chen [1,2,3,*]

[1] Institute of Atomic and Molecular Sciences, Academia Sinica, Taipei 106, Taiwan; hamzaqayyum02@yahoo.com
[2] Department of Physics, National Central University, Taoyuan 320, Taiwan
[3] Molecular Science and Technology Program, Taiwan International Graduate Program, Academia Sinica, Taipei 115, Taiwan
[4] Department of Mechanical Engineering, National Central University, Taoyuan 320, Taiwan; cjtseng@ncu.edu.tw (C.-J.T.); ck12332@gmail.com (T.-W.H.)
* Correspondence: sychen@ltl.iams.sinica.edu.tw; Tel.: +886-2-2366-8276

Academic Editors: Vincenzo Baglio and David Sebastián
Received: 29 September 2016 ; Accepted: 17 November 2016; Published: 22 November 2016

Abstract: The catalyst layers for polymer-electrolyte-membrane (PEM) fuel cells were fabricated by deposition of platinum directly onto the gas diffusion layer using pulsed laser deposition (PLD). This technique reduced the number of steps required to synthesize the catalyst layers and the amount of Pt loading required. PEM fuel cells with various Pt loadings for the cathode were investigated. With a cathode Pt loading of 100 $\mu g \cdot cm^{-2}$, the current density of a single cell reached 1205 $mA \cdot cm^{-2}$ at 0.6 V, which was close to that of a single cell using an E-TEK (trademark) Pt/C electrode with a cathode Pt loading of 400 $\mu g \cdot cm^{-2}$. Furthermore, for a PEM fuel cell with both electrodes prepared by PLD and a total anode and cathode Pt loading of 117 $\mu g \cdot cm^{-2}$, the overall Pt mass-specific power density at 0.6 V reached 7.43 $kW \cdot g^{-1}$, which was five times that of a fuel cell with E-TEK Pt/C electrodes. The high mass-specific power density was due to that a very thin nanoporous Pt layer was deposited directly onto the gas diffusion layer, which made good contact with the Nafion membrane and thus resulted in a low-resistance membrane electrode assembly.

Keywords: pulsed laser deposition; polymer-electrolyte-membrane fuel cell; catalyst; nanoparticle

1. Introduction

Recently, fuel cell technologies have received much attention due to growing concerns regarding the depletion of fossil fuels and climate change. The polymer-electrolyte-membrane (PEM) fuel cell is one of the most promising technologies. However, the high cost of PEM fuel cells has hindered its commercialization. The cost of a PEM fuel cell depends largely on the amount of Pt catalyst used. Therefore, the development of new methods to reduce Pt loading and achieve a higher Pt mass-specific power density (MSPD) is an active research area [1].

The preparation of membrane electrode assemblies (MEA) by directly depositing Pt onto a gas diffusion layer (GDL) is an efficient way because it results in a thin catalyst layer and good dispersion of Pt nanoparticles which translates into high Pt utilization. Various physical vapor deposition techniques such as high-power impulse magnetron sputtering [2], sputtering [3,4], e-beam evaporation [5], dual ion-beam assisted deposition [6] and pulsed laser deposition (PLD) [7,8] have been pursued to achieve high power density with lower Pt loading. Physical vapor deposition techniques offer benefits over chemical techniques of being a one-step process that leads to the formation of a thin film of nanoparticles. Some researchers have also deposited Pt onto the Nafion [9] or polytetrafluoroethylene (PTFE) sheets [10], but the overall cell performance was not as high as when Pt was deposited on GDL.

PLD has developed into an important deposition technique in the past few decades. One of the major advantages of PLD over other deposition techniques is its capability to generate high kinetic energy ions and atoms. It is also straightforward in generating nanoparticles by performing pulsed laser deposition in an ambient gas atmosphere, as a result of cooling of the ablation plume by the ambient gas [11]. By controlling the pressure of the ambient gas, one can control the size of deposited particles and the porosity of the grown film [12]. Cunningham et al. [7] reported that by using PLD in He atmosphere to deposit Pt onto GDL, with a very low Pt loading of 17 μg·cm^{-2}, the catalyst-loaded GDL could achieve a current density of 780 mA·cm^{-2} when used at the anode. In our earlier work [11], we had achieved a high current density of 1032 mA·cm^{-2} with an anode Pt loading of 13 μg·cm^{-2} by using PLD in Ar atmosphere. The better performance compared to the previous work could be ascribed to the smaller Pt nanoparticle size. Since the oxygen reduction reaction (ORR) at the cathode is very sluggish in nature compared to hydrogen oxidation reaction (HOR) at the anode, the cathode requires a much larger amount of catalyst compared to the anode. Therefore, it is generally recognized that the ORR at the cathode is the bottleneck and the reduction of Pt usage at the cathode holds the key to lowering the overall MSPD. Previously, only Mròz et al. [8] have reported using PLD to deposit Pt for the cathode of PEM fuel cell. They used PLD in vacuum to deposit Pt at a loading of 7 μg·cm^{-2} to achieve a maximum power density of 188 mW·cm^{-2} and a current density of 100 mA·cm^{-2} at 0.6 V. Although the Pt loading used was low, the power and current densities were too low to be practically useful. For instance, it is required that for vehicle operation a fuel cell should produce at least 1500 mA·cm^{-2} at 0.6 V [13]. The low power and current density values of Mròz et al. resulted from the use of PLD in vacuum. It led to the formation of a dense, low-porosity Pt layer instead of a nanoporous structure that ensures a high electrochemical surface area (ECSA) and good charge transfer and gas transport. For fabricating a high-performance MEA, it is paramount to control the size of nanoparticles and the porosity and thickness of the catalyst layer.

In this work, PLD in Ar atmosphere was used to deposit Pt nanoparticles onto GDL with a microporous layer for the cathode first and then for both electrodes for developing high-performance PEM fuel cells. Fuel cells using pulsed laser deposited Pt nanoparticles with four different Pt loadings (50, 75, 100 and 125 μg·cm^{-2}) as the cathode catalyst, and commercial E-TEK (trademark) Pt/C with a Pt loading of 200 μg·cm^{-2} as the anode catalyst, were characterized. The dependence of fuel cell performance on the cell outlet pressure was also investigated. The cells were also characterized by using electrochemical impedance spectroscopy. Finally, the fuel cell with the catalysts of both the anode and the cathode prepared by using PLD was characterized. With a low total Pt loading of 117 μg·cm^{-2}, the current density at 0.6 V reached 1450 mA·cm^{-2}, corresponding to a power density of 870 mW·cm^{-2}. This work demonstrated that PLD could be used to reduce the amount of Pt used while maintaining high current and power densities in PEM fuel cells for practical applications.

2. Results and Discussion

Figure 1 shows the transmission electron microscopy (TEM) images of Pt catalyst deposited on GDL with various Pt loadings. It was found that the average particle diameter increased from 2.7 nm to 5.0 nm with increasing Pt loading from 50 μg·cm^{-2} to 125 μg·cm^{-2}. This could be ascribed to the presence of free Pt atoms remaining when the ablation plume reaches the substrate. The free atoms in later laser shots could attach to the Pt nanoparticles deposited by earlier laser shots, thereby increasing the average particle size. This effect could also contribute to the reduction of mass-specific electrochemical surface area (MSECSA) with increasing Pt loading shown later. For the case of 100 μg·cm^{-2}, the particle diameter was 4.9 $\pm$ 0.4 nm. This number obtained from TEM measurement was consistent with that retrieved from X-ray diffraction measurement as reported previously [11].

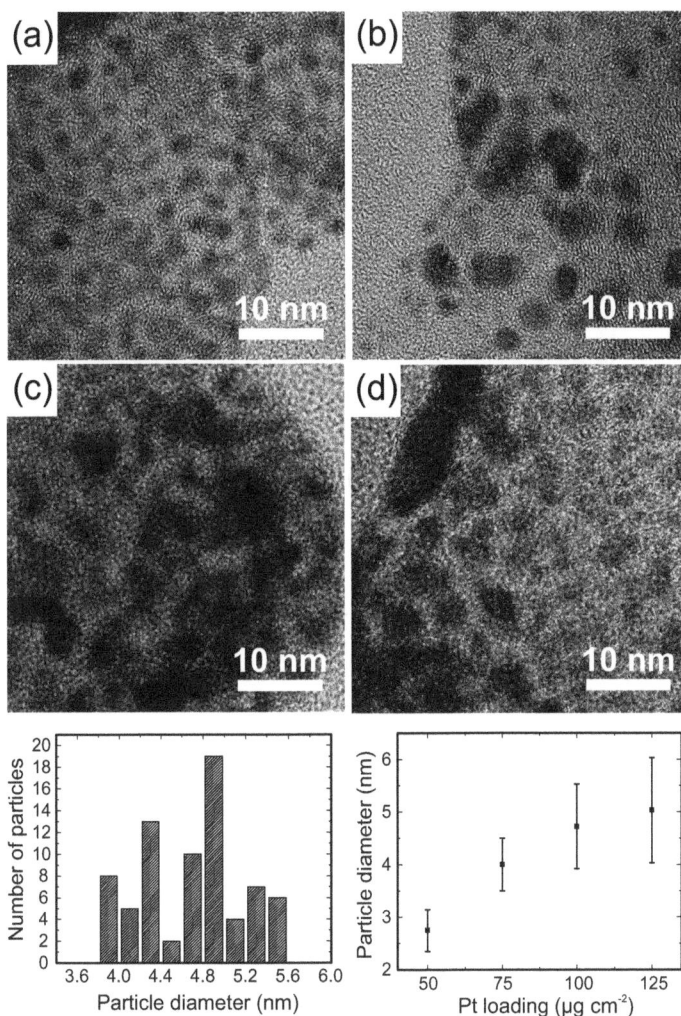

Figure 1. TEM images of Pt particles deposited by PLD at 107-Pa Ar pressure on a GDL with various Pt loadings: (**a**) 50 $\mu g \cdot cm^{-2}$; (**b**) 75 $\mu g \cdot cm^{-2}$; (**c**) 100 $\mu g \cdot cm^{-2}$ and (**d**) 125 $\mu g \cdot cm^{-2}$. The size distribution for 100 $\mu g \cdot cm^{-2}$ and the dependence of particle diameter on Pt loading are shown in the bottom panels. The error bar for each case indicates the standard deviation of particle size distribution.

Figure 2 shows the scanning electron microscopy (SEM) images of the surface of the GDLs with various Pt loadings. The deposited Pt did not form a flat film because of the high porosity of the GDL. With a Pt loading of 75 $\mu g \cdot cm^{-2}$, the deposited nanoparticles were uniformly dispersed on the surface of the underlying nanoporous support. With a Pt loading of 100 $\mu g \cdot cm^{-2}$, Pt nanoparticles aggregated to form small clusters, leading to decreased porosity as compared with the 75 $\mu g \cdot cm^{-2}$ sample. With a Pt loading of 125 $\mu g \cdot cm^{-2}$, aggregation was even more severe, resulting in large clusters. In addition, the voids between Pt nanoparticles became smaller, presumably due to the deposition of free atoms in the ablation plume [11]. The formation of large clusters and the decrease in porosity at higher Pt loadings resulted in reduction of specific surface area, which had a negative effect on fuel cell performance.

Figure 2. SEM images of Pt particles deposited by PLD at 107-Pa Ar pressure on a GDL with various Pt loadings: (**a**) GDL only; (**b**) 75 µg·cm^{-2}; (**c**) 100 µg·cm^{-2}; and (**d**) 125 µg·cm^{-2}.

Figure 3 shows the cyclic voltammograms (CV) recorded for PLD samples with various Pt loadings. The peaks of hydrogen adsorption/desorption in the potential region of 0 to 0.4 V vs. normal hydrogen electrode (NHE) can be seen clearly. The hydrogen desorption region was integrated to calculate ECSA. It was found that the ECSA increased from 1.118 to 1.631 m^2 when the Pt loading was increased from 50 to 125 µg·cm^{-2}, but with a trend of saturation. To relate the surface area of the catalyst with the mass-specific power density of the fuel cell, MSECSA was calculated as the ECSA divided by the Pt loading. The MSECSA was found to be 22.37, 17.32, 16.25, and 13.05 m^2·g^{-1} for 50, 75, 100, and 125 µg·cm^{-2} Pt loadings, respectively. A decrease in MSECSA with increasing loading was observed. In an ideal case when the Pt particles are distributed in single layer and their sizes are independent of Pt loading, the MSECSA should be constant for all Pt loadings. However, due to the increase of the degree of Pt particle aggregation and the decrease in porosity as shown in Figure 2, MSECSA decreased with increasing Pt loading. This result is consistent with the findings of Fabbri et al. [14]. They attributed the decrease in MSECSA to a transition of dispersed nanoparticles into aggregates and then to an extended layer with increasing Pt loading.

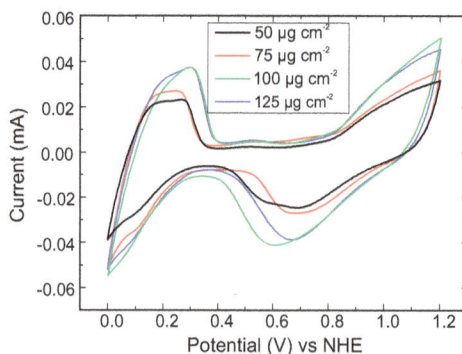

Figure 3. Cyclic voltammograms of PLD samples with various Pt loadings recorded in 0.5 M H$_2$SO$_4$ solution at a scan rate of 20 mV·s^{-1}.

Figure 4 shows the J-V and power density curves of PEM fuel cells with various Pt loadings on the cathode prepared using PLD. The anode was E-TEK Pt/C with a Pt loading of 200 $\mu g \cdot cm^{-2}$. The fuel cells with Pt loadings of 50 and 75 $\mu g \cdot cm^{-2}$ showed poor performance compared to the cell with commercial E-TEK Pt/C. This can be ascribed to insufficient Pt available for electrochemical reactions. Since the oxygen reduction reaction is sluggish in nature, the cathode requires a higher Pt content to attain a high reaction rate. With a Pt loading of 100 $\mu g \cdot cm^{-2}$ the current density at 0.6 V reached 1205 $mA \cdot cm^{-2}$, corresponding to a power density of 723 $mW \cdot cm^{-2}$ and a cathode MSPD of 7.3 $kW \cdot g^{-1}$. The cathode MSPD of the reference E-TEK sample at 0.6 V was 1.7 $kW \cdot g^{-1}$. The cathode MSPD at 0.6 V was by a factor of four higher than that of the reference sample using E-TEK Pt/C with a Pt loading of 400 $\mu g \cdot cm^{-2}$. The power densities at 0.6 V for various cases are shown in Table 1. The cathode prepared by PLD with a Pt loading of 100 $\mu g \cdot cm^{-2}$ had only one fourth of the Pt loading of the commercial E-TEK sample but showed similar power generation. The power density for the fuel cell with a cathode loading of 125 $\mu g \cdot cm^{-2}$ was close to that of the fuel cell with a cathode loading of 100 $\mu g \cdot cm^{-2}$. The saturation in power density with increasing Pt loading can be attributed to the saturation in ECSA (as discussed earlier in CV results). It may also result from mass transport limitation occurring in the catalyst layer due to the increased degree of aggregation of nanoparticles and the reduced porosity with increasing Pt loading (see Figure 2).

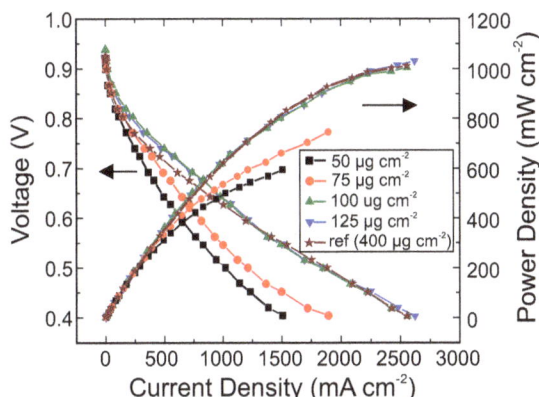

Figure 4. J-V and power density curves of PEM fuel cells having the cathode deposited by PLD with various Pt loadings. The performance of a PEM fuel cell using E-TEK Pt/C for the cathode with a Pt loading of 400 $\mu g \cdot cm^{-2}$ is shown for comparison. All data were measured at an outlet pressure of 100 kPa.

Table 1. Performance of PEM fuel cells using cathode catalyst prepared by PLD with various Pt loadings in comparison to that using E-TEK Pt/C with a Pt loading of 400 $\mu g \cdot cm^{-2}$.

Cathode Catalyst Deposition Method	Anode Pt Loading * ($\mu g \cdot cm^{-2}$)	Cathode Pt Loading ($\mu g \cdot cm^{-2}$)	Current Density at 0.6 V ($mA \cdot cm^{-2}$)	Power Density at 0.6 V ($mW \cdot cm^{-2}$)	Cathode MSPD ($kW \cdot g^{-1}$)
PLD	200	50	650	390	7.8
	200	75	820	490	6.5
	200	100	1205	723	7.3
	200	125	1216	729	5.8
E-TEK	200	400	1190	710	1.7

* The anodes of all samples use E-TEK Pt/C.

The resistance of a fuel cell is one of the key parameters that significantly affect the cell performance. To determine the magnitudes of the resistances, electrochemical impedance spectroscopy was used to measure the ohmic resistance (R_Ω) and charge transfer resistance (R_{ct}) of the cells with different Pt loadings. Figure 5 shows the in situ impedance response of the fuel cells in the form of Nyquist plots at 0.8 V for various Pt loadings. Only one semi-circle is observed in the Nyquist plot as the electrode process is dominated by the ORR at low current densities [15]. The high-frequency intercept on the real axis indicates the total ohmic resistance (R_Ω) of the fuel cell. The diameter of the arc is a measure of the charge transfer resistance of the ORR (R_{ct}) [15]. It can be seen that all samples had very similar ohmic resistance, which was expected since the deposition condition, membrane, GDL, and fuel cell hardware were the same in all of these cases. In contrast, the charge transfer resistance decreased with increasing Pt loading. This can be expected from the increase of ECSA with increasing Pt loading, which should result in larger total charge transfer rate and thus lower charge transfer resistance. The dependence of R_{ct} on Pt loading is consistent with that of fuel cell performance shown in Figure 4: the lower the resistance, the higher the fuel cell performance.

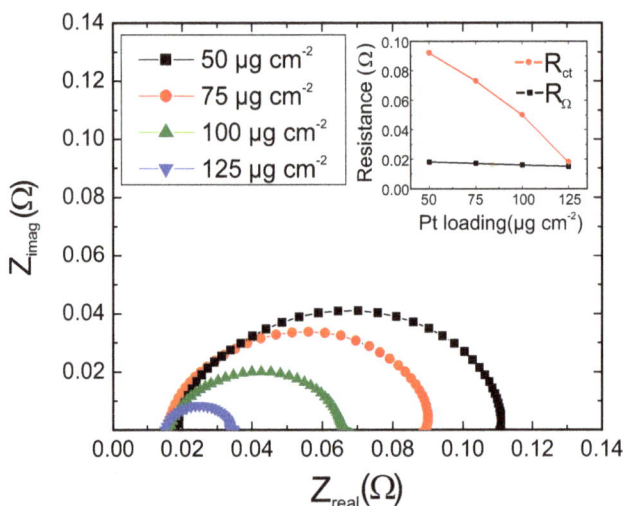

Figure 5. Electrochemical impedance spectroscopy curves of PEM fuel cells with various Pt loadings at the cathode at a cell voltage of 0.8 V. The inset shows the charge-transfer resistance and ohmic resistance as functions of Pt loading.

The cell outlet pressure is also an important parameter in the operation of a PEM fuel cell. Increasing the outlet pressure can positively affect the fuel cell performance by enhancing the ORR kinetics through increasing gas concentration and mass transport rate in the catalyst layer [16,17]. Figure 6 shows the J-V and power density curves of a PEM fuel cell with various outlet pressures. The cathode with a Pt loading of 100 μg·cm^{-2} was prepared by using PLD, and commercial E-TEK Pt/C electrode with a Pt loading of 200 μg·cm^{-2} was used for the anode. As can be seen, increasing outlet pressure raised the fuel cell performance. Figure 7 shows the current density at 0.6 V as a function of outlet pressure. The current density at 0.6 V increased with increasing outlet pressure. Figure 7 also shows R_{ct} measured with electrochemical impedance spectroscopy at various outlet pressures. R_{ct} decreased with increasing outlet pressure, indicating that an increase in outlet pressure can speed up the ORR kinetics and thus effectively enhance fuel cell performance.

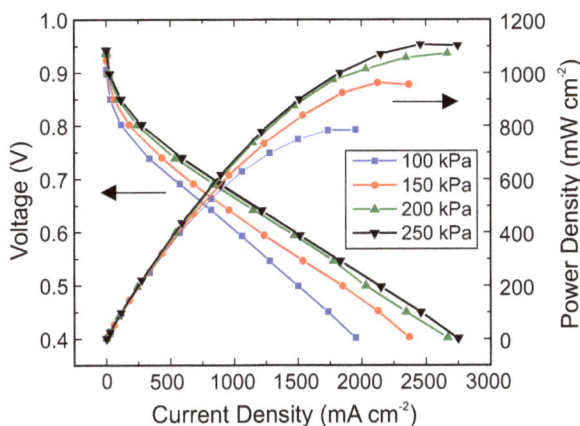

Figure 6. J-V and power density curves of a PEM fuel cell with various outlet pressures. The cathode was deposited by using PLD with a Pt loading of 100 $\mu g \cdot cm^{-2}$.

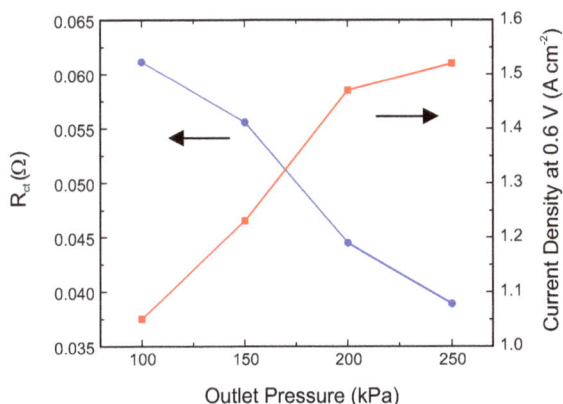

Figure 7. Current density at a cell voltage of 0.6 V and charge transfer resistance at a cell voltage of 0.8 V as functions of outlet pressure.

Figure 8 shows the J-V and power density curves of a PEM fuel cell having both electrodes prepared by PLD with an anode Pt loading of 17 $\mu g \cdot cm^{-2}$ and a cathode Pt loading of 100 $\mu g \cdot cm^{-2}$. The results for a cell using E-TEK Pt/C at an anode Pt loading of 200 $\mu g \cdot cm^{-2}$ and a cathode Pt loading of 400 $\mu g \cdot cm^{-2}$ under the same operating conditions are also shown for comparison. The current density at 0.6 V of the PLD-prepared fuel cell with an outlet pressure of 100 kPa was 1092 mA·cm^{-2}, which was about the same as that of a cell using E-TEK Pt/C with a total Pt loading of 600 $\mu g \cdot cm^{-2}$. This shows that electrodes fabricated by using PLD can achieve higher Pt utilization. The beneficial effect of increasing outlet pressure on fuel cell performance is also shown in Figure 8. At an outlet pressure of 200 kPa, the current density at 0.6 V was increased to 1450 mA·cm^{-2}. This corresponds to an overall MSPD at 0.6 V of 7.43 kW·g^{-1}, which is higher than that of a cell using E-TEK Pt/C by a factor of five.

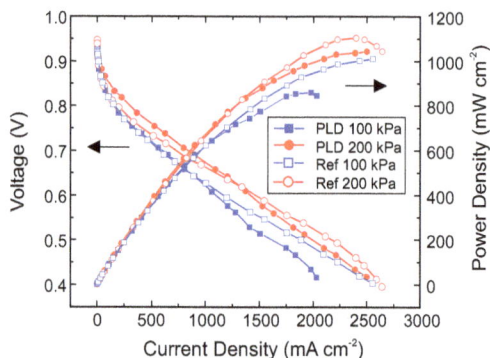

Figure 8. J-V and power density curves of a PEM fuel cell having both electrodes prepared by PLD with an anode Pt loading of 17 $\mu g \cdot cm^{-2}$ and a cathode Pt loading of 100 $\mu g \cdot cm^{-2}$ for various outlet pressures. The results using E-TEK Pt/C electrodes with an anode Pt loading of 200 $\mu g \cdot cm^{-2}$ and a cathode Pt loading of 400 $\mu g \cdot cm^{-2}$ under the same operating conditions are shown for comparison.

For comparison, Table 2 provides a partial summary of mass-specific power densities reported previously. Some previous works achieved high cathode mass-specific power density by using sputtering or PLD, corresponding to high Pt utilization [2,8], but their current density and power density are too low to be practically useful. The need to use a fuel cell with a larger area in order to attain the required current and power for practical application will raise the cost of other parts of fuel cells, occupy a larger space, and consume more power when installed on an automotive. To the best of our knowledge, the mass-specific power density obtained in the present work, using PLD, is the highest among all works that used pure Pt catalyst on this sort of carbon support with practically useful current density.

Table 2. Comparison of PEM fuel cells with catalyst fabricated by various deposition techniques. O_2 was used for the cathode in all cases.

Deposition Technique	Pt Loading Anode/Cathode ($\mu g \cdot cm^{-2}$)	Operating Condition (°C/kPa)	Current Density at 0.6 V (mA·cm^{-2})	Overall MSPD at 0.6 V (kW·g^{-1})	Reference
Pulsed electrodeposition	400/120	80/250	1400	1.61	[18]
Reactive spray deposition	75/75	80/180	1400	6.00	[19]
DC magnetron sputtering	50/100	70/100	800	3.28	[20]
Electrospray	40/120	70/200	1200	4.50	[21]
Sputtering	100/60	80/100	490	1.83	[4]
Impulse sputtering	20/20	70/300	400	6.00	[2]
Atomic layer deposition	180/500	80/100	1400	1.20	[22]
Pulsed laser deposition	-/7	80/240	100	-	[8]
Chemical vapor deposition	200/200	80/100	1100	1.65	[23]
This work	17/100	70/200	1450	7.43	

Besides the significant enhancement in the MSPD of the MEAs prepared by using PLD as compared to E-TEK Pt/C, from the prospective of mass production, the use of PLD to produce the catalyst layers is a single-step dry process in which Pt from its simplest form as a metal block is directly transferred onto GDL to form the end catalyst/support film. In contrast, the conventional ink process requires several wet chemical synthesis steps to reach the end product. In addition to being more tedious, a significant amount of Pt could be lost in these processes. Contrarily, with PLD, there is practically no loss of Pt [11]. In addition, the use of PLD allows for precise control over the size of Pt particles and the thickness of the deposited film, which is important in good manufacturing quality control for achieving high catalytic activity [24]. However, PLD has some potential drawbacks such as the requirement of a vacuum system and a longer process time, which have to be circumvented for mass production.

From a structural viewpoint, the catalyst layers prepared using PLD are at least 10 times thinner than that produced using the conventional ink process. The sub-micrometer thickness of the catalyst layer translates to a short ionic resistive path length because it takes a very short distance for protons to traverse between the Nafion membrane and the catalyst surface sites. In contrast, the thick catalyst layer produced with the conventional ink process requires the application of a substantial amount of ionomer to facilitate proton transport. It has been shown that the application of Nafion in the catalyst layer has a negative effect on the fuel cell performance due to an increase in the oxygen transport resistance. Since the oxygen transport resistance increases with increasing thickness of the catalyst layer and thus the Pt loading used [25–27], it becomes a significant limiting factor in the case of E-TEK Pt/C slurry. On the contrary, the small thickness of the catalyst layer prepared by PLD eliminates the need of applying an ionomer solution for improving proton conduction. This is because Pt particles deposited using PLD are all on the surface of the support and thus are all close to the Nafion membrane when assembled into an MEA. The polymer electrolyte could have good contact with the deposited Pt particles. The swelling of the Nafion membrane due to the humidified environment during the operation of fuel cell [28,29] may further enhance the contact between the Pt particles and the Nafion membrane. These lead to the high Pt utilization efficiency compared with that using E-TEK Pt/C slurry. A similar phenomenon was also observed for 3M nanostructured thin film catalyst [27].

Another critical issue for the catalyst of a fuel cell is its stability. In our earlier work, it was found that the PLD-produced catalyst with a low Pt loading of 17 $\mu g \cdot cm^{-2}$ exhibits much higher durability than E-TEK Pt/C. However, this characteristic may not be translatable to the case of high Pt loading such as 100 $\mu g \cdot cm^{-2}$ used in this work because, in the case of high Pt loading, the Pt nanoparticles form a porous Pt film on top of the carbon support, whereas in the case of low Pt loading, the Pt nanoparticles are dispersed on the carbon support as a very thin overlayer. To investigate the electrochemical stability of the catalyst with high Pt loading, an accelerated degradation test (ADT) was performed. Figure 9 shows the cyclic voltammograms of the PLD sample with 100 $\mu g \cdot cm^{-2}$ Pt loading before and after 5000 potential cycles. The inset shows the MSECSA as a function of cycling number. An MSECSA retention of 72% was observed after 5000 potential cycles, much better than the mere 11% MSECSA retention for the case of E-TEK Pt/C (shown in Ref. [11]). This indicates that the PLD-prepared electrode exhibits a much higher durability than E-TEK Pt/C for both the cases of high Pt loading and low Pt loading. The higher durability could be ascribed to the much higher degree of graphitization of the GDL used than the carbon black in E-TEK Pt/C [11].

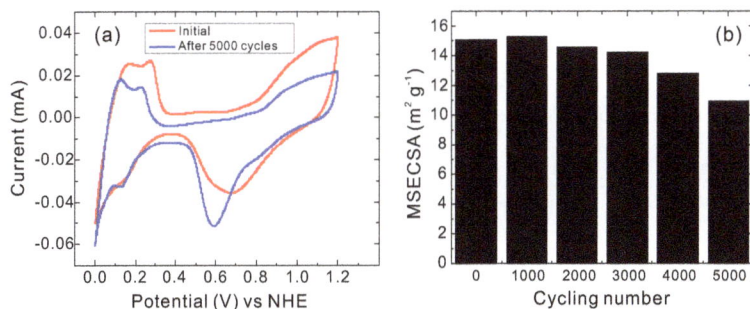

Figure 9. (**a**) cyclic voltammograms of PLD sample with 100 $\mu g \cdot cm^{-2}$ Pt loading before and after 5000 potential cycles; (**b**) MSECSA as a function of cycling number.

3. Materials and Methods

3.1. Catalyst Preparation

Pulsed laser deposition in Ar atmosphere was used to deposit Pt catalyst onto GDLs [30]. The PLD target used was a Pt disk (purity > 99.99%) with 10-mm diameter and 5-mm thickness. An Nd:YAG laser beam of 355-nm wavelength, 8-ns pulse duration, *p*-polarization, and 10-Hz repetition rate (PRO-350, Spectra-Physics, Santa Clara, CA, USA) was focused on the target with an incidence angle of 45°. The on-target beam size was 500 µm in clear aperture, and the peak laser fluence was 230 J·cm^{-2}. The substrate for coating was a 4 cm × 4 cm carbon GDL. It was a hydrophobized gas diffusion layer (5 wt. % PTFE) with a microporous layer on the top (SIGRACET GDL24BC, SGL CARBON GmbH, Meitingen, Germany). The substrate was located in the normal direction of the target, and the target-to-substrate distance was set at 4 cm. The target and the substrate was installed in a vacuum chamber. The chamber was pumped down and then backfilled with Ar gas before deposition. The Ar pressure was fixed at 107 Pa. A calibrated quartz microbalance (SQM-160, Sigma Instruments, Cranberry Township, PA, USA) was used to measure the deposition rate. The Pt loading on a GDL was varied by changing the number of laser shots fired. It was also confirmed by weighing the GDL before and after the deposition of Pt.

3.2. Electrochemical Measurements

For measuring the ECSA of the Pt catalyst, cyclic voltammetry was performed with a potentiostat (PGSTAT302, Metrohm Autolab, Utrecht, Netherlands) and then the hydrogen desorption region was integrated [31]. The ECSA was determined by using the formula:

$$ECSA = \frac{Q_H}{Qd_m},$$ (1)

where Q_H (in unit of mC) is the total charge obtained from the time integration of the hydrogen desorption peak in the CV curve, and Qd_m, which is 0.21 mC·cm^{-2}, is the areal density of electron charge in the hydrogen layer when a clean Pt surface is covered by a monolayer of hydrogen atoms with 100% surface coverage. A three-electrode setup using Ag/AgCl (saturated KCl) as the reference electrode and Pt wire as the counter electrode in 0.5 M H$_2$SO$_4$ solution was used. For this measurement, the PLD sample cut into 3 mm × 3 mm area was adhered to a glassy carbon electrode by using a carbon tape. The cyclic potentials were scanned between −0.2 and 1.0 V (vs. Ag/AgCl) at a rate of 20 mV·s^{-1} at room temperature. The electrolyte was purged with N$_2$ for 10 min before CV measurement. For investigating the electrochemical stability of the Pt-catalyst/support, an accelerated degradation test was performed by potential cycling in the potential region between the oxidation and reduction of Pt (0.5 V and 1.0 V versus Ag/AgCl, respectively) at a scan rate of 100 mV·s^{-1} [32].

3.3. Fabrication of MEAs and Measurement of Polarization Curves

In the first part of the experiment, electrodes made by using PLD were tested on the cathode of a single PEM fuel cell, while standard E-TEK Pt/C electrode of 200 µg·cm^{-2} Pt loading was used for the anode. The MEA with an active area of 3 cm × 3 cm was made by hot-pressing a GDL loaded with PLD Pt catalyst for the cathode, a Nafion XL membrane coated with E-TEK Pt/C on one side with a Pt loading of 200 µg·cm^{-2}, and an unloaded GDL at 1150 kg and 135 °C for 120 s. A PEM fuel cell was assembled by sandwiching the MEA with two flow-field plates which use metal foams coated with PTFE as flow distributors [33,34]. A standard fuel cell testing system was used for measuring the J-V curves of the PEM fuel cells [33]. In all fuel cell tests, pure H$_2$ and O$_2$ were used as fuel and oxidant, respectively, and they were both humidified to 100% relative humidity by using inline humidifers. The hydrogen (anode) flow rate and the oxygen (cathode) flow rate were both 450 sccm. The outlet pressures of the anode and the cathode were set to be the same and varied together. The temperatures of the cell and the humidifiers were all set to 70 °C for all of the measurements done in this work.

In the second part of the experiment, two different fuel cells were compared. The first one had both electrodes made by PLD with Pt loadings of 17 $\mu g \cdot cm^{-2}$ and 100 $\mu g \cdot cm^{-2}$ for the anode and the cathode, respectively. The second one had both electrodes using E-TEK Pt/C with Pt loadings of 200 $\mu g \cdot cm^{-2}$ and 400 $\mu g \cdot cm^{-2}$ for the anode and the cathode, respectively. Electrochemical impedance spectroscopy was carried out in situ with a potentiostat (PGSTAT 302) installed in the fuel cell testing system at a cell potential of 0.6 V with a modulation amplitude of 5 mV and a modulation frequency of 0.1–10,000 Hz.

3.4. Morphology Characterizations

The dispersion (spatial distribution) of Pt particles on GDL was investigated with scanning electron microscopy (SEM) (Nova NanoSEM 230, FEI, Hillsboro, OR, USA). The size distribution of Pt particles was characterized with transmission electron microscopy (TEM) (JEM-2100, JEOL, Peabody, MA, USA).

4. Conclusions

In summary, production of high-performance catalyst layer with low Pt loading was achieved by using PLD. The performance of fuel cells with PLD-prepared cathodes of various Pt loadings was investigated. The fuel cell with the highest MSPD was obtained with an anode Pt loading of 17 $\mu g \cdot cm^{-2}$ and a cathode Pt loading of 100 $\mu g \cdot cm^{-2}$. At an outlet pressure of 100 kPa and a cell temperature of 70 °C, the PEM fuel cell was able to deliver a power density of 674 $mW \cdot cm^{-2}$ at 0.6 V, which is comparable to that of a cell using commercial E-TEK Pt/C electrodes with a total Pt loading of 600 $\mu g \cdot cm^{-2}$. Moreover, at an outlet pressure of 200 kPa, the power density reached 870 $mW \cdot cm^{-2}$ at 0.6 V, corresponding to an overall MSPD of 7.43 $kW \cdot g^{-1}$. Such a high MSPD achieved with PLD-prepared catalyst could be ascribed to three factors: (1) the small size of the Pt nanoparticles provides a large mass-specific electrochemical surface area; (2) the very thin catalyst layer could provide shorter pathways for electron and proton transport; and (3) the good porosity of the catalyst layer could facilitate proton and gas transport.

The results reported here indicate that PLD could be a practically useful technique to deposit catalyst for PEM fuel cells. The performance of PEM fuel cells could be further raised by using other kinds of carbon support such as carbon nanotubes [35]. In addition, further reduction of Pt loading could be attained using an alloy of Pt [36,37], and the advantage offered by PLD should still hold for this case. Furthermore, the flow distributors used in this experiment were not optimised for this particular fuel cell active area yet, and thinner PEM can be used to reduce the proton transport resistance [33,34]. With the combination of pulsed laser deposition of Pt alloy, optimal carbon support, and optimal flow distributor and PEM, higher current density and power density could be achieved. Experiments are being conducted to implement these configurations.

Acknowledgments: The financial support from the Ministry of Science and Technology of Taiwan under contracts MOST-104-2112-M-001-003, MOST-104-2221-E-008-118-MY3, MOST-105-3113-F-008-003, and NSC-102-2923-E-008-002-MY3 are acknowledged.

Author Contributions: Szu-yuan Chen and Chung-Jen Tseng led the project together and contributed to this work equally. They conceived and designed the experiments. Hamza Qayyum and Ting-Wei Huang performed the experiments.

Conflicts of Interest: The authors declare no conflict of interest.

References

1. Brouzgou, A.; Song, S.Q.; Tsiakaras, P. Low and non-platinum electrocatalysts for PEMFCs: Current status, challenges and prospects. *Appl. Catal. B Environ.* **2012**, *127*, 371–388.
2. Cuynet, S.; Caillard, A.; Lecas, T.; Bigarrè, J.; Buvat, P.; Brault, P. Deposition of Pt inside fuel cell electrodes using high power impulse magnetron sputtering. *J. Phys. D Appl. Phys.* **2014**, *47*, 272001.

3. Cavarroc, M.; Ennadjaoui, A.; Mougenot, M.; Brault, P.; Escalier, R.; Tessier, Y.; Durand, J.; Roualdès, S.; Sauvage, T.; Coutanceau, C. Performance of plasma sputtered fuel cell electrodes with ultra-low Pt loadings. *Electrochem. Commun.* **2009**, *11*, 859–861.

4. Cogenli, M.S.; Mukerjee, S.; Yurtcan, A.B. Membrane electrode assembly with ultra low platinum loading for cathode electrode of PEM fuel cell by using sputter deposition. *Fuel Cells* **2015**, *15*, 288–297.

5. Raso, M.A.; Carrillo, I.; Mora, E.; Navarro, E.; Garcia, M.A.; Leo, T.J. Electrochemical study of platinum deposited by electron beam evaporation for application as fuel cell electrodes. *Int. J. Hydrog. Energy* **2014**, *39*, 5301–5308.

6. Saha, M.S.; Gullà, A.F.; Allen, R.J.; Mukerjee, S. High performance polymer electrolyte fuel cells with ultra-low Pt loading electrodes prepared by dual ion-beam assisted deposition. *Electrochim. Acta* **2006**, *51*, 4680–4692.

7. Cunningham, N.; Irissou, E.; Lefevre, M.; Denis, M.C.; Guay, D.; Dodelet, J.P. PEMFC anode with very low Pt loadings using pulsed laser deposition. *Electrochem. Solid-State Lett.* **2003**, *6*, A125–A128.

8. Mròz, W.; Budner, B.; Tokarz, W.; Piela, P.; Pawlowski, M.L.K. Ultra-low-loading pulsed-laser-deposited platinum catalyst films for polymer electrolyte membrane fuel cells. *J. Power Sources* **2015**, *273*, 885–893.

9. Hayre, R.O.; Lee, S.J.; Cha, S.W.; Prinz, F.B. A sharp peak in the performance of sputtered platinum fuel cells at ultra-low platinum loading. *J. Power Sources* **2002**, *109*, 483–493.

10. Nakakubo, T.; Shibata, M.; Yasuda, K. Membrane electrode assembly for proton exchange membrane fuel cells prepared by sputter deposition in air and transfer method. *J. Electrochem. Soc.* **2005**, *152*, A2316–A2322.

11. Huang, T.W.; Qayyum, H.; Lin, G.R.; Chen, S.Y.; Tseng, C.J. Production of high-performance and improved-durability Pt-catalyst/support for proton-exchange-membrane fuel cells with pulsed laser deposition. *J. Phys. D Appl. Phys.* **2016**, *49*, 255601.

12. Riabinina, D.; Irissou, E.; Drogoff, B.L.; Chaker, M.; Guay, D. Influence of pressure on the Pt nanoparticle growth modes during pulsed laser ablation. *J. Appl. Phys.* **2010**, *108*, 034322-1–034322-6, doi:10.1063/1.3463204.

13. Debe, M.K. Electrocatalyst approaches and challenges for automotive fuel cells. *Nature* **2012**, *486*, 43–51.

14. Fabbri, E.; Taylor, S.; Rabis, A.; Levecque, P.; Conrad, O.; Kotz, R.; Schmidt, T.J. The effect of platinum nanoparticle distribution on oxygen electroreduction activity and selectivity. *ChemCatChem* **2014**, *6*, 1410–1418.

15. Yuan, X.; Wang, H.; Sun, J.C.; Zhang, J. AC impedance technique in PEM fuel cell diagnosis-A review. *Int. J. Hydrog. Energy* **2007**, *32*, 4365–4380.

16. Zhang, J.; Song, C.; Zhang, J.; Baker, R.; Zhang, L. Understanding the effects of backpressure on PEM fuel cell reactions and performance. *J. Electroanal. Chem.* **2013**, *688*, 130–136.

17. Zhang, J.; Li, H.; Zhang, J. Effect of operating backpressure on PEM fuel cell performance. *ECS Trans.* **2009**, *19*, 65–76.

18. Onana, F.F.; Guillet, N.; AlMayouf, A.M. Modified pulse electrodeposition of Pt nanocatalyst as high-performance electrode for PEMFC. *J. Power Sources* **2014**, *271*, 401–405.

19. Yu, H.; Roller, J.M.; Mustain, W.E.; Maric, R. Influence of the ionomer/carbon ratio for low-Pt loading catalyst layer prepared by reactive spray deposition technology. *J. Power Sources* **2015**, *283*, 84–94.

20. Khan, A.; Nath, B.K.; Chutia, J. Nanopillar structured platinum with enhanced catalytic utilization for electrochemical reactions in PEMFC. *Electrochim. Acta* **2014**, *146*, 171–177.

21. Su, H.N.; Liao, S.J.; Shu, T.; Gao, H.L. Performance of an ultra-low platinum loading membrane electrode assembly prepared by a novel catalyst-sprayed membrane technique. *J. Power Sources* **2010**, *195*, 756–761.

22. Shu, T.; Dang, D.; Xu, D.W.; Chen, R.; Liao, S.J.; Hsieh, C.T.; Su, A.; Song, H.Y.; Du, L. High-performance MEA prepared by direct deposition of platinum on the gas diffusion layer using an atomic layer deposition technique. *Electrochim. Acta* **2015**, *177*, 168–173.

23. Yuan, Y.; Smith, J.A.; Goenaga, G.; Liu, D.J.; Luo, Z.; Liu, J. Platinum decorated aligned carbon nanotubes: electrocatalyst for improved performance of proton exchange membrane fuel cells. *J. Power Sources* **2011**, *196*, 6160–6167.

24. Hamel, C.; Garbarino, S.; Irissou, E.; Laplante, F.; Chaker, M.; Guay, D. Influence of the velocity of Pt ablated species on the structural and electrocatalytic properties of Pt thin films. *Int. J. Hydrog. Energy* **2010**, *35*, 8486–8493.

25. Greszler, T.A.; Caulk, D.; Sinha, P. The impact of platinum loading on oxygen transport resistance. *J. Electrochem. Soc.* **2012**, *159*, F831–F840.

26. Weber, A.Z.; Kusoglu, A. Unexplained transport resistances for low-loaded fuel-cell catalyst layers. *J. Mater. Chem. A* **2014**, *2*, 17207–17211.

27. Kongkanand, A.; Mathias, M.F. The priority and challenge of high-power performance of low-platinum proton-exchange membrane fuel cells. *J. Phys. Chem. Lett.* **2016**, *7*, 1127–1137.

28. Zhou, Y.; Lin, G.; Shih, A.J.; Hu, S.J. Assembly pressure and membrane swelling in PEM fuel cells. *J. Power Sources* **2009**, *192*, 544–551.

29. Bauer, F.; Denneler, S.; Willert-Porada, M. Influence of temperature and humidity on the mechanical properties of Nafion 117 polymer electrolyte membrane. *J. Polym. Sci. B Polym. Phys.* **2005**, *43*, 786–795.

30. Eason, R. *Pulsed Laser Deposition of Thin Films: Applications-Led Growth of Functional Materials*, 1st ed.; John Wiley and Sons, Inc.: Hoboken, NJ, USA, 2006.

31. Rodrìguez, J.M.D.; Meliàn, J.A.H.; Peña, J.P. Determination of the real surface area of Pt electrodes by hydrogen adsorption using cyclic voltammetry. *J. Chem. Educ.* **2000**, *77*, 1195–1197.

32. Wang, X.X.; Tan, Z.H.; Zeng, M.; Wang, J.N. Carbon nanocages: A new support material for Pt catalyst with remarkably high durability. *Sci. Rep.* **2014**, *4*, 4437.

33. Tsai, B.T.; Tseng, C.J.; Liu, Z.S.; Wang, C.H.; Lee, C.I.; Yang, C.C.; Lo, S.K. Effects of flow field design on the performance of a PEM fuel cell with metal foam as the flow distributor. *Int. J. Hydrog. Energy* **2012**, *37*, 13060–13066.

34. Tseng, C.J.; Tsai, B.T.; Liu, Z.S.; Cheng, T.C.; Chang, W.C.; Lo, S.K. A PEM fuel cell with metal foam as flow distributor. *Energy Convers. Manag.* **2012**, *62*, 14–21.

35. Murata, S.; Imanishi, M.; Hasegawa, S.; Namba, R. Vertically aligned carbon nanotube electrodes for high current density operating proton exchange membrane fuel cells. *J. Power Sources* **2014**, *253*, 104–113.

36. Tseng, C.J.; Lo, S.T.; Lo, S.C.; Chu, P.P. Characterization of Pt-Cu binary catalysts for oxygen reduction for fuel cell applications. *Mater. Chem. Phys.* **2006**, *100*, 385–390.

37. Su, B.J.; Wang, K.W.; Cheng, T.C.; Tseng, C.J. Preparation of PtSn/C electrocatalysts with improved activity and durability toward oxygen reduction reaction by alcohol-reduction process. *Mater. Chem. Phys.* **2012**, *135*, 395–400.

catalysts

MDPI

Article

Hydrothermal Method Using DMF as a Reducing Agent for the Fabrication of PdAg Nanochain Catalysts towards Ethanol Electrooxidation

Yue Feng, Ke Zhang, Bo Yan, Shumin Li and Yukou Du *

College of Chemistry, Chemical Engineering and Materials Science, Soochow University, Suzhou 215123, China; fengyueyue183@163.com (Y.F.); zhangke910222@163.com (K.Z.); boyan623@outlook.com (B.Y.); lishumin329@126.com (S.L.)
* Correspondence: duyk@suda.edu.cn; Tel./Fax: +86-512-6588-0089

Academic Editor: Vincenzo Baglio
Received: 21 June 2016; Accepted: 12 July 2016; Published: 15 July 2016

Abstract: In this article, we developed a facile one-step hydrothermal method using dimethyl formamide (DMF) as a reducing agent for the fabrication of PdAg catalyst. The scanning electron microscope (SEM) and transmission electron microscopy (TEM) images have shown that the as-synthesized PdAg catalyst had a nanochain structure. The energy-dispersive X-ray analyzer (EDX) spectrum presented the actual molar ratio of Pd and Ag in the PdAg alloy. Traditional electrochemical measurements, such as cyclic voltammetry (CV), chronoamperometry (CA) and electrochemical impedance spectrometry (EIS), were performed using a CHI 760D electrochemical analyzer to characterize the electrochemical properties of the as-synthesized catalyst. The results have shown that the PdAg catalyst with a nanochain structure displays higher catalytic activity and stability than pure Pd and commercial Pd/C catalysts.

Keywords: hydrothermal; PdAg; nanochain; ethanol oxidation

1. Introduction

Nowadays, direct ethanol fuel cells (DEFCs) as a promising clean energy have attracted great interest for their low cost and toxicity [1,2]. Moreover, DEFCs act as a power source because of their high energy density and low emissions, and they can be produced in a sustainable way [3]. However, there are still some limitations to realize the commercialization of DEFC, for example the high cost of catalysts, and the poor activity and stability of catalysts. To solve these problems, nanostructure electrocatalysts have been well designed to improve the catalytic activity and durability for ethanol electrooxidation [4–6].

As is known, Pt and Pt-based catalysts, such as PtSn [7], PtAu [8] and PtNi [9], with high activity have been widely investigated by researchers for ethanol electrooxidation in the past few decades. However, the high price and limited resources of Pt may hinder the application of Pt in the field of fuel cells. Pd is an active material for ethanol electrooxidation in alkaline media, which is less expensive and more active than Pt and Pt-based catalysts [10–12]. Thus, Pd and Pd-based catalysts have attracted much attention as effective anode catalysts for DEFCs. According to previous studies, it has been found that the particle size, shape, and surface compositions of the catalysts have great impact on the catalytic properties of the catalysts [13,14]. To enhance the catalytic activity and reduce the cost, another transition metal can be introduced into the Pd catalyst. In addition, to effectively obtain Pd-based catalysts with large surface areas and abundant active sites, it is important to manipulate the morphology and size of the as-prepared catalysts. Up to now, lots of methods have been attempted to achieve Pd-based catalysts with various kinds of shapes and sizes, for example nanocube PdAu catalysts [15], porous PdPt catalysts [16], and core shell structure PdCu catalysts [17], which have been employed in the field of DEFCs. Pd combined with another transition metal can modify the electronic structure of the Pd catalyst and reduce the experimental cost [18,19]. However, PdAg catalysts for ethanol oxidation in alkaline media with a nanochain structure are rarely reported, and the addition of Ag can not only reduce the cost of the experiment, but can also enhance the catalytic activity and durability of Pd catalysts.

In this article, we synthesized nanochain-structure PdAg catalysts via a one-step hydrothermal method, using dimethyl formamide (DMF) as a reducing agent. The as-prepared nanochain PdAg catalysts can significantly promote the catalytic activity towards the ethanol oxidation reaction in alkaline media. For comparison, the activity and stability of commercial Pd/C towards ethanol electrooxidation was also studied by cyclic voltammetry (CV) and chronoamperometry (CA) measurements. The structure, morphologies and compositions of the as-synthesized catalysts were characterized by SEM, TEM, and XRD.

2. Results and Discussion

The morphology of the as-synthesized PdAg catalyst was characterized by SEM. As shown in Figure 1, nanochain-structure PdAg catalysts were synthesized by the hydrothermal method using DMF as a reducing agent. It can be seen that the nanochain-structure PdAg was constituted by many irregular nanoparticles. The small-size irregular nanoparticles may increase the catalytic active surface of the as-synthesized catalyst, which may further enhance the catalytic activity of the catalyst for ethanol oxidation in alkaline media. Furthermore, EDX analysis was also carried out to detect the chemical composition of the as-synthesized PdAg catalyst, which was shown in Figure 1B. Apparently, the peak of the Si element corresponded to the substrate. The EDX spectrum data confirms that the molar ratio between Pd and Ag is close to 1.1:1, which is nearly consistent with the rate of charge.

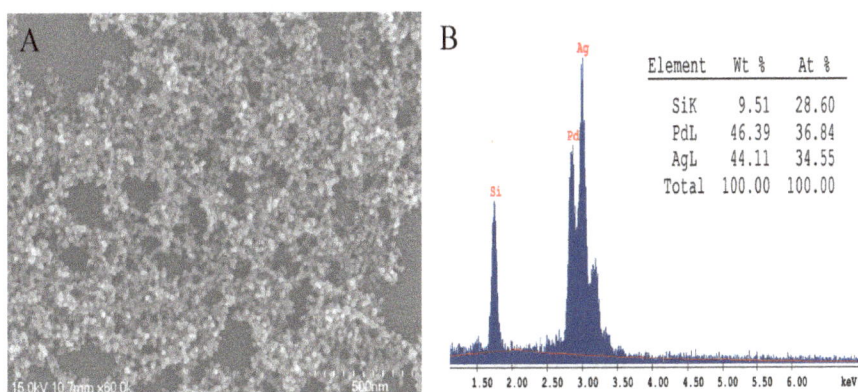

Element	Wt %	At %
SiK	9.51	28.60
PdL	46.39	36.84
AgL	44.11	34.55
Total	100.00	100.00

Figure 1. SEM (scanning electron microscope) image of PdAg catalyst (**A**); EDX (energy-dispersive X-ray analyzer) spectrum of PdAg catalyst (**B**).

In order to further analyze the structure and morphology of the PdAg catalyst, TEM images with different magnifications were displayed in Figure 2. As known, the DMF organic solvent is a powerful reducing agent to reduce Ag^+. The reaction mechanism is as the following equation [20]: $HCONMe_2 + 2Ag^+ + H_2O \rightarrow 2Ag + Me_2NCOOH + 2H^+$. Ag is more active than Pd, so it is easy for DMF solution to reduce Pd^{2+} to Pd^0 nanoparticles. The presence of the Br^- ions enables us to etch the as-synthesized nanoparticles to form the irregular nanoparticles [21]. The introduction of (polyvinylpyrrolidone) PVP in this experiment is to prevent the irregular particles from agglomerating together, because the as-synthesized PdAg particles can coordinate with N or O elements in PVP, which can generate a covered layer on the surface of the PdAg catalyst [22]. In addition, protonated carbonyl groups provided PVP with abundant positive charges, and the electrostatic repulsion effect can disperse the as-synthesized irregular PdAg particles. The PVP polymer chains connected to each other easily to form the nanochain-structure PdAg catalyst [23].

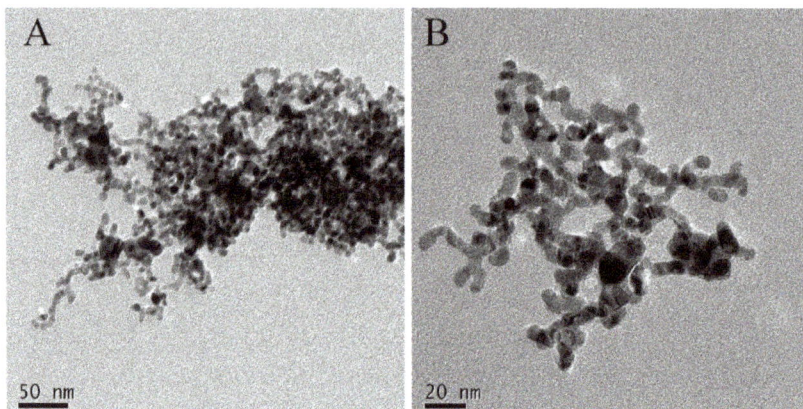

Figure 2. TEM images of the as-synthesized PdAg catalysts with different magnifications.

The structural features of the as-synthesized PdAg and Pd catalysts were characterized by XRD, and the XRD profiles are presented in Figure 3. Four typical diffraction peaks at around 39.4°, 45.9°, 67.1° and 81.3°, corresponding to the (111), (200), (220), and (311) lattice planes of Pd, can be evidently observed in Figure 3 [24,25]. It is worth noticing that there are some slight shifts of diffraction peaks on the PdAg catalyst compared to the pure Pd catalyst. This phenomenon suggests that the addition of Ag can cause the lattice contraction of the Pd catalyst and the formation of the PdAg alloys [26]. Moreover, the diffraction peaks of Ag or Ag oxides cannot be observed, which can further confirm that the PdAg alloys are synthesized.

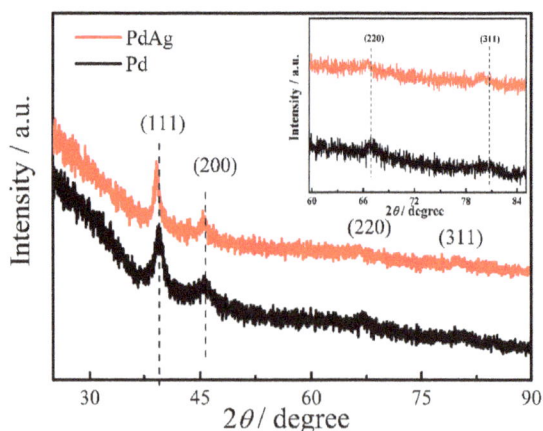

Figure 3. XRD (X-ray diffraction) patterns of as-synthesized PdAg and Pd catalysts.

CV curves of the as-prepared PdAg, Pd and commercial Pd/C catalysts, shown in Figure 4A, were obtained using cyclic voltammetry measurement in 1.0 M KOH from −0.8 to 0.3 V. It can be observed in Figure 4A that the typical cathodic peaks of the as-synthesized catalysts at around -0.4 V are related to the reduction of Pd oxide [27]. The reduction peak of the PdAg catalyst was higher and shifted to negative potential compared with the pure Pd and commercial Pd/C catalysts, which may be ascribed to the simultaneous reduction of PdO and Ag oxides. The electrochemical surface area (ECSA) is usually applied to assess the electrochemical active sites of the catalysts, which can be deduced to be 402.1 cm$^2 \cdot$mg^{-1}, 328.2 cm$^2 \cdot$mg^{-1} and 241.8 cm$^2 \cdot$mg^{-1} according to the following Equation (1) [27,28]:

$$ECSA = \frac{Q}{0.43 \times [Pd]} \tag{1}$$

The value 0.43 (mC$\cdot$cm^{-2}) is a constant charge value assumed for the reduction of Pd oxide on the surface of the catalysts; Q in mC can be calculated according to the integration of the area under the reduction peak of the CVs in Figure 4A; $[Pd]$ in mg represents the mass of Pd loading on the surface of the (glassy carbon electrode) GCE. The results of the ECSA are shown in Table 1. It can be found that the ECSA of the as-synthesized PdAg catalyst is remarkably higher among all these catalysts, which indicates that the addition of Ag can enhance the electrochemical active area of the Pd catalyst.

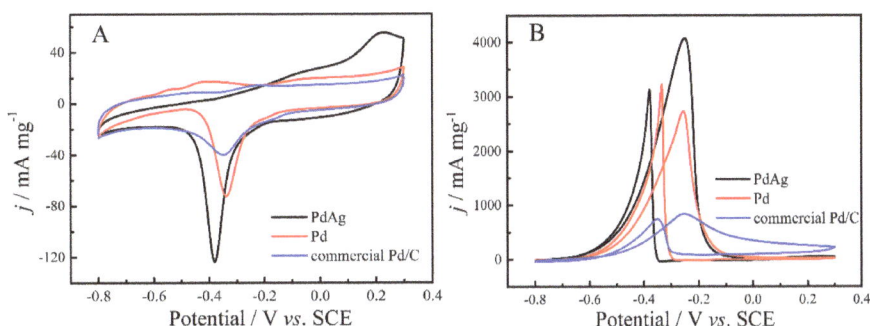

Figure 4. Cyclic voltammograms of the as-prepared PdAg, Pd and commercial Pd/C in (**A**) 1.0 M KOH; (**B**) 1.0 M C$_2$H$_5$OH + 1.0 M KOH solution, at a scan rate of 50 mV$\cdot$s^{-1}.

Table 1. Comparison of properties among the as-prepared catalysts.

Catalysts	ECSA (Electrochemical Surface Area) (cm$^2 \cdot$mg^{-1})	I_f (mA$\cdot$mg^{-1})	I_b (mA$\cdot$mg^{-1})	I_f/I_b
PdAg	402.1	4098.2	3137.7	1.3
Pd	328.2	2735.3	3207.8	0.9
Commercial Pd/C	241.8	827.2	756.9	1.1

The electrochemical catalytic activity of all these catalysts toward ethanol oxidation in alkaline media was further investigated by testing the catalysts in 1.0 M C$_2$H$_5$OH and 1.0 M KOH solution. The mass normalization activity was usually used to evaluate the electrocatalytic activity of the as-prepared catalysts. In Figure 4B, the forward anodic peak current density of the PdAg catalyst is about 4098.2 mA$\cdot$mg^{-1}, which is higher than the pure Pd catalyst (2735.3 mA$\cdot$mg^{-1}) and commercial Pd/C catalyst (827.2 mA$\cdot$mg^{-1}). This result may suggest that the introduction of Ag to the Pd catalyst promotes the activity of the Pd catalyst. Moreover, the ratio of the forward peak current density to the backward peak current density (I_f/I_b) can be calculated to evaluate the tolerance of the catalysts to the poison species on the surface of the as-prepared working electrodes [29–31]. As seen in Table 1, the I_f/I_b value of the PdAg catalyst is about 1.3, higher than the pure Pd (0.9) and commercial Pd/C (1.1) catalyst. This result indicates that the addition of Ag to the Pd can enhance the poison tolerance of the Pd catalyst.

To further investigate the stability of the as-prepared catalysts, the chronoamperometric measurement was carried out in 1.0 M KOH + 1.0 M C_2H_5OH solution at −0.3 V. As can be learned from Figure 5, the current density of the as-prepared PdAg and pure Pd catalysts decayed rapidly in the initial period and gradually remained stable, which is ascribed to the formation of intermediate carbonaceous species (such as CO_{ads}) on the surface of catalysts during ethanol electrooxidation [32]. After 3600 s, the current density of the PdAg catalyst is higher than that of the pure Pd and commercial Pd/C catalysts, suggesting that the PdAg catalyst had better long-term stability for ethanol electrooxidation.

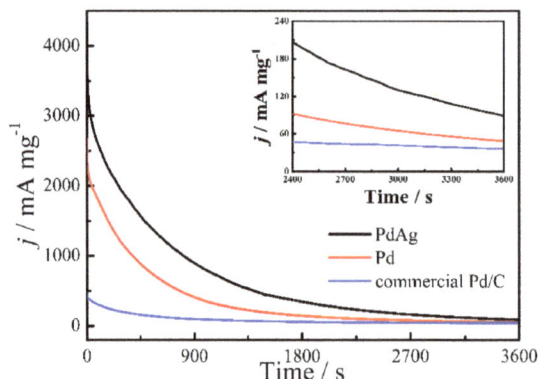

Figure 5. Chronoamperograms of the as-prepared catalysts for 1.0 M C_2H_5OH on 1.0 M KOH solution at −0.3 V.

EIS spectra of the PdAg catalyst are recorded in Figure 6, which can be used to assess the kinetics of the PdAg catalyst for ethanol oxidation reactions. In Figure 6A, the diameter of the impedance arc (DIA) on the PdAg catalyst decreased with the potential increasing from −0.6 to −0.3 V, which indicates that the ethanol oxidation rate increased owing to the fact that the carbonaceous intermediates generated from ethanol dehydrogenation were removed by oxidation [33]. As the potential continued to increase, the impedance arc reversed to the second quadrant at −0.25 V and −0.20 V, because of the removal of the intermediate species (such as CO_{ads}) generated on the surface of the catalyst and the recovery of the catalytic active sites. At the potential of −0.15 V, the impedance arc turned back to the first quadrant with a large diameter, which is mainly due to the formation of Pd oxides under the high electrode potential [34].

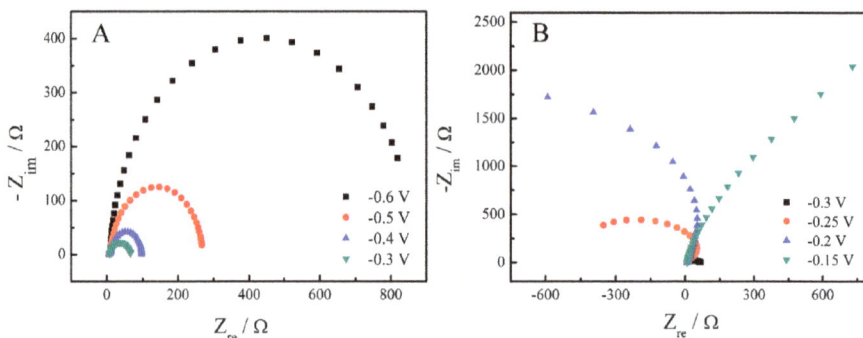

Figure 6. Nyquist plots of C_2H_5OH electrooxidation on the PdAg catalyst in 1.0 M C_2H_5OH + 1.0 M KOH solution with potential range of −0.6 V to −0.3 V (**A**), and −0.3 V to −0.15 V (**B**).

The PdAg-, Pd- and commercial Pd/C-modified GCE were investigated by EIS in 1.0 M KOH and 1.0 M C_2H_5OH at −0.3 V, and the Nyquist plots are displayed in Figure 7. As seen in Figure 7, all these catalysts exhibit a semicircle at high frequencies. The DIA of the PdAg catalyst is smaller than that of the pure Pd and commercial Pd/C catalysts, indicating the smaller charge-transfer resistance (Rct). This result also demonstrates that the electron transfer rate on PdAg is faster than that on pure Pd and commercial Pd/C catalysts, which is consistent with the higher catalytic activity of PdAg catalyst.

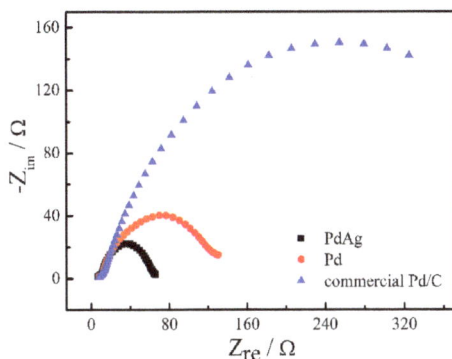

Figure 7. Nyquist plots of PdAg, Pd and commercial Pd/C for ethanol oxidation in 1.0 M C_2H_5OH + 1.0 M KOH solution with potential of −0.3 V.

3. Materials and Methods

3.1. Materials and Instruments

Palladium acetate ($Pd(C_2H_3O_2)_2$) was purchased from J&K Scientific Co., Ltd, Beijing, China, silver nitrate ($AgNO_3$), dimethyl formamide (DMF), NaBr, C_2H_5OH, KOH, and polyvinylpyrrolidone (PVP, K30) were all supplied by Sinopharm Chemical Reagent Co., Ltd, Shanghai, China, using without further purification. Commercial Pd/C catalyst (containing 20% Pd) were obtained from Shanghai Hesen Electric Co., Ltd, Shanghai, China. Doubly distilled water was used in the whole experiment.

The electrocatalytic experiments were performed using a CHI 760D electrochemical analyzer (CH Instrument, Inc, Shanghai, China). The three-electrode setup was made up of a saturated calomel electrode (SCE), a Pt wire and a modified glassy carbon electrode (GCE, diameter of 3 mm). The SCE, Pt wire and GCE worked as reference, counter electrode and working electrode, respectively. Before testing, the GCE was polished with α-Al_2O_3 powders (Tianjin AIDA hengsheng Science-Technology Development Co., Ltd., Tianjin, China, 0.05 m), washed with doubly distilled water and ethanol, then dried in the oven at 60 °C. Cyclic voltammetry (CV) measurement was used to analyze the activity of the as-prepared catalysts for ethanol electrooxidation, while the amperometric i-t curves obtained under −0.3 V was used to detect the long-term stability of as-prepared catalysts.

3.2. Synthesis of Nanochain PdAg Catalyst

The nanochain structure PdAg catalyst was prepared via hydrothermal method at 160 °C, using DMF as a reducing agent. The preparation procedure is as following: 7 mL DMF was added into a Teflon autoclave, 60 mg PVP, 20 mg NaBr, 3 mg $AgNO_3$ and 3 mL doubly distilled water were all added to the DMF solution, and the mixture was stirred under room temperature for 10 min. An orange mixture was formed gradually. Then, the mixture was heated to 160 °C from room temperature, keeping for 6 h and cooling down to room temperature naturally. Finally, the black solid product was obtained by centrifugation and thoroughly washed with water and ethanol several times.

For comparison, pure Pd catalyst was also prepared under the same condition according to the above preparation steps.

3.3. Preparation of Pd and PdAg Modified Electrodes

The working electrodes modified with Pd and PdAg catalysts were prepared as following: Firstly, the obtained black precipitates were dispersed in 9 mL doubly distilled water and 1 mL C_2H_5OH. After ultrasonic homogenization, the catalysts suspension was obtained. Then, 10 L catalysts "ink" was dripped onto the surface of the pretreated GCE. Finally, the modified GCE was dried at 60 °C in the oven. The mass loading of Pd for both the PdAg catalyst and pure Pd catalyst coated on the GCE is 2.37 mg.

3.4. Characterization

X-ray diffraction (XRD) (Philips, X'Pert-Pro MRD, Amsterdam, Netherland) profiles of the as-synthesized catalysts were obtained at 40 kV and 30 mA with Cu Kα radiation (λ = 1.54056 Å), using glass slide as substrate, which are used to analyze the crystalline structure of the as-prepared catalyst. Transmission electron microscopy (TEM) images were taken on a TECNAI-G20 electron microscope (FEI Tecnai G2 F20 S-TWIN TMP, Hongkong, China), which are operated at 200 kV. The scanning electron microscope (SEM) (Hitachi Corporation, S-4700, Tokyo, Japan) equipped with energy-dispersive X-ray analyzer (EDX) (Hitachi Corporation, Tokyo, Japan) was applied to characterize the morphology and composition of the as-synthesized catalysts. The electrochemical impedance spectrometry (EIS) spectra were measured in 1.0 M C_2H_5OH alkaline solution between 1 and 10^5 Hz, and the AC voltage amplitude is 5.0 mV.

4. Conclusions

In conclusion, the PdAg catalyst was prepared by a facile one-step hydrothermal method at 160 °C, using DMF solution as a reducing agent. The SEM, TEM and XRD characterization results have confirmed that the as-prepared catalyst is PdAg alloy and it has a nanochain structure. The electrochemical results have demonstrated that the catalytic activity and stability of the obtained PdAg catalyst for ethanol electrooxidation in alkaline media were remarkably enhanced because of the special structure of the catalyst and the introduction of the Ag element to the Pd catalyst.

Supplementary Materials: The following are available online at www.mdpi.com/2073-4344/6/7/103/s1, Figure S1: Cyclic voltammograms of PdAg catalysts with different molar ratio in 1.0 M C_2H_5OH/1.0 M KOH solution with scan rate of 50 mV s^{-1}. Figure S2: SEM image of PdAg catalyst prepared with NaBr (A), without NaBr (B). Figure S3: Typical N_2 adsorption-desorption isotherm of the as-prepared PdAg catalyst.

Acknowledgments: This work was financially supported by the Priority Academic Program Development of Jiangsu Higher Education Institutions (PAPD), the National Natural Science Foundation of China (Grant No. 51373111) and State and Local Joint Engineering Laboratory for Novel Functional Polymeric Materials.

Author Contributions: Yue Feng, Ke Zhang and Yukou Du conceived and designed the experiments; Yue Feng performed the experiments; Yue Feng, Bo Yan and Shumin Li analyzed the data; Yue Feng wrote the paper.

Conflicts of Interest: The authors declare no conflict of interest.

References

1. Guo, Y.; Zheng, Y.; Huang, M. Enhanced activity of PtSn/C anodic electrocatalyst prepared by formic acid reduction for direct ethanol fuel cells. *Electrochim. Acta* **2008**, *53*, 3102–3108. [CrossRef]
2. Li, Y.S.; Zhao, T.S.; Liang, Z.X. Performance of alkaline electrolyte-membrane-based direct ethanol fuel cells. *J. Power Sources* **2009**, *187*, 387–392. [CrossRef]
3. Antolini, E. Catalysts for direct ethanol fuel cells. *J. Power Sources* **2007**, *170*, 1–12. [CrossRef]

4. Bambagioni, V.; Bianchini, C.; Marchionni, A.; Filippi, J.; Vizza, F.; Teddy, J.; Serp, P.; Zhiani, M. Pd and Pt–Ru anode electrocatalysts supported on multi-walled carbon nanotubes and their use in passive and active direct alcohol fuel cells with an anion-exchange membrane (alcohol = methanol, ethanol, glycerol). *J. Power Sources* **2009**, *190*, 241–251. [CrossRef]

5. Zhu, C.; Guo, S.; Dong, S. PdM (M = Pt, Au) Bimetallic Alloy Nanowires with Enhanced Electrocatalytic Activity for Electro-oxidation of Small Molecules. *Adv. Mater.* **2012**, *24*, 2326–2331. [CrossRef] [PubMed]

6. Peng, C.; Hu, Y.; Liu, M.; Zheng, Y. Hollow raspberry-like PdAg alloy nanospheres: High electrocatalytic activity for ethanol oxidation in alkaline media. *J. Power Sources* **2015**, *278*, 69–75. [CrossRef]

7. Feng, Y.; Wang, C.; Bin, D.; Zhai, C.; Ren, F.; Yang, P.; Du, Y. One-pot Synthesis of PtSn Bimetallic Composites and Their Application as Highly Active Catalysts for Ethanol Electrooxidation. *ChemPlusChem* **2016**, *81*, 93–99. [CrossRef]

8. Cheng, F.; Dai, X.; Wang, H.; Jiang, S.P.; Zhang, M.; Xu, C. Synergistic effect of Pd–Au bimetallic surfaces in Au-covered Pd nanowires studied for ethanol oxidation. *Electrochim. Acta* **2010**, *55*, 2295–2298. [CrossRef]

9. Soundararajan, D.; Park, J.H.; Kim, K.H.; Ko, J.M. Pt–Ni alloy nanoparticles supported on CNF as catalyst for direct ethanol fuel cells. *Current Appl. Phys.* **2012**, *12*, 854–859. [CrossRef]

10. Xu, C.; Tian, Z.; Shen, P.; Jiang, S.P. Oxide (CeO_2, NiO, Co_3O_4 and Mn_3O_4)-promoted Pd/C electrocatalysts for alcohol electrooxidation in alkaline media. *Electrochim. Acta* **2008**, *53*, 2610–2618. [CrossRef]

11. Shen, P.K.; Xu, C. Alcohol oxidation on nanocrystalline oxide Pd/C promoted electrocatalysts. *Electrochem. Commun.* **2006**, *8*, 184–188. [CrossRef]

12. Wang, Z.; Hu, F.; Shen, P.K. Carbonized porous anodic alumina as electrocatalyst support for alcohol oxidation. *Electrochem. Commun.* **2006**, *8*, 1764–1768. [CrossRef]

13. Hong, W.; Wang, J.; Wang, E. Synthesis of porous PdAg nanoparticles with enhanced electrocatalytic activity. *Electrochem. Commun.* **2014**, *40*, 63–66. [CrossRef]

14. Hong, J.W.; Kim, D.; Lee, Y.W.; Kim, M.; Kang, S.W.; Han, S.W. Atomic-distribution-dependent electrocatalytic activity of Au-Pd bimetallic nanocrystals. *Angew. Chem.* **2011**, *50*, 8876–8880. [CrossRef] [PubMed]

15. Zhong, J.; Bin, D.; Feng, Y.; Zhang, K.; Wang, J.; Wang, C.; Guo, J.; Yang, P.; Du, Y. Synthesis and high electrocatalytic activity of Au-decorated Pd heterogeneous nanocube catalysts for ethanol electro-oxidation in alkaline media. *Catal. Sci. Technol.* **2016**, *6*, 5397–5404. [CrossRef]

16. Zhu, C.; Guo, S.; Dong, S. Rapid, general synthesis of PdPt bimetallic alloy nanosponges and their enhanced catalytic performance for ethanol/methanol electrooxidation in an alkaline medium. *Chem. Eur. J.* **2013**, *19*, 1104–1111. [CrossRef] [PubMed]

17. Cai, J.; Zeng, Y.; Guo, Y. Copper@palladium–copper core–shell nanospheres as a highly effective electrocatalyst for ethanol electro-oxidation in alkaline media. *J. Power Sources* **2014**, *270*, 257–261. [CrossRef]

18. Li, G.; Jiang, L.; Jiang, Q.; Wang, S.; Sun, G. Preparation and characterization of Pd_xAg_y/C electrocatalysts for ethanol electrooxidation reaction in alkaline media. *Electrochim. Acta* **2011**, *56*, 7703–7711. [CrossRef]

19. Zhu, L.D.; Zhao, T.S.; Xu, J.B.; Liang, Z.X. Preparation and characterization of carbon-supported sub-monolayer palladium decorated gold nanoparticles for the electro-oxidation of ethanol in alkaline media. *J. Power Sources* **2009**, *187*, 80–84. [CrossRef]

20. Pastoriza-Santos, I.; Liz-Marzán, L.M. Synthesis of Silver Nanoprisms in DMF. *Nano Lett.* **2002**, *2*, 903–905. [CrossRef]

21. Wiley, B.J.; Xiong, Y.; Li, Z.Y.; Yin, Y.; Xia, Y. Right Bipyramids of Silver: A New Shape Derived from Single Twinned Seeds. *Nano Letters* **2006**, *6*, 765–768. [CrossRef] [PubMed]

22. Wang, H.; Qiao, X.; Chen, J.; Wang, X.; Ding, S. Mechanisms of PVP in the preparation of silver nanoparticles. *Mater. Chem. Phys.* **2005**, *94*, 449–453. [CrossRef]

23. Lan, F.; Wang, D.; Lu, S.; Zhang, J.; Liang, D.; Peng, S.; Liu, Y.; Xiang, Y. Ultra-low loading Pt decorated coral-like Pd nanochain networks with enhanced activity and stability towards formic acid electrooxidation. *J. Mater. Chem. A* **2013**, *1*, 1548–1552. [CrossRef]

24. Jana, R.; Subbarao, U.; Peter, S.C. Ultrafast synthesis of flower-like ordered Pd_3Pb nanocrystals with superior electrocatalytic activities towards oxidation of formic acid and ethanol. *J. Power Sources* **2016**, *301*, 160–169. [CrossRef]

25. Wang, M.; Zhang, W.; Wang, J.; Wexler, D.; Poynton, S.D.; Slade, R.C.; Liu, H.; Winther-Jensen, B.; Kerr, R.; Shi, D.; Chen, J. PdNi hollow nanoparticles for improved electrocatalytic oxygen reduction in alkaline environments. *ACS Appl. Mater. Inter.* **2013**, *5*, 12708–12715. [CrossRef] [PubMed]

26. Colmati, F.; Antolini, E.; Gonzalez, E.R. Ethanol oxidation on a carbon-supported Pt$_{75}$Sn$_{25}$ electrocatalyst prepared by reduction with formic acid: Effect of thermal treatment. *Appl. Catal.* **2007**, *73*, 106–115. [CrossRef]
27. Bin, D.; Ren, F.; Wang, Y.; Zhai, C.; Wang, C.; Guo, J.; Yang, P.; Du, Y. Pd-nanoparticle-supported, PDDA-functionalized graphene as a promising catalyst for alcohol oxidation. *Chem. Asian J.* **2015**, *10*, 667–673. [CrossRef] [PubMed]
28. Yang, J.; Xie, Y.; Wang, R.; Jiang, B.; Tian, C.; Mu, G.; Yin, J.; Wang, B.; Fu, H. Synergistic effect of tungsten carbide and palladium on graphene for promoted ethanol electrooxidation. *ACS Appl. Mater. Inter.* **2013**, *5*, 6571–6579. [CrossRef] [PubMed]
29. Yao, Z.; Yue, R.; Zhai, C.; Jiang, F.; Wang, H.; Du, Y.; Wang, C.; Yang, P. Electrochemical layer-by-layer fabrication of a novel three-dimensional Pt/graphene/carbon fiber electrode and its improved catalytic performance for methanol electrooxidation in alkaline medium. *Int. J. Hydrogen Energy* **2013**, *38*, 6368–6376. [CrossRef]
30. Bai, L.; Zhu, H.; Thrasher, J.S.; Street, S.C. Synthesis and electrocatalytic activity of photoreduced platinum nanoparticles in a poly(ethylenimine) matrix. *ACS Appl. Mater. Inter.* **2009**, *1*, 2304–2311. [CrossRef] [PubMed]
31. Lin, Y.; Cui, X. Platinum/Carbon Nanotube Nanocomposite Synthesized in Supercritical Fluid as Electrocatalysts for Low-Temperature Fuel Cells. *J. Phys. Chem. B* **2005**, *109*, 14410–14415. [CrossRef] [PubMed]
32. Dutta, A.; Mahapatra, S.S.; Datta, J. High performance PtPdAu nano-catalyst for ethanol oxidation in alkaline media for fuel cell applications. *Int. J. Hydrogen Energy* **2011**, *36*, 14898–14906. [CrossRef]
33. Wang, C.; Wang, H.; Zhai, C.; Ren, F.; Zhu, M.; Yang, P.; Du, Y. Three-dimensional Au0.5/reduced graphene oxide/Au0.5/reduced graphene oxide/carbon fiber electrode and its high catalytic performance toward ethanol electrooxidation in alkaline media. *J. Mater. Chem. A* **2015**, *3*, 4389–4398. [CrossRef]
34. Zhou, W.; Du, Y.; Ren, F.; Wang, C.; Xu, J.; Yang, P. High efficient electrocatalytic oxidation of methanol on Pt/polyindoles composite catalysts. *Int. J. Hydrogen Energy* **2010**, *35*, 3270–3279. [CrossRef]

catalysts

MDPI

Article

Oxygen Reduction Electrocatalysts Based on Coupled Iron Nitride Nanoparticles with Nitrogen-Doped Carbon

Min Jung Park [1,2,†], Jin Hee Lee [1,†], K. P. S. S. Hembram [3], Kwang-Ryeol Lee [3], Sang Soo Han [3], Chang Won Yoon [1,4], Suk-Woo Nam [1,2,5] and Jin Young Kim [1,*]

1   Fuel Cell Research Center, Korea Institute of Science and Technology, Seongbuk-gu, Seoul 02792, Korea; mjpark@kist.re.kr (M.J.P.); sinhwa1030@gmail.com (J.H.L.); cwyoon@kist.re.kr (C.W.Y.); swn@kist.re.kr (S.-W.N.)
2   Department of Energy and Environmental Engineering, Korea University of Science and Technology, Daejeon 34113, Korea
3   Computational Science Research Center, Korea Institute of Science and Technology (KIST), Seongbuk-gu, Seoul 02792, Korea; hembramhembram@gmail.com (K.P.S.S.H.); krlee@kist.re.kr (K.-R.L.); sangsoo@kist.re.kr (S.S.H.)
4   Department of Clean Energy and Chemical Engineering, Korea University of Science and Technology, Daejeon 34113, Korea
5   Green School, Korea University, 145, Anam-ro, Seongbuk-gu, Seoul 136-701, Korea
*   Correspondence: jinykim@kist.re.kr; Tel.: +82-2-958-5294
†   These authors contributed equally to this work.

Academic Editors: Vincenzo Baglio and David Sebastián
Received: 14 April 2016; Accepted: 6 June 2016; Published: 15 June 2016

Abstract: Aimed at developing a highly active and stable non-precious metal electrocatalyst for oxygen reduction reaction (ORR), a novel Fe_xN_y/NC nanocomposite—that is composed of highly dispersed iron nitride nanoparticles supported on nitrogen-doped carbon (NC)—was prepared by pyrolyzing carbon black with an iron-containing precursor in an NH_3 atmosphere. The influence of the various synthetic parameters such as the Fe precursor, Fe content, pyrolysis temperature and pyrolysis time on ORR performance of the prepared iron nitride nanoparticles was investigated. The formed phases were determined by experimental and simulated X-ray diffraction (XRD) of numerous iron nitride species. We found that Fe_3N phase creates superactive non-metallic catalytic sites for ORR that are more active than those of the constituents. The optimized Fe_3N/NC nanocomposite exhibited excellent ORR activity and a direct four-electron pathway in alkaline solution. Furthermore, the hybrid material showed outstanding catalytic durability in alkaline electrolyte, even after 4,000 potential cycles.

Keywords: oxygen reduction reaction; non-precious metal electrocatalysts; Iron nitride; nitrogen-doped carbon

1. Introduction

Over the last few decades, low temperature polymer electrolyte membrane fuel cells (PEMFCs) have been recognized as a promising substitute to the current fossil fuel based power generation system [1,2]. Despite extensive efforts to improve performance of PEMFCs, enhancement of sluggish kinetics of oxygen reduction reaction (ORR) has remained a major challenge. Thus far, Pt derived materials have been considered state of the art electrocatalysts for ORR. [3,4] Nevertheless, the high price and instability of Pt catalysts in various fuel cell operation conditions limit practical applications of PEMFCs.

Development of efficient non-precious metal electrocatalysts is thus a topic of great interest in order to commercialize PEMFCs. However, although performance of such catalysts is continuously increasing, the active species responsible for the activation and reduction of adsorbed oxygen molecules are not fully understood yet.

In recent years, it is suggested that that novel materials substituting for Pt are non-precious Fe or Co metals embedded within an *N*-doped carbon (NC) matrix. Inexpensive Fe and/or Co constitute the catalytically active center, while NC materials provide enhanced electron mobility and oxygen adsorption capacity, which in turn, improve electrocatalytic activity as well as durability [5]. Iron carbide (Fe_xC_y) [6–9], iron oxide (Fe_xO_y) [10,11], and iron bonded with multiple nitrogen atoms [12] have been suggested as plausible active structures for ORR. However, the complexities of creating the desired bond configuration and characterization of the specific Fe species, has resulted in difficulties improving the materials overall catalytic activity in the electrocatalysts.

Herein, we report a facile, one-step synthetic route that leads to a novel precious-metal-free ORR electrocatalyst composed of iron nitride (Fe_xN_y) nanoparticles supported on NC (Fe_xN_y/NC). Various analytical techniques were employed to elucidate catalytically active sites responsible for the ORR performance. The strong correlation between the Fe_xN_y phases and ORR activity was proven by experimental and theoretical studies. For electrocatalytic performance, we show that Fe_3N/NC hybrid material exhibits a high ORR activity (half-wave potential ~0.77 V) and low HO_2^- (<5%) yield in alkaline medium. The prepared electrocatalysts exhibited a direct four-electron ORR pathway and superior stability in an alkaline medium. The proposed synthetic route is low-cost and scalable, providing a feasible method for the development of highly efficient non-precious metal electrocatalysts.

2. Results and Discussions

The Fe_xN_y/NC catalyst was synthesized through high-temperature pyrolysis of the precursor containing the mixture of the Fe salt and carbon black in an NH_3 atmosphere (Scheme 1). The catalyst structure contains a few ten nanometer-sized Fe_xN_y nanoparticles and NCs with sizes in the range of hundreds of nanometers. Most of the Fe_xN_y nanoparticles are well dispersed on the NCs. Importantly, the composites showed high inertness under harsh electrochemical conditions, as well as high surface area (1110 m^2g^{-1}) and electrical conductivity (0.112 S· cm^{-1}), which are comparable to those of commercial carbon black products (surface area: 1350 m^2g^{-1}, and electrical conductivity: 0.092 S· cm^{-1}) and beneficial for a number of electrochemical applications. Moreover, the synthetic approach allowed for the low-cost, large-scale production of Fe_xN_y/NC electrocatalysts. The structural phase of Fe_xN_y prepared by heat-treatment has been reported to depend on preparation conditions. [13–15] Therefore, in order to characterize high-performance Fe_xN_y/NC electrocatalysts, a series of synthetic parameters that could influence the ORR activity, including the types of Fe salts used, Fe content, heat-treatment temperature, and pyrolysis time were investigated. ORR behaviors of all the catalysts prepared under varying synthetic parameters were compared in 0.1 M KOH solution (Figure 1).

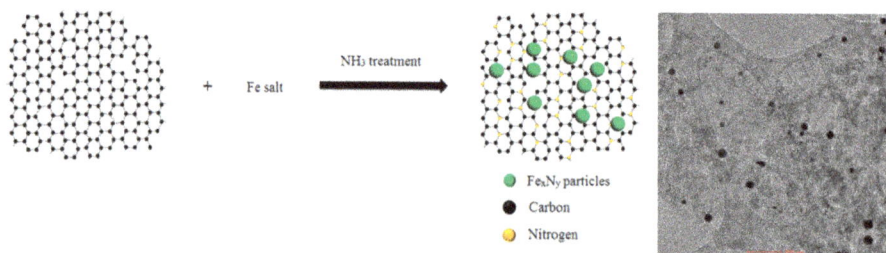

Scheme 1. Illustration of Fe_xN_y/nanocomposite (NC) synthesis.

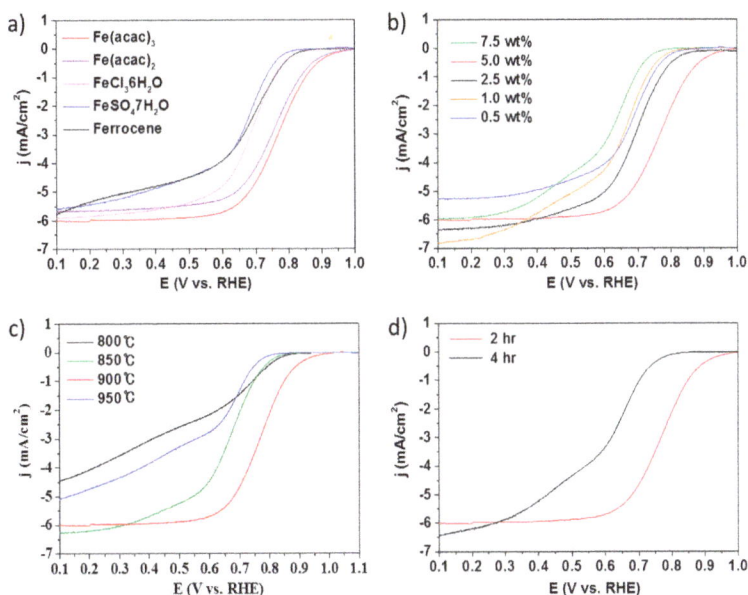

Figure 1. Linear sweep voltammetry curves of oxygen reduction reaction (ORR) in 0.1 M KOH catalyzed by Fe$_x$N$_y$/NC prepared by altering synthetic conditions. (**a**) Fe source, (**b**) Fe contents, (**c**) pyrolysis temperature, and (**d**) pyrolysis time.

For the standard synthesis, Fe(acac)$_3$ and carbon black mixture were pyrolyzed at 900 °C for 2 h to give 5 wt. % Fe contents of the catalyst. To test the influence of the aforementioned parameters of the Fe$_x$N$_y$/NC catalysts, synthetic conditions were altered one by one from the standard condition. The catalysts without Fe and/or nitrogen were also prepared for the comparison. The counter anion of Fe precursors prominently influenced the ORR activity of the final catalysts (Figure 1a). The catalyst synthesized using Fe(acac)$_3$ showed the highest ORR activity among the five Fe precursors tested. The higher catalytic performance of the catalyst prepared using Fe(acac)$_3$, as opposed to Fe(acac)$_2$, indicates that Fe(III) is a more suitable starting material for the catalyst preparation compared to Fe(II). Half-wave potential reached to 0.77 V which is comparable to the most active non-noble metal based ORR catalysts so far [16,17], as well as commercially available Pt/C catalysts [18] The carbon itself did not show good ORR activity and waving curve indicates 2 electron reduction is dominant (Figure S1, black curve). The catalyst without Fe followed 4 electron pathway, but half-wave potential and limiting current were far lower than Fe-N-C catalyst (Figure S1, blue curve).

It was found that proper amounts of Fe in the catalysts are also a critical factor affecting catalyst performance. Catalytic activity initially increased as the amount of Fe increased, possibly due to the increase of active site density, however the activity significantly diminished when Fe was employed over 5 wt. % (Figure 1b).

Likewise, there was an optimal temperature for heat-treatment. Pyrolysis temperature has been well understood to be a key factor to achieve proper ORR catalysts in several reports [19,20]. In these attempts, pyrolysis at 900 °C yielded a superior ORR catalyst than any other pyrolysis temperature, and both higher and lower temperatures were detrimental to catalytic activity (Figure 1c). Furthermore, an extended heat-treatment time of 4 h was found to be harmful (Figure 1d). These observations indicate that optimal temperature and duration upon heating considerably affect the formation of active catalytic species in Fe$_x$N$_y$/NC electrocatalysts.

To determine the cause of activity difference of the Fe_xN_y/NC electrocatalysts obtained under the studied synthetic conditions, X-ray diffraction (XRD) analysis was performed for every catalyst tested. Interestingly, distinct XRD patterns were observed from the catalysts (Figure 2). The characteristic XRD peaks of Fe_3N phase [21] centered at 42.8° (sharp peak), 40.7°, 37.4°, 56.5°, 67.6°, and 75.6° appeared in the catalysts synthesized using $Fe(acac)_3$ or $Fe(acac)_2$ under the standard preparation conditions, *i.e.*, 5 wt. % Fe and pyrolysis at 900 °C for 2 h. Theoretical calculation of several Fe_xN_y complexes were carried out to identify the most probable structure with the lowest energy (Figure S2). Of the various phases of Fe_xN_y, the only stable phases obtained were those of FeN (F4-3m) and Fe_3N (P6$_3$22). All the others phases were metastable, and decomposed to FeN and Fe_3N. A similar trend from the theoretical calculation of the XRD-spectra was observed [22]. Selecting the most optimized structures of FeN, Fe_2N, Fe_3N and Fe_4N, their XRD spectra were simulated and compared with the experimental results as shown in Figure 3.

With regard to the Fe_xN_y/NC composites, the combined ORR electrocatalytic test results and XRD analysis revealed that the best-performing active sites contain an Fe_3N (P6$_3$22) phase on the NCs. The Fe_3N phase was not formed in the catalysts prepared using $FeCl_3$, $FeSO_4$, or ferrocene (Figure 2a), which showed lower ORR activity. These results signify that the Fe_3N phase coupled with NC is the principle structure responsible for efficient ORR in this catalytic system. Analogous results appeared as well in other catalysts prepared under control of other parameters, such as weight percent of Fe, pyrolysis temperature, and time. All the less active catalysts contained less pronounced, or did not contain any, Fe_3N XRD structure in their respective XRD spectra (Figure 2b–d). The catalysts with lower amounts of Fe (*i.e.*, less than 5% in this study) did not show distinct XRD peaks, indicating that the Fe species were present in an amorphous phase (Figure 2b). In addition, higher amounts (*i.e.*, higher than 5% in this study) of Fe contents resulted in the formation of yet another crystal phase that corresponded to metallic Fe (Figure 2b, green).

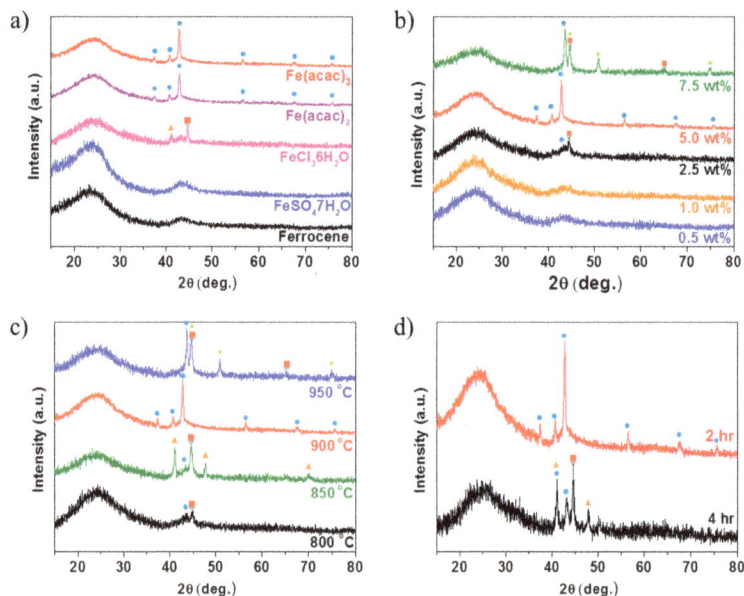

Figure 2. X-ray diffraction (XRD) spectra of Fe-N-C catalysts prepared by altering synthetic conditions. (**a**) Fe source, (**b**) Fe contents, (**c**) pyrolysis temperature, and (**d**) pyrolysis time. The symbols denote specific Fe species. ■ denotes metallic Fe (JCPDS # 87-0722), ★ denotes FeN (JCPDS # 50-1087), ● denotes Fe_3N (JCPDS # 86-0232), and ▲ denotes Fe_4N (JCPDS # 86-0231).

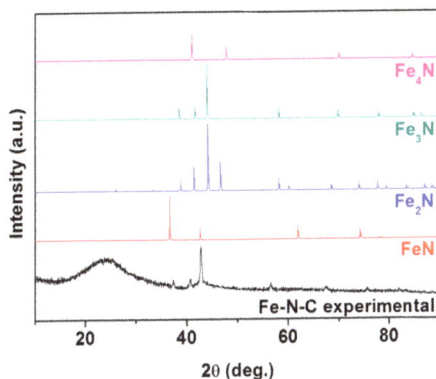

Figure 3. XRD patterns from the experiment compared with the theoretical generated pattern.

The effect of pyrolysis temperature the catalysts ORR activity can also be explained by the presence, or lack thereof, of the Fe_3N phase. The Fe_3N crystal phase was only found in the catalysts treated at 900 °C. There were both Fe_3N and metallic Fe species in the catalysts heated at 800 °C, and Fe_4N and metallic Fe species appeared in the catalysts pyrolyzed at 850 °C and 950 °C, respectively (Figure 2c). Interestingly, the catalyst containing higher Fe contents of 7.5 wt. % and prepared at the higher temperature of 950 °C, possessed an identical metallic Fe species. It is assumed that higher amounts of Fe, combined with high heating temperature, induce an aggregation of Fe nanoparticles to give less active, metallic Fe. Longer reaction time also induced the phase changes of Fe_3N to the mixture of Fe_4N and metallic Fe (Figure 2d). The strong correlation between the existence of Fe_3N phase and ORR performance show that Fe_3N is the essential active species for superior ORR performance of iron nitride electrocatalysts.

The presence of Fe_3N was likewise observed in transmission electron microscope (TEM) analysis, which showed the characteristic d-spacing (2.05 nm) of <111> peak of Fe_3N phase matched with the experimental d-spacing (2.10 nm) observed from HRTEM images (Figure 4a). Nanoparticles with an average size of 23.7 nm were supported on the NC support (Figure 4b). The nanoparticles were determined to be Fe_3N by the length of 0.21 nm of the lattice distance and strong ring pattern corresponding to the (111) phase of Fe_3N in a selected area electron diffraction (SAED) (Figure 4c,d).

In addition, fine distribution of N atoms over the carbon support was evident by EDS mapping (Figure 5). Along with well dispersed N over the carbon support, a higher population of N atoms near Fe atoms was observed, which again showed the favorable bonding between N and Fe to form active catalytic components.

The measured value of the Tafel slope of the single phase Fe_3N on NC catalyst obtained from pyrolysis at 900 °C was much lower than the corresponding values of the samples containing other iron nitride phases prepared at other conditions: 850 °C (61 mV), 950 °C (65 mV) and 800 °C (69 mV) (Figure 6a). This attests to the effect of the Fe_3N species as showing better catalytic performance. Further, the catalytic activity of ORR is closely related to electron transfer mechanism during the reaction [23–25]. Thus, the number of electrons transferred per one mole of oxygen and yield of HO_2^- byproduct were determined by rotational ring disk electrode (RRDE) measurement using the most active Fe_xN_y/NC catalyst. The number of transferred electrons was determined to be higher than 3.9 throughout the whole potential range (Figure 6b). This selective and efficient catalytic process supports the high activity of our Fe_3N/NC catalyst. Formation of HO_2^- was also well inhibited and the amounts did not exceed 5%. Accelerated durability test (ADT) demonstrated the superior stability of the iron nitride catalyst that showed slight negative shift after 4,000 repeated potential cycles (Figure 6c).

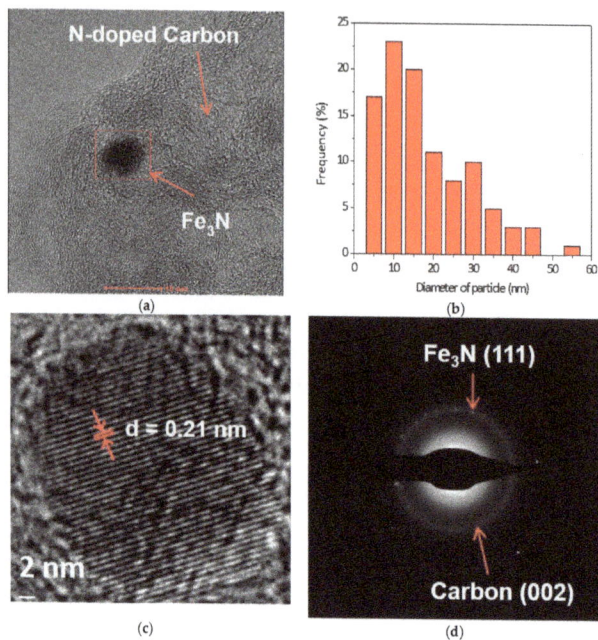

Figure 4. (**a**) Transmission electron microscope (TEM) image of Fe-N-C catalyst, (**b**) particle size distribution of Fe$_3$N nanoparticles in Fe-N-C catalyst, (**c**) high resolution (HR)-TEM image of Fe$_3$N nanoparticle, and (**d**) selected area electron diffraction (SAED) image of Fe-N-C catalyst.

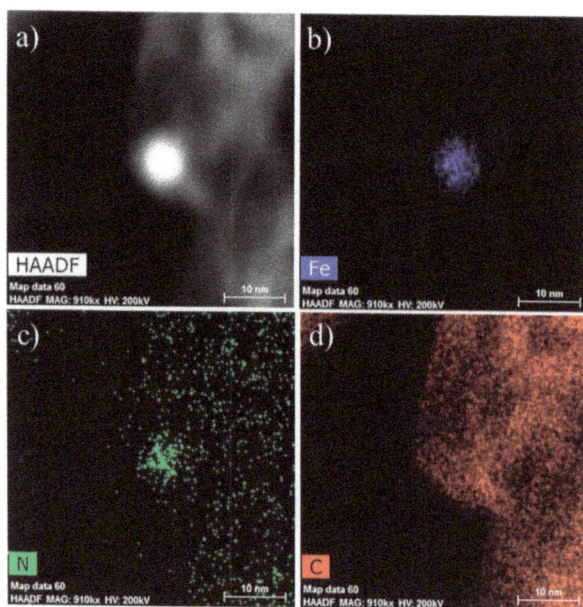

Figure 5. (**a**) High angle annular dark field-scanning transmission electron microscope (HAADF-STEM) image of Fe-N-C catalyst, electron energy loss spectroscopy (EELS) mapping images of Fe-N-C catalyst (**b**) Fe, (**c**) N, and (**d**) C.

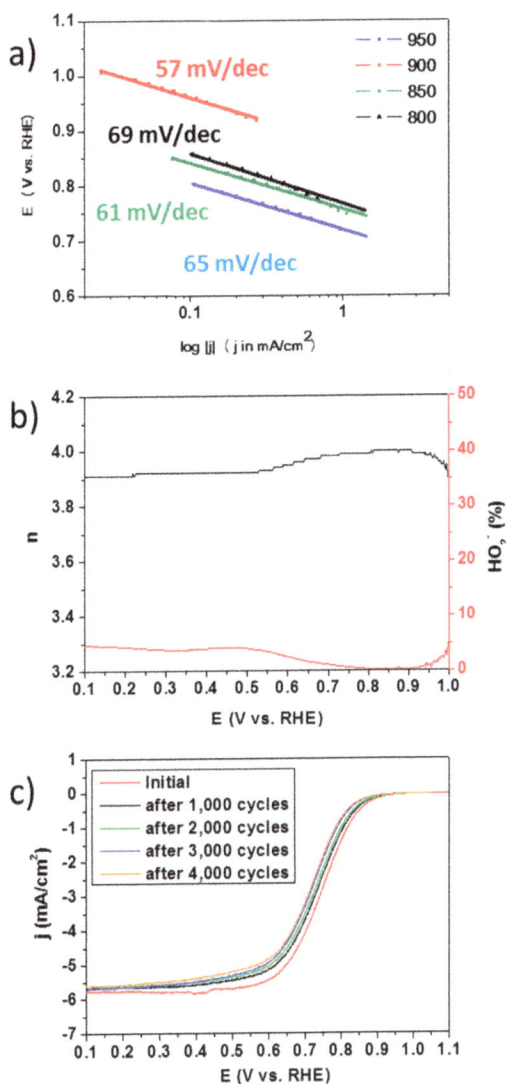

Figure 6. (**a**) Tafel slopes of the Fe-N-C catalysts prepared by different pyrolysis temperature, (**b**) number of electron transferred and HO_2^- yield during the ORR catalyzed by Fe-N-C and (**c**) accelerated durability testing (ADT) of Fe-N-C; initial, after 1,000, 2,000, 3,000, and 4,000 potential cycles in O_2-saturated 0.1 M KOH.

3. Experimental

3.1. Synthesis of Fe_xN_y/NC Catalyst

All of the chemicals were purchased from Sigma-Aldrich (St. Louis, MO, USA) and used without further purification throughout the entire synthetic procedure. A mixture of Carbon black (Ketjen, Chicago, IL, USA) and Fe salt (iron(III) acetylacetonate, Fe(acac)₃, 5 wt. %, Sigma-Aldrich, St. Louis, MO, USA) was ground in a mortar. For the other Fe sources, desired amounts of Fe salts such as iron(II) acetylacetonate (Fe(acac)$_2$, Sigma-Aldrich, St. Louis, MO, USA), Iron(III) chloride hexahydrate (FeCl$_3$·6H$_2$O, 97%, Sigma-Aldrich, St. Louis, MO, USA), Iron(II) sulfate heptahydrate (FeSO$_4$·7H$_2$O, 99%, Sigma-Aldrich, St. Louis, MO, USA), and ferrocene (Fe(C$_5$H$_5$)$_2$, 98%, Sigma-Aldrich, St. Louis, MO, USA) were employed, instead of Fe(acac)$_3$. The nominal amount was kept at 5.0 wt. % theoretical loading. Subsequently, the powder was annealed at 900 °C for 2 h with a ramping rate of 5 °C·min^{-1} under NH$_3$ atmosphere. Fe loading, reaction temperature, and pyrolysis time were altered for the optimization of synthetic conditions.

3.2. Characterization of Catalysts

Morphology of the as-synthesized catalyst was characterized by high resolution transmission electron microscope (HRTEM, FEI Tecnai F20, FEI, Hillsboro, OR, USA) and high angle annular dark field-scanning transmission electron microscope (HAADF-STEM, FEI Titan at 200 Kv, FEI, Hillsboro, OR, USA). Energy dispersive spectrum (EDS) mapping was carried out using a Talos TEM (FEI; Talos F200X 80-200, FEI, Hillsboro, OR, USA) microscope equipped with X-FEG and super-X EDS system with four silicon drift detectors (SDDs) (Bruker, Bermen, Germany) to determine atomic distribution of each element.

3.3. Electrochemical Measurements

An Autolab PGSTAT20 potentiostat (Metrohm Autolab B.V., Kanaalweg, Utrecht, The Netherlands) was used to carry out all electrochemical measurements. The experiments were conducted in a standard three-electrode cell and a rotating disk electrode (RDE) system (Eco Chemie B.V., Utrecht, Utrecht, The Netherlands). A saturated calomel electrode (SCE) and Pt wire were selected as the reference and counter electrode, respectively. All potentials were calibrated using a reversible hydrogen electrode (RHE) and reported with respect to the RHE. A glassy carbon (GC) RDE (5 mm in diameter) acted as the working electrode. A rotating ring-disk electrode (RRDE, 5.61 mm in diameter) was conducted using an AFMSRCE advanced electrochemical system (Pine Instrument Co., Grove City, PA, USA). A three-electrode cell system was employed incorporating a rotating GC disk (disk area: 0.2475 cm^2) and a Pt ring electrode (ring area: 0.1886 cm^2) after loading the electrocatalyst as the working electrode. The suspension of the catalytic ink was prepared by ultrasonically dispersing Fe-N-C catalyst in a mixture of Nafion (5 wt. %, Sigma-Aldrich, St. Louis, MO, USA) and 2-propanol. Then, the amount of the ink slurry 0.464 mg·cm^{-2}geo of Fe-N-C was loaded onto the GC electrode followed by air-drying. To measure the current of the ORR, linear sweep voltammetry (LSV) was conducted in an O$_2$ saturated 0.1 M KOH solution with a potential range of 0.05 to 1.1 V (*vs.* RHE) and positive scanning rate of 5 mV·s^{-1} at 1600 rpm. The durability of the prepared catalyst was examined by repeating 4,000 potential cycles of cyclic voltammetry (CV) between 0.6 V and 1.0 V (*vs.* RHE) at a scan rate of 100 mV·s^{-1} in O$_2$-saturated 0.1 M KOH solution. The peroxide percentage (%HO$_2^-$) and electron transfer number (*n*) were evaluated based on the following equations:

$$n = \frac{4 \times I_D}{I_D + \frac{I_R}{N}} \tag{1}$$

$$\%HO_2^- = \frac{200 \times \frac{I_R}{N}}{\left(\frac{I_R}{N}\right) + I_D} \tag{2}$$

In equation (1), n is the number of electrons transferred per oxygen molecule; IR and ID are the ring and disk current, respectively. The manufacturer's value of N (ring collection efficiency) is 37%.

The electrochemical measurements on RDE and RRDE greatly depend on the electrode conditions such as the structure, porosity, and geometry of the catalyst layer [26–29]. We tried to minimize those effects by strictly maintaining the standard procedure for electrode preparation. Thus, we suppose that those effects do not alter our catalytic activity comparison.

4. Conclusion

We reported the essential role of Fe_3N as the active site for ORR in Fe_xN_y/NC catalysts. Several synthetic variables were optimized to obtain a highly active ORR catalyst, and to reveal the relationship between ORR activity and the density of Fe_3N sites. The identical patterns in simulated XRD and measured XRD patterns verified the presence of Fe_3N phase in the active catalysts. TEM images and SAED patterns further demonstrated the existence of small sized Fe_3N nanoparticles on the N doped carbon material. Numerous parameters such as the type of Fe precursor, Fe contents in the catalyst, pyrolysis temperature, and pyrolysis time affected the formation of the Fe_3N nanoparticles. ORR activity of Fe_xN_y/NC catalyst was proven to be closely related to the Fe_3N phase. The optimized catalyst exhibited excellent ORR activity and durability in alkaline solution. Our effort to identify the active species of iron nitride catalyst for ORR may bring a new opportunity to develop more efficient electrocatalysts and establish a PEMFC based power generation system as well.

Supplementary Materials: The following are available online at www.mdpi.com/2073-4344/6/6/86/s1, Figure S1. Linear sweep Voltammetry curves of ORR in 0.1 M KOH catalyzed by Fe-N-C, C, and N-C., Figure S2. Calculated formation energy for different Fe_xN_y compounds.

Acknowledgments: This research was supported by the Technology Development Program to Solve Climate Changes of the Natinal Research Foundation (NRF) funded by the Ministry of Science, ICT, & Future Planning (NRF-2015M1A2A2056690). This work was supported by the Korean Government through the New and Renewable Energy Core Technology Program of the Korea Institute of Energy Technology Evaluation and Planning (KETEP) funded by MOTIE (No.20133030011320), and the National Research Foundation of Korea Grant funded by MSIP (2016, University-Institute cooperation program). This work was also financially supported by KIST through Institutional Project (2E25411).

Author Contributions: Min Jung Park and Jin Hee Lee contributed equally to this work. They prepared the catalysts and evaluated the electrocatalytic activity of the catalysts. K. P. S. S. Hembram and Sang Soo Han carried out the theoretical calculations and analysis of Fe_xN_y compounds. Chang Won Yoon and Suk-Woo Nam contributed the interpretation of the analytical data and writing the manuscript. Jin Young Kim provided the concept of this research and managed all the experimental and writing process as the corresponding author.

Conflicts of Interest: The authors declare no conflict of interest.

References

1. Brian, C.H.S.; Angelika, H. Materials for fuel-cell technologies. *Nature* **2001**, *414*, 345–352.
2. Perry, M.L.; Fuller, T.F. A historical perspective of fuel cell technology in the 20th century. *J. Electrochem. Soc.* **2002**, *149*, S59–S67. [CrossRef]
3. Nie, Y.; Li, L.; Wei, Z. Recent advancements in Pt and Pt-free catalysts for oxygen reduction reaction. *Chem. Soc. Rev.* **2015**, *44*, 2168–2201. [CrossRef] [PubMed]
4. Wang, Y.J.; Zhao, N.; Fang, B.; Li, H.; Bi, X.T.; Wang, H. Carbon-supported Pt-based alloy electrocatalysts for the oxygen reduction reaction in polymer electrolyte membrane fuel cells: Particle size, shape, and composition manipulation and their impact to activity. *Chem. Rev.* **2015**, *115*, 3433–3467. [CrossRef] [PubMed]
5. Proietti, E.; Jaouen, F.; Lefevre, M.; Larouche, N.; Tian, J.; Herranz, J.; Dodelet, J.P. Iron-based cathode catalyst with enhanced power density in polymer electrolyte membrane fuel cells. *Nat. Commun.* **2011**, *2*, 416. [CrossRef] [PubMed]
6. Hu, Y.; Jensen, J.O.; Zhang, W.; Cleemann, L.N.; Xing, W.; Bjerrum, N.J.; Li, Q. Hollow spheres of iron carbide nanoparticles encased in graphitic layers as oxygen reduction catalysts. *Angew. Chem. Int. Ed.* **2014**, *53*, 3675–3679. [CrossRef] [PubMed]
7. Jang-Soo, L.; Gi, S.P.; Sun, T.K.; Meilin, L.; Jaephil, C. A highly efficient electrocatalyst for the oxygen reduction reaction; *n*-doped ketjenblack incorporated into Fe/Fe_3C-functionalized melamine foam. *Angew. Chem. Int. Ed.* **2013**, *52*, 1026–1030.
8. Wen, Z.; Ci, S.; Zhang, F.; Feng, X.; Cui, S.; Mao, S.; Luo, S.; He, Z.; Chen, J. Nitrogen-enriched core-shell structured Fe/Fe_3C-C nanorods as advanced electrocatalysts for oxygen reduction reaction. *Adv. Mater.* **2012**, *24*, 1399–1404. [CrossRef] [PubMed]

9. Xiao, M.; Zhu, J.; Feng, L.; Liu, C.; Xing, W. Meso/macroporous nitrogen-doped carbon architectures with iron carbide encapsulated in graphitic layers as an efficient and robust catalyst for the oxygen reduction reaction in both acidic and alkaline solutions. *Adv. Mater.* **2015**, *27*, 2521–2527. [CrossRef] [PubMed]

10. Jang, B.; Bong, S.; Woo, S.; Park, S.K.; Ha, J.; Choi, E.; Piao, Y. Facile synthesis of one-dimensional iron-oxide/carbon hybrid nanostructures as electrocatalysts for oxygen reduction reaction in alkaline media. *J. Nanosci. Nanotechnolo.* **2014**, *14*, 8852–8857. [CrossRef]

11. Ma, Y.; Wang, H.; Key, J.; Linkov, V.; Ji, S.; Mao, X.; Wang, Q.; Wang, R. Ultrafine iron oxide nanoparticles supported on n-doped carbon black as an oxygen reduction reaction catalyst. *Int. J. Hydrogen Energy* **2014**, *39*, 14777–14782. [CrossRef]

12. Tylus, U.; Jia, Q.; Strickland, K.; Ramaswamy, N.; Serov, A.; Atanassov, P.; Mukerjee, S. Elucidating oxygen reduction active sites in pyrolyzed metal-nitrogen coordinated non-precious-metal electrocatalyst systems. *J. Phys. Chem. C* **2014**, *118*, 8999–9008. [CrossRef] [PubMed]

13. Palaniselvam, T.; Kannan, R.; Kurungot, S. Facile construction of non-precious iron nitride-doped carbon nanofibers as cathode electrocatalysts for proton exchange membrane fuel cells. *Chem. Commun.* **2011**, *47*, 2910–2912. [CrossRef] [PubMed]

14. Yin, H.; Zhang, C.; Liu, F.; Hou, Y. Hybrid of iron nitride and nitrogen-doped graphene aerogel as synergistic catalyst for oxygen reduction reaction. *Adv. Funct. Mater.* **2014**, *24*, 2930–2937. [CrossRef]

15. Zhang, J.; He, D.; Su, H.; Chen, X.; Pan, M.; Mu, S. Porous polyaniline-derived FeN_xC/C catalysts with high activity and stability towards oxygen reduction reaction using ferric chloride both as an oxidant and iron source. *J. Mater. Chem. A* **2014**, *2*, 1242–1246. [CrossRef]

16. Michel, L.; Eric, P.; Frédéric, J.; Jean-Pol, D. Iron-based catalysts with improved oxygen reduction activity in polymer electrolyte fuel cells. *Science* **2009**, *324*, 71–74.

17. Wu, G.; More, K.L.; Johnston, C.M.; Zelenay, P. High-performance electrocatalysts for oxygen reduction derived from polyaniline, iron, and cobalt. *Science* **2011**, *332*, 443–447. [CrossRef] [PubMed]

18. Ganesan, S.; Leonard, N.; Barton, S.C. Impact of transition metal on nitrogen retention and activity of iron-nitrogen-carbon oxygen reduction catalysts. *Phys. Chem. Chem. Phys.* **2014**, *16*, 4576–4585. [CrossRef] [PubMed]

19. Hung, T.F.; Tu, M.H.; Tsai, C.W.; Chen, C.J.; Liu, R.S.; Liu, W.R.; Lo, M.Y. Influence of pyrolysis temperature on oxygen reduction reaction activity of carbon-incorporating iron nitride/nitrogen-doped graphene nanosheets catalyst. *Int. J. Hydrogen Energy* **2013**, *38*, 3956–3962. [CrossRef]

20. Peng, H.; Mo, Z.; Liao, S.; Liang, H.; Yang, L.; Luo, F.; Song, H.; Zhong, Y.; Zhang, B. High performance Fe- and n-doped carbon catalyst with graphene structure for oxygen reduction. *Sci. Rep.* **2013**. [CrossRef]

21. Schnepp, Z.; Thomas, M.; Glatzel, S.; Schlichte, K.; Palkovits, R.; Giordano, C. One pot route to sponge-like Fe_3N nanostructures. *J. Mater. Chem.* **2011**, *21*, 17760–17764. [CrossRef]

22. Jain, A.; Ong, S.P.; Hautier, G.; Chen, W.; Richards, W.D.; Dacek, S.; Cholia, S.; Gunter, D.; Skinner, D.; Ceder, G.; *et al.* Commentary: The materials project: A materials genome approach to accelerating materials innovation. *APL Mater.* **2013**, *1*, 011002. [CrossRef]

23. Kattel, S.; Atanassov, P.; Kiefer, B. A density functional theory study of oxygen reduction reaction on non-PGM Fe-N_x-C electrocatalysts. *Phys. Chem. Chem. Phys.* **2014**, *16*, 13800–13806. [CrossRef] [PubMed]

24. Kramm, U.I.; Herranz, J.; Larouche, N.; Arruda, T.M.; Lefevre, M.; Jaouen, F.; Bogdanoff, P.; Fiechter, S.; Abs-Wurmbach, I.; Mukerjee, S.; *et al.* Structure of the catalytic sites in Fe/N/C-catalysts for O_2-reduction in PEM fuel cells. *Phys. Chem. Chem. Phys.* **2012**, *14*, 11673–11688. [CrossRef] [PubMed]

25. Szakacs, C.E.; Lefevre, M.; Kramm, U.I.; Dodelet, J.P.; Vidal, F. A density functional theory study of catalytic sites for oxygen reduction in Fe/N/C catalysts used in H_2/O_2 fuel cells. *Phys. Chem. Chem. Phys.* **2014**, *16*, 13654–13661. [CrossRef] [PubMed]

26. Masa, J.; Batchelor-McAuley, C.; Schuhmann, W.; Compton, R.G. Koutecky-levich analysis applied to nanoparticle modified rotating disk electrodes: Electrocatalysis or misinterpretation. *Nano. Res.* **2013**, *7*, 71–78. [CrossRef]

27. Ward, K.R.; Gara, M.; Lawrence, N.S.; Hartshorne, R.S.; Compton, R.G. Nanoparticle modified electrodes can show an apparent increase in electrode kinetics due solely to altered surface geometry: The effective electrochemical rate constant for non-flat and non-uniform electrode surfaces. *J. Electroanal. Chem.* **2013**, *695*, 1–9. [CrossRef]

28. Gara, M.; Ward, K.R.; Compton, R.G. Nanomaterial modified electrodes: Evaluating oxygen reduction catalysts. *Nanoscale* **2013**, *5*, 7304–7311. [CrossRef] [PubMed]
29. Ward, K.R.; Compton, R.G. Quantifying the apparent 'catalytic' effect of porous electrode surfaces. *J. Electroanal. Chem.* **2014**, *724*, 43–47. [CrossRef]

MDPI AG

St. Alban-Anlage 66

4052 Basel, Switzerland

Tel. +41 61 683 77 34

Fax +41 61 302 89 18

http://www.mdpi.com

Catalysts Editorial Office

E-mail: catalysts@mdpi.com

http://www.mdpi.com/journal/catalysts

www.ingramcontent.com/pod-product-compliance
Lightning Source LLC
Chambersburg PA
CBHW051847210326
41597CB00033B/5811